3rd Annual International Wafer-Level Packaging Conference (IWLPC 2006)

San Jose, California, USA
1 - 3 November 2006

ISBN: 978-1-5108-3302-9

Printed from e-media with permission by:

Curran Associates, Inc.
57 Morehouse Lane
Red Hook, NY 12571

Some format issues inherent in the e-media version may also appear in this print version.

Copyright© (2006) by Surface Mount Technology Association (SMTA)
All rights reserved.

Printed by Curran Associates, Inc. (2017)

For permission requests, please contact Surface Mount Technology Association (SMTA)
at the address below.

Surface Mount Technology Association (SMTA)
6600 City West Parkway
Eden Prairie, MN 55344
USA

Phone: (952) 920-7682
Fax: (952) 926-1819

smta@smta.org

Additional copies of this publication are available from:

Curran Associates, Inc.
57 Morehouse Lane
Red Hook, NY 12571 USA
Phone: 845-758-0400
Fax: 845-758-2633
Email: curran@proceedings.com
Web: www.proceedings.com

3rd Annual International Wafer-Level Packaging Conference (IWLPC 2006)

San Jose, California, USA
1 - 3 November 2006

TABLE OF CONTENTS

Opening Session: Business and Marketing Issues of Wafer-Level and IC Packaging

The Expansion of Wafer-Level Packaging: Challenges and Opportunities.....1
E. Jan Vardaman, *TechSearch International, Inc.*

The IC Packaging World and Its Latest Developments.....3
Sandra L. Winkler, *Electronic Trend Publications*

WAFER-LEVEL PACKAGING AND PROCESS MATERIALS TRACK

Session 1

RF Crosstalk Suppression Based on Wafer-Level Packaging Concept.....5
S.M. Sinaga, A. Polyakov, M. Bartek, and J.N. Burghartz, *Delft Institute of Microelectronics and Submicron Technology*

Wafer Level Stacking of 8 to 10 Dice per mm for Consumer Products – Wireless Die-on-Die "WDoD".....10
Christian Val, Ph.D. and Pascal Couderc, Ph.D., *3D PLUS*

Squeegee Influence on Bump Metrics for Stencil Printed Wafers.....16
Jeff Schake, *DEK USA Inc.* and Guy Burgess, *Flip Chip International, LLC*

Session 3

Fabrication of Tapered Through-Vias on (100) Silicon for Wafer-Level Packaging.....22
Huang Shuang Wu and Chia Yong Poo, *Micron Semiconductor Asia Pte Ltd.*

Lithography-Grade Controlled Expansion Substrates for Wafer Level Packaging.....28
Greg Rudd and Bob Cronk, *SMI (Spectra-Mat, Inc.)*

Session 5

Su-8 Bonding for Transparent Packaging.....32
C. Brubaker, T. Matthias and M. Wimplinger, *EV Group*

Metrology for Ultra-Thin Wafer and Die Strength Characterization and Related Edge Damage and Modeling Challenges 39
David Liu, Anwei Liu, Michael I. Current, Wojtek J. Walecki, and Ann Koo, *Frontier Semiconductor*

Utilization of Die Attach Adhesives in Wafer Level Assembly of Cavity Packages for Image Sensors.....42
G. Humpston and M. Nystrom, *Tessera Inc.;* S. Kanagavel, M. Previti, and M. Wilson, *Cookson Electronics Inc.*

Session 7

BCB Wafer Bonding with Electrical Interconnects.....48
Praveen Pandojirao-S, Rachita Dewan, Dan O. Popa, and J.-C. Chiao, *Automation & Robotics Research Institute, University of Texas at Arlington*

Wafer-to-Wafer and Chip-to-Wafer Integration Schemes for Systems-in-a-Package and 3D Interconnects.....54
Thorsten Matthias, Stefan Pargfrieder, Herwig Kirchberger, Markus Wimplinger, and Paul Lindner, *EV Group*

Pattern Effects on Electroplated Copper Pillars.....61
Arthur Keigler, Bill Wu, Jim Zhang, and Zhenqiu Liu, *NEXX Systems, Inc.*

Session 9

C4NP – Data for Fine Pitch to CSP Flip Chip Solder Bumping.....65
Eric Laine and Klaus Ruhmer, *SUSS MicroTec, Inc.;* Luc Belanger and Michel Turgeon, *IBM Canada Ltd.;* Eric Perfecto, Hai Longworth, and David Hawken, *IBM Microelectronics*

An Integrated Deep Silicon Etch/Directional Physical Vapor Deposition Process for Through-Wafer Via Applications.....72
G. Reynolds, C. Constantine, S. Lai, K. Mackenzie, R. Westerman, D. Johnson, C. Johnson, and R. Benz, *Oerlikon USA, Inc.;* J-B Chevrier, *Oerlikon France;* J. Weichart, S. Kadlec, M. Elghazzali, H. Hirscher, and H. Auer, *OC Oerlikon Balzers Limited*

Study of Ni-P/Pd/Au as a Final Finish for Wafer.....78
Kazuki Yoshikawa, Toshiaki Shibata, Masayuki Kiso, and Shigeo Hashimoto, *C. Uyemura & Company, Ltd.;* Don Gudeczauskas, *Uyemura International Corporation*

Session 11

Advanced Plasma Processing Techniques for Improving Descum and Other WLP Process Performance.....82
Scott D. Szymanski, *March Plasma Systems*

Using the 2D Macro CD Metrology Package to Measure CD Lines.....87
Rajiv Roy, Matt Wilson, and Chris Hawes, *Rudolph Technologies*

Surface Cleaning Flip Chip Wafers for Test and Assembly Improvements.....91
Terence Collier, *CVInc.*

Non Lithographic Microcell Plating for Integrated Passives and RDL.....98
P. Möller and M. Fredenberg, *Replisaurus Technologies AB;* P. Leisner, *Acreo AB;* M. Ostling, *Royal Institute of Technology, IMIT*

3D, STACKED AND NOVEL DIE PACKAGING TRACK

Session 2

Assembling Optical Devices Utilizing Wafer Level Technology and Chip on Board Process to Enable Higher Yields and Reduced Costs.....102
Yehudit Dagan, Giles Humpston, and Michael J. Nystrom, *Tessera Inc.*

Hybrid Wafer-Level Packaging for RF-MEMS Applications.....106
J. Iannacci, R. Gaddi, and A. Gnudi, *ARCES-DEIS Università di Bologna;* J. Iannacci, M. Bartek, J. Tian, S. Sosin, and A. Akhnoukh, *HiTeC-DIMES, Delft University of Technology*

UTCP: 60µm Thick Bendable Chip Package.....114
W. Christiaens, B. Vandevelde, E. Bosman, and J. Vanfleteren, *IMEC/TFCG Microsystems*

Copper Panel Fabrication and Stacking Concept for VLP FB DIMMS.....120
Peter Salmon, *Peter C. Salmon, LLC*

Session 4

Single Wafer Bumping.....128
Yixiang Xie, Qiang Fu, and Solomon Basame, *Surfect Technologies, Inc.*

SiP – Identifying Issues for Stacked (3D) Multichip Packaging Adoption.....131
Larry Gilg, *Die Products Consortium*

Challenges in Flip Chip Die Sorting, Handling and Inspection.....135
Gerald Steinwasser, *Mühlbauer, Inc.*

Session 6

Overview of MEMS Wafer Level Processes and Patents.....140
Ken Gilleo, Ph.D., *ET-Trends LLC*

Effects of Plasma Pretreatment on Flip Chip and CSP Substrate Level Assembly Yield and Reliability.....147
Daniel Baldwin, Ph.D., *Georgia Institute of Technology;* Paul Houston and Brian Lewis, *Engent, Inc.*

Embedded IC Polyimide Multi-Layer Substrate.....157
M. Okamoto, S. Ito, S. Okude, T. Suzuki, O. Nakao, T. Ito, and R. Yamauchi, Ph.D., *Fujikura Ltd.*

Stair-Step IC Packages for Low Cost and High Performance.....161
Joseph Fjelstad, *SiliconPipe, Inc.*

Session 8

Advanced Package Prototyping Using Nano-Particle Silver Printed Interconnects.....167
Sungchul Joo and Daniel F. Baldwin, Ph.D., *Georgia Institute of Technology*

DRIE with High Rate and Uniformity for MEMS and WLP.....174
Leslie Lea, *Surface Technology Systems plc*

Aerosol-Jet Printing for 3-D Interconnects, Flexible Substrates and Embedded Passives.....179
Martin Hedges, *Neotech Services MTP;* Mike Kardos, Bruce King, and Mike Renn, *Optomec Inc.*

Session 10

Study on Adhesion of Dicing Die Attach Two-in-One Film for 3-D Stack Packaging.....184
Shijian Luo, Ph.D. and Tom Jiang, Ph.D., *Micron Technology Inc.*

Placing Wafer Level Devices in a High Speed Workflow.....189
Gheorghe Pascariu, *Hover-Davis Inc.*

Session 12

Methodology for Stacking of Power Semiconductors for the Harsh Automotive Environment.....193
Todd P. Oman, *Delphi Corporation*

White Ring Defect Formation in Lead-Free Wafer Level Packaging.....199
Kimberly D. Pollard, Ph.D., Raymond Chan, Ph.D., and Diane Scheele, *Dynaloy, LLC*

Formation of Lead-Free Microbumps by Electroplating for Flip-Chip and WLP Applications.....203
R. Kiumi, F. Kuriyama, and N. Saito, *Precision Machinery Co., Ebara Corp.*

Poster Session

Wafer-Level Packaging: Effective Cost Reduction with Wafer Bonding......210
Thorsten Matthias, Markus Wimplinger, and Paul Lindner, *EV Group*

THE EXPANSION OF WAFER LEVEL PACKAGING: CHALLENGES AND OPPORTUNITIES

E. Jan Vardaman
TechSearch International, Inc.
Austin, TX, USA
jan@techsearchinc.com

ABSTRACT

Shipments of wafer level packages (WLPs) have exceeded expectations, both in the number of units shipping and size and complexity of the devices. Once relegated to the few I/O range, wafer level packages are routinely shipping with more than 100 I/Os. They type of devices shipping in WLPs are expanding from integrated passives and analog devices, to a variety of integrated circuits including RF and memory. While many of the early parts were on 6-inch wafers, companies have migrated to 8-inch wafers and some companies will use 12-inch wafers in the future. What difficulties will be encountered with the expansion of WLPs? What are the opportunities for companies providing fabrication services, equipment, and materials? This paper examines the current status and provides projections for the future.

Key words: WLP, flip chip.

INTRODUCTION

Wafer level packages (WLPs) are fully packaged before dicing and include bumped die that are not packaged or underfilled due to thermal stress management concerns. Typically the ball or bump is a larger diameter and has a larger pitch than found in flip chip bump applications. Often the WLP uses a preformed solder ball. Table 1 compares a typical WLP with a flip chip bumped device. Some WLPs are found inside packages such as system-in-packages (SiPs).

Characteristic	Flip chip	WLP
Die size (mm)	< 24 x 24	< 5 x 5
I/O count	8 to 1,000s	4 to 100
Minimum I/O pitch (µm)	> 150	400
Bump diameter (µm)	< 150	250 – 500
Bump height (µm)	100	180 – 400
Package height (mm)	0.4 – 0.75	0.5 – 1.2
Underfilled for stress relief	Yes	No
Requires high accuracy placement equipment	Yes	No
All packaging is done in wafer level format	n/a	Yes

Source: TechSearch International, Inc., adapted from Amkor.
Table 1. WLP versus Flip Chip

DEVICES SHIPPING IN WAFER LEVEL PACKAGES

An increasing number of devices, including analog, digital, and memory, are shipping in the ultimate form of a chip size package—a wafer-level CSP. Many of these devices historically shipped in QFN or SOIC packages. WLPs are replacing these packages for many companies for form factor and performance reasons. Some companies have also indicated that for larger wafer sizes, WLPs offer the potential for lower cost. WLPs have typically been used for low pin count (≤100 I/O) applications, including analog devices such as power amplifiers, battery management devices, MOSFETs, image sensors, controllers, memory, and integrated passives[i]. Philips uses WLPs for its integrated passives such as transistors, diodes, resistors, and capacitors, in order to provide better performance in less space. Companies supplying MOSFETs in WLPs report up to one-third the area and the same performance versus a standard SOIC package.

Many companies plan to use WLPs for higher pin count applications, including analog parts with larger die sizes. This will increase the number of wafers to be processed, as well as the unit volumes. The memory die is one example of a large die whose adoption significantly increases the number of wafers. Several DRAM manufacturers have plans to offer WLPs for high-speed DRAM memory products, but a conservative demand forecast has been adopted for this transition because of concerns with testing, die shrink, board-level solder joint reliability, and handling.

Casio and Oki Electric's WLPs with copper posts are shipping in production. Shinko Electric assembles power regulators and analog devices in WLPs. Ricoh supplies voltage regulators and detectors in wafer level packages using Japanese subcontractors. National Semiconductor's micro SMD package is used for an increasing number of analog devices such as voltage regulators, controllers, and op amp devices. Makers of analog devices such as Analog and Intersil will supply many future products in WLPs. ShellCase has been purchased by Tessera. Xintec in Taiwan, Sanyo in Japan, and China Wafer Level CSP, Ltd. in Suzhou, China have all licensed the ShellCase process and package image sensors. Fairchild, Vishay Siliconix, and International Rectifier sell MOSFETs in WLPs. Integrated passives are supplied in WLPs by AVX, California Micro Devices (CMD), STMicroelectronics, On Semiconductor, Philips, and others. New WLP processes are being been introduced by companies such as Furukawa and Schott.

APPLICATIONS FOR WAFER LEVEL PACKAGES

WLPs can be found in applications such as mobile phones, laptop computers, handheld products such as personal digital assistants, and consumer products such as watches, MP3 players, cameras, and camcorders. Watch modules are one of the early applications with models from Casio containing microcontrollers and flash memory package in WLPs. These packaged are typically underfilled in order to pass the product drop test.

One of the major drivers for the adoption of WLPs in portable products is form factor, and mobile phones increasingly contain WLPs. Demand for greater functionality in smaller spaces is driving the adoption of WLPs in mobile phones faster than in any other segment of the market. Motorola's RAZR V3, the thinnest phone of its time, contained at least 14 wafer level packages (WLPs). The height of the wafer level packages ranged from 0.27 mm to 0.5 mm, offering the lowest profile possible[ii]. An increased number of companies are using small, low profile WLPs in their mobile phone products. Many phones also contain CMOS image sensors packaged in WLPs using the ShellCase technology.

CHALLENGES

All types of wafer level packages are shipping today. Devices packaged as WLPs have transitioned from 150mm or smaller wafers to 200mm wafers, and some 300mm wafers will be used in the future for applications including DRAM. As company transition to 300mm wafers, challenges include capacity for 300mm wafer level packaging, testing, backgrinding large wafers, the need for underfill, and reliability of finer pitch parts (≤0.4 mm).

The majority of WLPs currently shipping use a redistribution bump process, although for larger die sizes many companies are looking for alternative methods to provide improved reliability for larger die sizes. Reliability test requirements vary by application and by company. One company requires the WLP to pass thermal 750 thermal cycles at -40°C to 125°C. Another company has temperature cycle requirements of 1,000 hours of -55 to 125°C. For yet another companies 500 cycles may be sufficient.

In order to pass the required thermal cycles, some companies underfill WLPs. For example, Samsung underfills large die size memory in order to pass thermal cycle test of 1,000 hours, but an underfill-free WLP is under development[iii]. Panasonic's P901iS mobile phone contains many WLPs. One is a 5mm x 5mm 120 I/O part with 0.4mm pitch and the part is underfilled.

A number of companies have developed materials for underfilling WLPs. Some of these materials are wafer-applied materials that require different curing and initiating agents and must be able to withstand the elevated temperatures of Pb-free solder materials. The materials must also allow assembly without voiding and avoid outgassing and entrapment problems.

New material sets may be required to meet WLP reliability requirements without the use of underfill materials. Because the WLP attaches the die directly to the circuit board, only a solder ball and a thin polymer layer separate the silicon die from the product board in final assembly. This allows stresses to the die and the interface system during thermal conditioning and mechanical shock that often exceed the fracture strength of the various packaging materials[iv].

OPPORTUNITIES

As the price of wafer level packaging declines, the use of the technology will expand into higher volumes and a greater number of applications. Companies that offer wafer level packaging services should benefit from the strong growth of more than 25 percent CAGR from 2005 to 2010. Because so many parts ship in tape and reel, companies that manufacturer this type of equipment should also see opportunities for growth. For some devices such as memory, testing at the wafer level remains a concern, but much progress has been made. The future for wafer level packages is bright.

REFERENCES

---

[i] Flip Chip and WLP: 2006 Market Update and Technology Developments, TechSearch International, October 2006.
[ii] "Analysis of Motorola RAZR V3," TPSS Japan, December 2005.
[iii] S.Y. Jeong, "Integrity of Solder Joint after Mechanical Shock in 300mm Wafer Level Fabricated Packaging," *2006 KGD Packaging & Test Workshop*, September 10-13, 2006.
[iv] J. Hunt, et al., "WLCSP Materials: Science and Alchemy," *International Symposium and Exhibition on Advanced Packaging Materials*, March 15-17, 2006.

THE IC PACKAGING WORLD AND ITS LATEST DEVELOPMENTS

Sandra L. Winkler
Electronic Trend Publications
San Jose, CA, USA
slwinkler@electronictrendpubs.com

ABSTRACT

The worldwide integrated circuit (IC) market is steadily growing, and innovations at the packaging end continue to occur, which allow for greater innovation for end products. A number of technologies are coming to the foreground, including through-silicon vias in stacked packages, increased integration such as SiPs, embedded active devices in substrates, and larger-than-die-sized WLPs.

Key Words: IC packaging market, through-silicon vias, increased integration, larger-than-die-sized WLP

THE WORLDWIDE IC PACKAGING MARKET

The worldwide IC market is seeing healthy growth, with units growing at a compound annual growth rate of 10.6 percent through 2010. The industry has always had its cycles. Approximately every three years there is an inventory correction which is viewed as a downturn in the industry. Some corrections are more severe than others. But the industry is growing steadily as an overall trend. IC-laden products, such as cell phones and game consoles, are growing in popularity. Other products such as automobiles are incorporating more and more ICs within each vehicle sold. Thus, even when the industry experienced its worst downturn in 2001, it was able to recover without new applications coming on the market. Thus the market should be considered healthy and robust, even with its occasional downturns.

Within the IC packaging market, a variety of package styles are assembled each year, from the older DIPs and SOs to the newer stacked packages, SiPs (system-in-package), and wafer level packages (WLPs). SOs, which include SOPs, SOICs, SOJs, TSOPs, TSSOPs, and the like, comprise more than 40 percent of the total IC packaging market; adding SOTs increases that to more than 50 percent. Thus, contract package assemblers need to offer SOs within their package line-up, as there will always be a large demand for these packages, regardless of industry cycle.

The more sophisticated IC packages, such as PGAs, BGAs, FBGAs, stacked packages, and SiPs garnish more revenue and profit per package. They are very desirable products for contract package assemblers. An increasing number of package assemblers are offering stacked packages and the like. Thus it is prudent to offer a variety of package styles, to be able to ride out whatever cycles the industry has to offer and to keep the assembly lines running.

STACKED PACKAGE TRENDS

Stacked packages come in a variety of styles. The most popular ones include placing two or more bare die vertically on a common substrate, known as a die stack, or placing two or more fully packaged and tested devices vertically in a package stack, otherwise known as package-on-package. The common theme being that the die are vertically stacked and ultimately packaged as a single unit.

A variety of interconnection styles exist for stacked packages, the most popular being wire bonded. Reverse wire bonding was invented for stacked packages to prevent wire wash, where the wires touch and short out each other. Increasingly flip chip is being incorporated, generally for the lower die, thus increasing the speed and performance of the overall package, while reducing size. Another form of interconnect style, through-silicon vias, is gathering interest. This method provides the ultimate in size, speed, and performance, without the parasitics associated with wire bonds. Vias are created in the silicon of the lower die which electrically connect the upper die to the substrate below the lower die. Thus the die must be designed to accommodate keep-out areas for these vias. This can increase the size of the die, even if ever so slightly. But the most important thing is that the semiconductor manufacturers need to design the wafers so that they either incorporate the vias initially, or allow for room to add them later. Getting this to happen may be the largest roadblock preventing acceptance to this technology. Manufacturers want to create all the wafers one way for manufacturing efficiency, which is why WLPs have redistribution layers to move the bond pads from the periphery. But the vertically-integrated Freescale recently designed a WLP which can incorporate through-silicon vias for stacked WLPs. This may be the impetus needed to get this technology going. The much-established wire bonding, however, will remain the primary interconnection method for the foreseeable future.

INCREASED INTEGRATION

Portable electronics, particularly cellular telephones, have exploded in popularity. Much of this popularity stems from the fact that these products are so small and lightweight (the Motorola brick was not anywhere near as popular as the cellular phones today). A variety of package options have allowed for the shrinkage of these phones, such as stacked packages and SiPs. SoC, or system-on-chip, is a front-end operation which also provides for increased integration. SoC generally requires more time to create than an SiP, thus is less reactive to changes in the market place. But all of these options, front- and back-end, increase the level of integration, which makes for an even smaller, lighter, and higher performance product, with additional functionality able to be placed in the unit. This is what many consumers are demanding. Cell phone manufacturers are responding, and are incorporating as many features as possible in the smallest, lightest form factors, by using the technologies just discussed.

The latest integration announced by Nokia is to place an SoC within an SiP, and to integrate an active IC within the substrate as well. This involves all levels of the manufacturing process to create the ultimate in integration. Nokia is not the only organization thinking about integrating an active device within the package substrate. The Electronics and Optoelectronics Research Laboratories (EOL) of Taiwan's Industrial Technology Research Institute (ITRI) is developing a Chip-in-Substrate Package (CiSP) which embeds an active IC within the substrate. Casio Computer Co., Ltd. is developing an embedded wafer level package (EWLP) which embeds chips which are initially packaged as WLPs in a LGA (land grid array) format. Integrating passive devices in substrates has been occurring for some time. Integrating an active device is a new twist that other companies have tried, but have not successfully moved to a production environment.

LARGER THAN DIE-SIZED WLP
WLPs are die-sized packages, with the package formed on an uncut wafer, so that all the dice on that wafer are packaged in this manner. A draw back to this technology is that it limits the number of I/O on the package to what can be fit on the underside of these small dice, and be routed out on a PCB. While there are plenty of these applications around—voltage regulators being the largest WLP market—by expanding the number of I/O on a WLP, the market for WLPs can also be expanded.

Two companies as of this date have devised methods of creating a larger-than-die-sized WLP. Freescale Semiconductor has created its redistributed chip packaging (RCP) technology, and Infineon Technologies, Nitto Denko, and Apic Yamada Corporation have devised a molded reconfigured wafer level package (WLP). The beginning step in creating such a WLP is to dice the wafer and transfer the good die to another surface (call it a panel or dummy wafer) in which the die are spaced further apart. Molding compound must fill in the gaps between the initial silicon die, and then wafer processing can occur to build up the layers which constitute the package. Freescale Semiconductor takes this a step further, which allows for vias to be incorporated within a die so to as allow for stacking two WLPs into a stacked package, using through-silicon via technology.

CONCLUSIONS
Innovations at the IC packaging level are key to furthering the advancement of the IC world, which will bring about more desirable and variable end products. Numerous industries benefit from these innovations, from portable electronics for the consumer to medical devices that will save lives. Innovations are still occurring, even in today's cost-cutting world.

RF CROSSTALK SUPPRESSION BASED ON
WAFER-LEVEL PACKAGING CONCEPT

S.M. Sinaga, A. Polyakov, M. Bartek, and J.N. Burghartz
Laboratory of High Frequency Technology and Components
Delft Institute of Microelectronics and Submicron Technology
Delft University of Technology
Delft, the Netherlands

ABSTRACT

In this paper, new through-substrate trenching schemes for radio-frequency (RF) substrate crosstalk suppression are introduced. Wafer-level chip-scale packaging (WLCSP) technology, in which the active silicon substrate is bonded to a low-loss supporting substrate for maintaining mechanical integrity, is employed. Air-filled trenches are primarily useful at low RF, at which they effectively block low-frequency parasitic signals. At high RF the crosstalk signals propagate more easily across such trench barriers, which simply feature a locally minimized permittivity. Providing a grounded metal within the trench provides a means of draining the signal to ground besides blocking it. At 10GHz such a grounded trench gives an isolation that is ~40dB better than that of a control reference structure, while with the air trench only ~10dB improvement is achieved. The effectiveness of the grounded trench depends on the minimization of the parasitic impedance between the metal in the trench and the physical ground potential.

Key words: Isolation, substrate coupling, wafer-level packaging, trench.

INTRODUCTION

The growth of the communication technology in recent years has led to a great demand for a single-chip radio frequency integrated circuits (RFIC) with highly increased complexity and operating frequency. One of the drawbacks, yet playing a key role, is that a mixed-signal chip is very prone to RF crosstalk due to the presence of resistive and capacitive paths in the silicon substrate. As it is of great interest, various attempts have been made in order to suppress the substrate crosstalk. These attempts include various schemes of isolation and/or strategies which are technology and layout dependent. Substrate crosstalk suppression techniques are in general divided into two. One is to attempt to block potential RF crosstalk signals and the other is to drain the crosstalk signal to ground. Some techniques, however, consider both aspects.

Silicon-on-insulator (SOI) technology was used by many such as described in [1]-[8]. In [1], low-resistivity silicon was used instead of high-resistivity silicon. There, it was proven that by using low-resistivity Si substrate instead of the high resistivity the substrate crosstalk can be suppressed as much as 30dB at 10GHz. The low-resistivity Si substrate provides low resistive path to drain out the substrate parasitic current to the ground. Many also created additional features such as buried ground planes, guard rings, trench, faraday cages, and etc, to suppress the substrate crosstalk. Guard ring is one of the most common isolation schemes used in commercial ICs since it is easy to implement and was proven to be effective [3], [7]. The use of triple-well technology in CMOS process shown in [9] achieved an additional isolation of 20dB at 10GHz. It also has been shown that metal-filled vias in the substrate can be arranged to form a grounded Faraday cage, by which undesired RF signals can be absorbed [6], [10], and [11]. Such Faraday cages consisting of metal vias are, however, not likely very effective in crosstalk suppression. Furthermore, the thin substrate required [11] may lead to mechanical instability [12]. Another crosstalk suppression technique is the trench isolation. Trench isolation has been widely used for quite some time [4], [13], and [14]. However, these trenches were limited in terms of depth. A through-substrate trench was proposed in [15] and [16] by using porous silicon as the crosstalk barrier. Comparisons of various isolation methods and strategies were discussed in [17] and [18]. Very good theories about substrate crosstalk can be found in [19] and [20].

In this paper, we propose a new isolation technique that utilizes the features of emerging WLCSP to form a silicon wafer structure featuring a fully enclosing metallized trench through the entire silicon mounted onto a low-loss supporting substrate [21], thus overcoming both the electrical and mechanical limitations. Preliminary results of this work were presented in [22]. New results on grounded-metallized trench as new isolation scheme are presented here.

The outline of this paper is as follows: in section II, the fabrication flow acquired is described. In this section, step-by-step fabrication sequence is schematically explained. Section III addresses the test structures design issues. In particular, this section describes the types of test structures designed and fabricated. Section IV is devoted to experimental results and analysis. Finally, conclusions are drawn in section V.

FABRICATION FLOW

The fabrication sequence is schematically shown in Fig. 1. Test samples were fabricated on a low-resistivity (2-5 Ω-cm) 4-inch p-type silicon wafers. The substrate taps (ohmic contacts) were defined by ion implantation (B+ 40keV) through a photo-resist mask. A 300nm thermally grown SiO$_2$ layer, a photolithographic mask step to define via contacts, and a 600 nm 99% Al /1% Si metal layer were used to complete the front-side processing (Fig.1a). After the completion of the front-side processing, one wafer was selected as the measurement reference, and the remaining wafers were bonded to AF45 glass carriers (ε_r = 6.2, tan δ = 9x10^4). The bonding was done using ~10μm-thick Benzocyclobutene (BCB: ε_r = 2.7, tan δ = 8x10^4) layer that was spun on the processed silicon wafers (Fig.1b). Silicon and glass wafers were brought into contact under vacuum condition and then bonded at an elevated pressure of 2 Bar. Finally, BCB polymerization was carried out at 300°C for 3 hours (Fig.1c). Afterwards, the bonded glass-silicon wafers were turned upside down and then thinned down to ~50μm by using 33% TMAOH solution at 80°C (Fig.1d). A 500μm thick PECVD SixNy layer was deposited onto the exposed (back-side) silicon surface. A trench and contact pattern was then defined in that SixNy layer (Fig.1e). The exposed silicon was selectively removed by anisotropic wet etching in 33% TMAOH solution. The silicon separation was therefore reduced to 20μm instead of the 90μm mask dimension. Next, the remaining SixNy layer was removed by means of plasma etching and a 500μm PECVD SiO2 isolation layer was deposited. Photoresist spray coating was applied in order to provide a controlled resist application over the resulting excessive wafer topography and to form well-defined contact windows to the front side metallization (Fig.1f) [23]. Due to considerably large size of contact windows, there was no DOF (Depth on Focus) problem found. The backside metallization was formed by using 10/40nm Ti/TiN barrier and 1.5μm thick sputtered copper layers, an electrodeposited photoresist mask, and wet etching (Fig.1g).

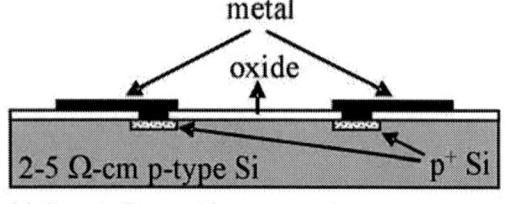

(a) Step 1: Front-side processed

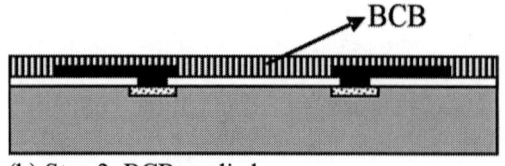

(b) Step 2: BCB applied

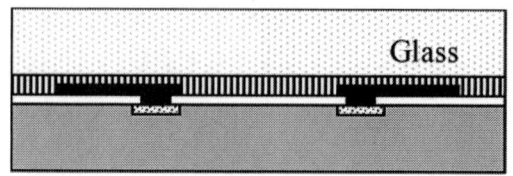

(c) Step 3: Glass bonded

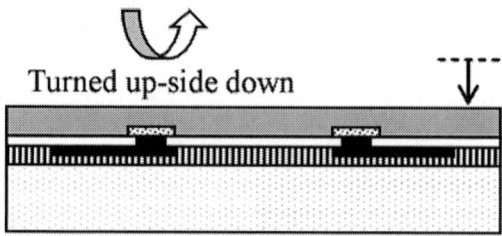

(d) Step 4: Silicon thinned down to 50μm

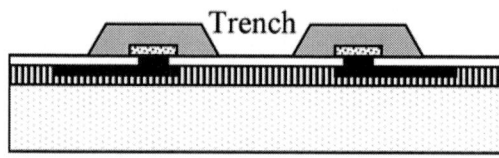

(e) Step 5: Silicon etched

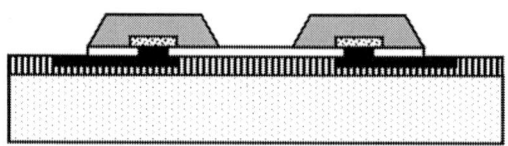

(f) Step 6: Oxide removed

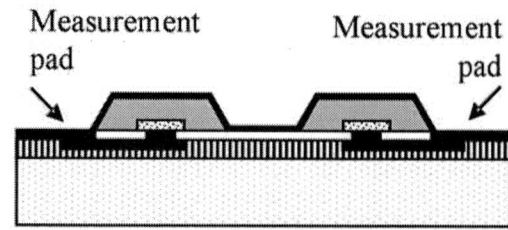

(g) Step 7: Back-side metallization

Figure 1. Test structures fabrication flow.

TEST STRUCTURES DESIGN AND MEASUREMENT STRATEGY

The reference test structure, namely the control device (#1), consists of two substrate contacts (100μm x 100μm each) which represent the noise injector and noise receiver (see Fig. 1 and Fig. 2). The noise injector and the noise receiver are denoted as port 1 and port 2, respectively. These two ports are separated by 100μm distance. The silicon substrate is grounded through substrate contacts located on the frontside (not visible from the figure). The contact pads area for the measurement were made accessible during the back-side metallization step (see Fig. 1g and Fig. 2a). The test structures were designated for on-wafer two port measurement using GSG probes. As clearly seen from Fig. 2a, probes are landed on the back-side contact pad for measurement (see Fig. 1g). The first isolation test structure

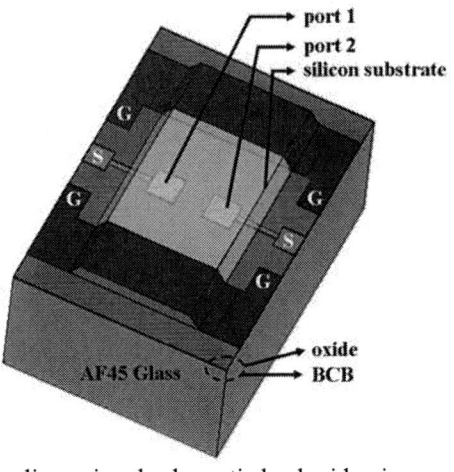

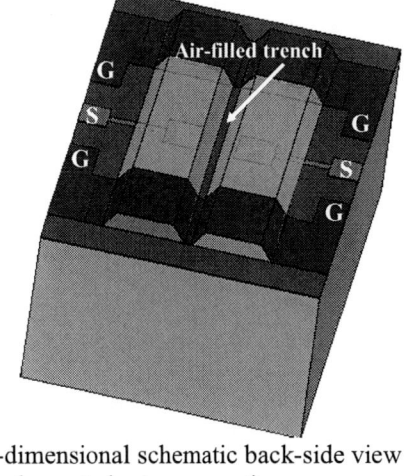

(a) Three-dimensional schematic back-side view (silicon is shown to be transparent)

(a) Three-dimensional schematic back-side view (silicon is shown to be transparent)

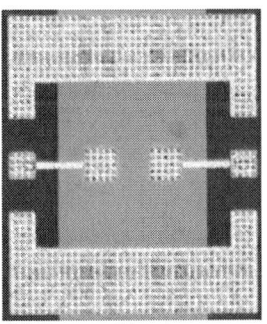
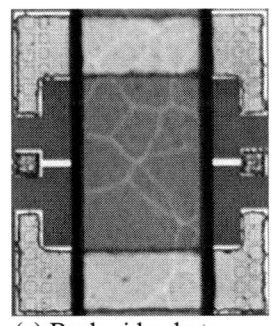

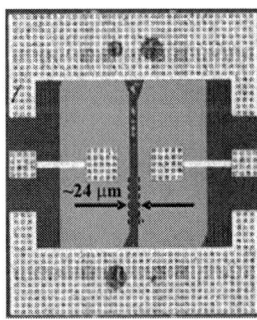
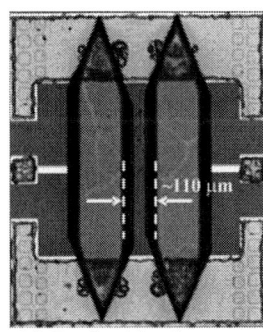

(b) Front-side photo   (c) Back-side photo

(b) Front-side photo   (c) Back-side photo

FIGURE 2. Reference test structure (#1) consists of two 100µm x 100µm p$^+$ substrate contacts, denoted as port 1 and port 2, separated by 100µm distance.

FIGURE 3. Air-filled trench isolation test structure (#2).

is open (air-filled) trench isolation structure (#2). In addition to the control device (#1), open trench was formed from the back-side to divide the silicon island into two regions to create device #2 (see Fig. 3). The trench width shown in the front-side view as depicted in Fig. 3b is different from the back-side view as depicted in Fig. 3c. This is due to the angle created during the etching [24]. The second isolation scheme is the grounded-metallized trench which is illustrated in Fig 4a. In this work, the trench width is kept at one dimension (see Fig. 3), while the metallization width is varied (see Fig. 4).

S-parameter measurement were carried out with a HP8510c network analyzer within the frequency range from 50MHz to 40GHz. GSG picoprobes with 200µm pitch was used for the RF measurement. The scattering parameter of the two-port measurement, S_{21}, provides a quantitative measure of the degree of the isolation [17]. The lower the magnitude of S_{21} the better is the isolation, i.e., the lower is the crosstalk.

RESULTS AND DISCUSSIONS

The control device (#1) in Fig. 2 exhibited an apparent isolation of ~25dB at low frequencies that resulted from the spacing of the probes and from the silicon resistivity chosen

(#1 in Fig. 5). This is clearly understood that at low frequencies, silicon substrate behaves resistively [19]. At frequencies over 10GHz the isolation started to weaken by increasing capacitive currents through the silicon, indicated by a 20dB/dec increase in $|S_{21}|$ [25]. The formation of an open (air-filled) trench that divided the silicon island in between the probes (Fig.2) yielded a considerable improvement in isolation at low frequencies ($|S_{21}|$ ~50 dB, i.e. a 25dB improvement, at 1GHz; #2 in Fig.5). At frequencies above 10GHz, however, the level of isolation approaches that of the control device #1. The residual improvement in isolation related only to the lower dielectric constant of air as compared to silicon within the trench (ε_r,Air = 1 vs. ε_r,Si = 11.9), leading to a higher 'electrical' spacing of the probes for #2 compared to #1. With a grounded metal formed within the trench (#3 in Fig.4 and Fig.5), however, the isolation was greatly improved both at high and low RF. At 10 GHz the relative increase in isolation was as much as ~30dB at 10GHz and ~20dB at 40GHz. At low frequencies the isolation was close to the noise floor of the measurement set up (see Fig.5). Fundamentally, i.e. by assuming an ideal ground in the trench, the isolation should be minimized at all frequencies. The fact that there is finite impedance between the metal in the trench and the connection to the physical ground, i.e. the ground of the GSG probes, leads to the apparent decrease of isolation towards higher frequencies. An extension of the grounded metal to the regions outside of the trench (300μm metal width) lead to a further improvement in isolation (~40 dB from #1 at 10GHz), yet with a similar frequency dependence (#4 in Fig.5).

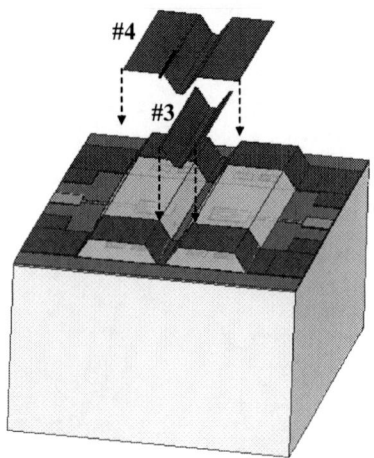

(a) Three-dimensional schematic back-side view (silicon is shown to be transparent)

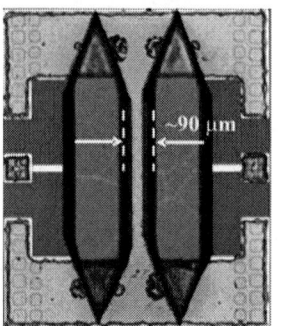
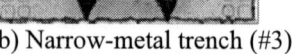
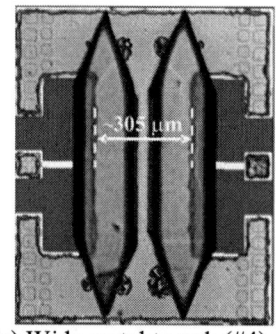

(b) Narrow-metal trench (#3). (c) Wide-metal trench (#4).

FIGURE 4. Grounded-metallized trench isolation test structure. The width of the trench is the same as in #2 (see Fig. 3).

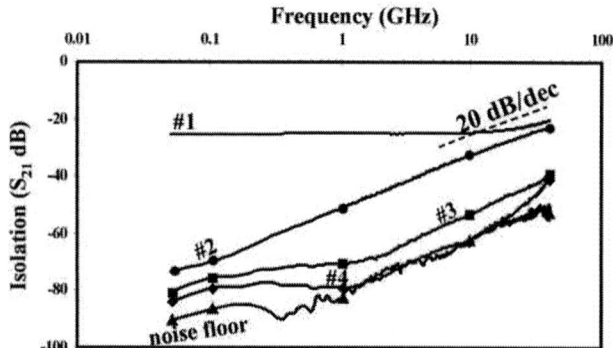

FIGURE 5. Isolation test structures measurement results.

CONCLUSION AND FUTURE WORKS

The experimental results have demonstrated that an enclosing fully metallized grounded trench is an excellent means of suppressing RF crosstalk through a conductive silicon substrate. Compared to an open, air-filled trench, by which a crosstalk signal can only effectively be blocked at low RF, the grounded metal in the trench provides a crosstalk drainage path to ground that also ensures isolation at high RF. The feasible degree of high-RF isolation depends on the parasitic impedance between each point in the metallized trench and the physical ground at the test probes at the silicon wafer surface. That impedance needs to be minimized, meaning that in practical circuit layout sufficient repetitions of ground connection should be provided and the metal should be sufficiently thick.

ACKNOWLEDGMENT

Financial support of this work by Philips Semiconductor and Philips Research, as part of the Philips Associated Center at DIMES (PACD), is acknowledged.

REFERENCES

[1] Ankarcrona, L. Vestling, K.-H. Eklund, and J. Olsson, "Low Resistivity SOI for Substrate Crosstalk Reduction," IEEE Trans. Electron Devices, vol. 52, no. 8, pp. 1920–1922, 2005.

[2] J. S. Hamel, S. Stefanou, M. Bain, B. M. Amstrong, and H. S. Gamble, "Substrate Crosstalk Suppression Capability of Silicon-On-Insulator Substrates with Buried Ground Planes (GPSOI)," IEEE Microwave Guided Wave Lett., vol. 10, no. 4, pp. 134–135, 2000.

[3] M. Kumar, Y. Tan, and J. K. O. Sin, "Novel Isolation Structures for TFSOI Technology," IEEE Electron Device Lett., vol. 22, no. 9, pp. 435–437, 2001.

[4] S. Maeda, Y. Wada, K. Yamamoto, H. Komurasaki, T. Matsumoto, Y. Hirano, T. Iwamatsu, Y. Yamaguchi, T. Ipposhi, K. Ueda, K. Mashiko, S. Maegawa, and M. Inuishi, "Feasibility of 0.18μm SOI CMOS Technology using Hybrid Trench Isolation with High Resistivity Substrate for Embedded RF/Analog Applications," IEEE Trans. Electron Devices, vol. 48, no. 9, pp. 2065–2073, 2001.

[5] M.-P. Raskin, A. Viviani, D. Flandre, and J.-P. Colinge, "Substrate Crosstalk Reduction using SOI Technology," IEEE Trans. Electron Devices, vol. 44, no. 12, pp. 2252–2261, 1997.

[6] S. Stefanou, J. S. Hamel, P. Baine, M. Bain, B. M. Armstrong, H. S. Gamble, M. Kraft, and H. A. Kemhadjian, "Ultralow Silicon Substrate Noise Crosstalk using Metal Faraday Cages in an SOI Technology," IEEE Trans. Electron Devices, vol. 51, no. 3, pp. 486–491, 2004.

[7] Y. Hiraoka, S. Matsumoto, and T. Sakai, "New Substrate-Crosstalk Reduction Structure using SOI Substrate," in Proc. IEEE International SOI Conference, Oct. 2001, pp. 107–108.

[8] Viviani, J. P. Raskin, D. Flandre, J. P. Colinge, and D. Vanhoenacker, "Extended Study of crosstalk in SOI-

SIMOX Substrates," in Proc. IEEE International Electron Devices Meeting, Dec. 1995, pp. 713–716.

[9] T. Blalack, Y. Leclercq, and C. P. Yue, "On-chip RF Isolation Techniques," in Proc. IEEE 2002 Bipolar/BiCMOS Circuits and Technology Meeting, Oct. 2002, pp. 205–211.

[10] K. Chong, X. Zhang, K.-N. Tu, D. Huang, M.-C. Chang, and Y.-H. Xie, "Three-dimensional substrate impedance engineering based on p^-/p^+ Si substrate for mixed-signal system-on-chip (SoC)," IEEE Trans. Electron Devices, vol. 52, no. 11, pp. 2440–2446, 2005.

[11] J. H. Wu, J. Scholvin, J. A. del Alamo, and K. A. Jenkins, "A Faraday Cage Isolation Structure for Substrate Crosstalk Suppression," IEEE Microwave Wireless Compon. Lett., vol. 11, no. 10, pp. 410–412, 2001.

[12] Polyakov, M. Bartek, and J. N. Burghartz, "Mechanical Stability and Handling-Induced Failure of Micromachined Wafers for RF Applications," in Digest of papers. 2001 Topical Meeting on Silicon Monolithic Integrated Circuits in RF Systems, 2001, pp. 102–109.

[13] S. Wane, D. Bajon, H. Baudrand, C. Biard, J. Langanay, and P. Gamand, "Effects of Buried Layers Doping Rate on Substrate Noise Coupling: Efficiency of Deep Trench Techniques to Improve Isolation Capability," in Digest of papers. 2004 IEEE Radio Frequency Integrated Circuits (RFIC) Symposium, 2004, pp. 179–182.

[14] D. Szmyd, L. Gambus, and W. Wilbanks, "Strategies and Test Structures for Improving Isolation Between Circuit Blocks," in Proc. IEEE 2002 International Conference on Microelectronic Test Structures, 2002, pp. 89–93.

[15] H.-S. Kim, K. A. Jenkins, and Y.-H. Xie, "Effective Crosstalk Isolation Through p^+ Si Substrates with Semi-Insulating Porous Si," IEEE Electron Device Lett., vol. 23, no. 3, pp. 160–162, 2002.

[16] K. Joardar, "A Simple Approach to Modeling Crosstalk in Integrated Circuits," IEEE J. Solid-State Circuits, vol. 29, no. 10, pp. 1212–1219, 1994.

[17] K. A. Jenkins, "Substrate Coupling Noise Issues in Silicon Technology," in Digest of papers. 2004 Topical Meeting on Silicon Monolithic Integrated Circuits in RF Systems, 2004, pp. 91–94.

[18] X. Aragonès, J. L. González, and A. Rubio, Analysis and Solutions for Switching Noise Coupling in Mixed-Signal ICs. Dordrecht, the Netherlands: Kluwer Academic Publishers, 1999.

[19] S. Donnay and G. Gielen, Substrate Noise Coupling in Mixed-Signal ASICs. Dordrecht, the Netherlands: Kluwer Academic Publishers, 2003.

[20] Polyakov, S. Sinaga, P. M. Mendes, M. Bartek, J.H. Correia, and J.N. Burghartz, "High-resistivity Polycrystalline Silicon as RF Substrate in Wafer-Level Packaging," Electronics Letters, vol. 41, no. 2, pp. 100–101, 2005.

[21] S.M. Sinaga, A. Polyakov, M. Bartek, and J. N. Burghartz, "Circuit Partitioning and RF Isolation by Through-Substrate Trenches," in Proc. IEEE Electronic Components and Technology, 2004, pp. 1519–1523.

[22] N. P. Pham, E. Boellaard, J. N. Burghartz, and P. M. Sarro, "Photoresist Coating Methods for The Integration of Novel 3-D RF Microstructures," Journal of Microelectromechanical Systems, vol. 13, no. 3, pp. 491–499, 2004.

[23] N.P. Pham, "Silicon Micromachining for RF Technology," Ph.D. dissertation, Delft University of Technology, Delft, the Netherlands, May 2003.

[24] M. Pfost and H.-M. Rein, "Modeling and Measurement of Substrate Coupling in Si-Bipolar IC's up to 40GHz," IEEE J. Solid-State Circuits, vol. 33, no. 4, pp. 582–591, 1998.

WAFER LEVEL STACKING OF 8 TO 10 DICE PER MM FOR CONSUMER PRODUCTS – WIRELESS DIE-ON-DIE "WDoD"

Christian Val, Ph.D. and Pascal Couderc, Ph.D.
3D PLUS
Buc, France
cval@3d-plus.com, pcouderc@3d-plus.com

INTRODUCTION

We know that what drives the packaging of dice for consumer products are:
- Volume/weight decrease (X, Y – AXIS), size, Thin (Z)
- Cost: reduced packaging cost and System level costs
- Ease of integration for Systems in Package (logic/memory, logic/analog, logic/RF, analog/MEMS).

Considering the State of the Art, several technological approaches have been made.

Table 1 represents both the various 3-D stacking families related to 3-D interconnection techniques.

It can first be noted that what is named Package-on-Package (PoP) consists in fact in stacking 2 non-standard 3-D modules which we named Proprietary Package. We therefore have a Proprietary Package-on-Proprietary Package rather than a Standard Package-on Standard Package. This constitutes the difference between a one-source product and a multi-source product.

The PoP interconnection is carried out with balls, which, on the one hand increases the package surface and on the other hand poses the general problem of a second reflow of the balls when the customer carries out the surface mounting of this module.

Also, the fact that it is not moulded will allow the external elements (Flux, dust, humidity) to later create corrosions and leakage currents.

3D PLUS approach, which we named Standard Package-on-Standard Package (SPoSP) allows to avoid these inconveniences. The customer can for instance use standard memories, packaged in available and low-cost TSOP packages; Their stacking with bare dice such as microcontrollers, ASIC, etc is carried out as per our 3-D standard technique.

On table 1 in the "Bare die" column, we can see the standard approach ("Chip-on-Chip") for which two 3-D modules can also be stacked through wiring interconnection.

These Chip-on-Chip techniques have the big advantage of using the Back-End equipments, which are available worldwide. However, they have the important disadvantage to use area because of the wires. Also, we are faced with all the problems linked to the multi chip modules i.e. manufacturing global yields. As a matter of fact, if no KGD are used (generally not available and very expensive), the yield begins to lower beyond 3 or 4-stacked dice.

We use our 3-D standard technique, which we adapted to the Wafer Level Process technologies.

We named this technology "Wireless Die-on-Die ("WDoD") to show that it is wireless and that consequently the area of the "WDoD" 3-D module is almost identical to the area of the larger dice.

In order to avoid the well-known yield problems, we use the concept of the Known Good rebuilt wafer, i.e. we rebuild a wafer only with the good-tested dice.

The stacking of these rebuilt wafers with our mature 3-D technique does not present any particular problem.

As a summary, the criteria which have been retained when we launched this important European programme in 2001 are the following:
. Use of multisource standard wafers
. Use of non-Thinned wafers
. Use of RLC passive components, for example Philips PICS technology
. Stacking of 8 to 10 levels per mm
. Electrical test of each level prior to stacking thanks to the test on the "rebuilt wafer"
. This process is totally collective from A to Z steps.

PROCESS

This process has been developed in the European Walpack project with ST Micro, CEA/LETI, THALES, IBS, WSI, etc..). It was then simplified with Philips Semiconductors, to make it collective from A to Z. It can be noticed that this Wafer Level Packaging type of approach necessitates material and know-how which is between front-end and back-end.

Naturally the front-end part does not necessitate an equal high-quality cleanroom as for the fabrication of the dice, since the width of the conductors for the pads redistribution is rather around 10 to 20 microns than sub micronics. This is why a new paradygm has appeared : Mid-End. This industrial approach corresponds exactly to our "WDoD".

The "WDoD" technology is compatible with the stacking of heterogeneous components such as HF dice, MEMS, etc.

For these 2 last applications, a variant consists in using a silicon substrate or at least a substrate which can be wet-etched (silicon, glass…).

Steps 1, 2, 8 to 12 are presented on Figure 1, the other steps are part of a patent currently under registration. They will be presented orally during the conference.
The standard variant is presented hereafter:

The approach proposed by 3D-PLUS consists in reconstituting a wafer from heterogeneous chips (DSP, XRAM, logic glue and passives), with the active sides placed in the same plane, embedded in a resin which gives the wafer its mechanical consistency. Components pads are then redistributed following 3-D architecture design rules. Finally, the wafer is diced and back-side thinned before being tested. The Known Good re-built Chips are then stacked and 3D interconnected by means of the patented 3D-PLUS process .

The process presented in Figure 2 includes rebuilt wafer, pad redistribution, back-side thinning, testing, stacking and 3D interconnection.

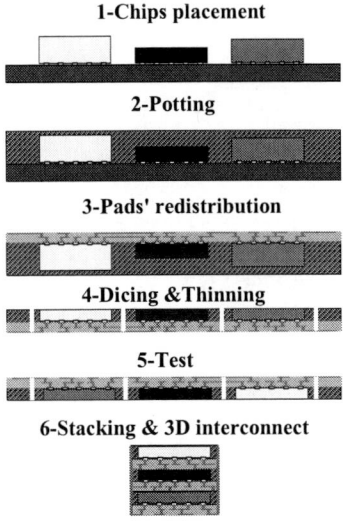

1-Chips placement

2-Potting

3-Pads' redistribution

4-Dicing &Thinning

5-Test

6-Stacking & 3D interconnect

Fig.2 Technical process for "re-built wafer"

The feasibility of this "WDoD" process started in 2001 on 100 m diametres. We are currently working on 200 mm diametres chip placement and potting.

Chip placement and potting
Wafer with Memory dice has been rebuilt (fig.3). The chips have been placed on an adhesive tape to hold them during potting. This step is crucial because at the end of the process the placement accuracy of the dice must be compatible with the pad redistribution. The displacement induced by curing has to be controlled and the warp of the wafer has to be limited. A great deal of work has been performed to select, assess and qualify the materials.

The equipement used for the placement has been developed by Delvotec.

Fig. 3 Re-built wafer with Memory

Slight displacement induced by potting has been observed. The average placement of the dice in several rebuilt wafers after potting is -0.3 µm with a standard deviation of 7.5 µm which is compliant with redistribution of 50 x 50 µm pads.

Pad redistribution
Two processes were assessed: "Cu/BCB" process " (Figure 4) and "laminated film/Cu.

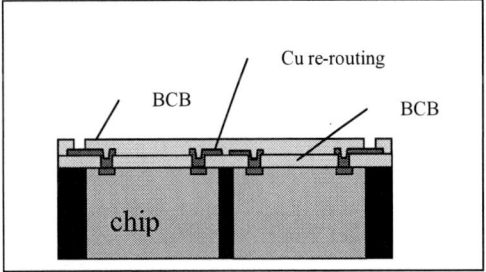

Fig 4 pad redistribution on Re-built Wafer

Cu/BCB
The pad redistribution process with Cu and BCB on silicon wafer was adapted to the re-built wafer (fig.5). The material properties of the epoxy were taken into account during the process and finally the pad redistribution process was performed by the CEA-LETI. The process includes:

. Spin coating and BCB paterning
. BCB curing
. Descum
. Seed conductive layer is deposited
. photo resist is spin coated on the re-built wafer
. Cu is electroplated through the resist
. The photo resist is removed.
. The open seed layer is etched.
. Cleaning
. Spin coating and BCB paterning
. BCB curing
. Descum.

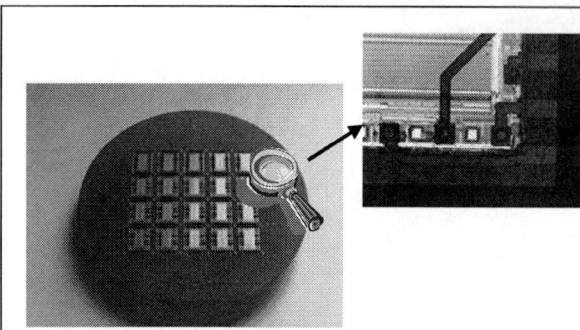

Fig 5 photo of pad redistribution with Cu/BCB on demonstrator

Laminated film/Cu

Compared to a Cu/BCB process, the film/Cu process requires more fabrication steps as the dielectric film is not photosensitive: prior to Cu processing, laser drilling of the film is necessary to process the contact vias down to the die pads. Such a process requires an additional prior metal deposition over the die pads to strengthen the metal structure, particularly at the end of the laser ablation process when the beam stops on the metallized layer, preventing the laser fluence from damaging the pad surface.

The work performed consisted in simplifying the process by reducing the number of steps by replacing the adhesive film through the use of a spin coated resin as gluing interface, and by reduction of the dielectric film thickness.

The main advantage of this approach is that we can separate the manufacturing of the laminated film and that of the rebuilt wafer.

The potential advantage of this approach for complex circuits lays in the manufacturing global yield (Figure 6).

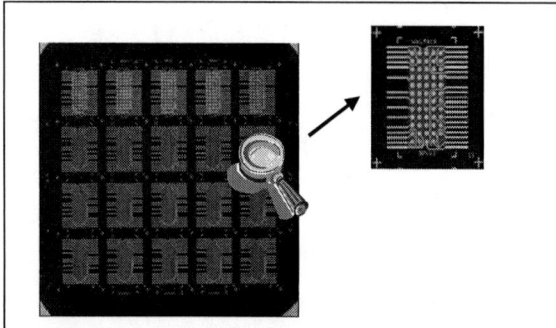

Fig 6 Photo of pad redistribution with film/Cu on demonstrator

Testing has been performed by probing the re-built Chips at wafer scale by the CEA-LETI. Pad redistribution with yield of 99% has been reached for both processes which is very encouraging results.

Dicing & Thinning

A preliminary process was developed. It involved coarse grinding following by fine polishing. Developments on the coarse grinding were performed to process heterogeneous material such as silicon and resin at the same time. For the same reason, mechanical polishing was selected. In order to reduce any failure risk during dicing, dicing by thinning method was evaluated and qualified. Re-built wafer thicknesses of 125µm are currently achieved.

The 100 µm thickness has not posed any yield problem so far. This thickness allows to reach 10 levels per mm.

Stacking & 3D Interconnect

Only Known Good Re-built Chips are stacked. The stacking principle is as follows:

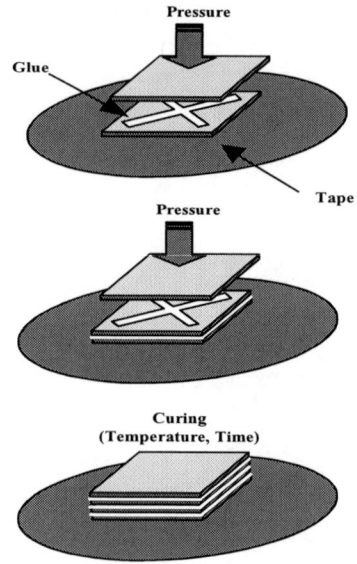

Fig. 7 Stacking principle

Liquid adhesive is dispensed on top of the first level placed by Pick & Place equipment on an adhesive tape (fig.7). The next level is then placed on top of the previous level.

This process enables the thickness of the adhesive joint between the levels to be limited. The feasibility of this process has been demonstrated and validated.

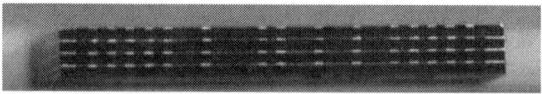

Fig 8. Copper tracks on the side of 4 stacked modules after dicing

3D interconnect of the module is performed, after dicing of the edge of the module to reveal the copper tracks (fig.8), using the fully qualified 3D-PLUS technology. The sides of the module are metallized before removing the metal between the tracks by laser in order to redistribute the pads.

Electrical tests

These 3-D modules have external connections either in LGA (Land Grid Array) or in BGA (Ball Grid Array). The electrical test is carried out with standard sockets for LGA or BGA.

APPLICATIONS
Memory modules
This technique represents a very important technological advance for the stacking of rapid memories (SRAM, DDR2, DDR3, etc), thanks to, on the one hand the deletion of wires (wireless technique) and on the other hand the deletion of PBC or in-between flex.

Figure 9 presents a 2-Gb memory module made of 8 SRAM, the total thickness of which is 1 mm.

Fig. 9 Demonstrator

The market for very large volumes is naturally the market of FLASH memories for all portable equipments (iPod, mobile phone, MP3, micro cards...). As a matter of fact, the comparison in volume (area and height) with the CoC gives an advantage to the "WDoD" comprised between 20 and 50 %.

Additionally, our approach since we started the 3-D interconnection, which has always consisted in testing each level prior to stacking, appears to be more and more valuable for a stacking over 4 memories.

- SiP
The 2 main families are totally "WDoD" compatible.
. Integration of passive elements, either above IC techniques (Camelia European project with 3D PLUS) or with capacitors embedded in a dedicated or passive substrate.

This type of substrate is built industrially by Philips (PICS technology) and perfectly compatible with the "WDoD" technique.

. With heterogeneous components of HF dice or MEMS type ; Here again, the silicon substrate is very interesting since these components are extremely sensitive to hermeticity and stress.

The approach presented allows to be at the same time zero stress and hermetic, despite the use of resin for the encapsulation, thanks to the use of the "twin cavities approach".

NEW DEVELOPMENTS FOR LARGE DIAMETER "REBUILT WAFER"
The lowering of manufacturing costs demands to place the maximum of "rebuilt components", i.e. components surrounded with resin on their 4 sides, per "rebuilt wafer". For 10-12 inches "rebuilt wafers", the placement accuracy of the dice could become critical, this is why a variant to our "WDoD" consists in using a temporary substrate (silicon, glass) ; some of the process steps are presented on Figure 1. The temporary substrate allows to mount the components with the Flip-Chip or the Stud Bump techniques.

The placement accuracy is naturally perfect and independent from the resin which surrounds the components.

The choice of this variant rather than the one without temporary substrate depends on the volumes to be manufactured and on the pad minimal pitch on the dice.

Some applications are being carried out for the stacking of 4 to 8 FLASH memories with or without microcontroller.

CONCLUSION
The approach which we have used for "WDoD" is similar to the one which we have used as of 1990 when we launched the 3-D interonnection, i.e.:
- use of standard components and available off-the-shelf.
-    test   of   each   level   prior   to   stacking.
The use of "WDoD" level process has not fundamentally changed the approach. : in fact, we did not choose the same direction as the one which is currently dominant, i.e. :
. use of non-thinned wafers.
. use of standard wafers, which means non modified (no interconnection through holes in the silicon).
. Building of Known Good Rebuilt wafers in order not to be faced again with the problems of multi chips modules and Wafer Scale Integration, which have failed.

As far as these large volume productions are concerned, 3D PLUS works as a fabless and as IP provider.

We have already identified a foundry and are looking for some others.

Kind of 3-D Technology / 3-D Interconnection Technique	Package – in – Package (PiP)	Proprietary Packaged - On - Proprietary Packaged (PPoPP)	Standard Package - On – Package (SPoP)	Main Features
Using **Wire Bond**				• Very Low Profile • XY >> Die Area
Using **Solder Balls**		**PoP** **PPoPP** 		• Low Profile • XY >> Die Area
Using **Metallization "Bus Metal"**			**SPoP** 	• Low Profile • XY ≤ Body of the Plastic Package
	Wireless Die-on-Die (WDoD) 			• Very Low Profile • XY ≈ Die Area

Table 1 – System-In-Package (SiP)

1 - Silicon Carrier with redistribution
   conductors

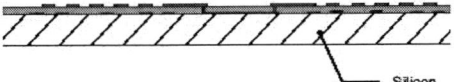

8 - Stacking of the Known Good rebuilt wafer.

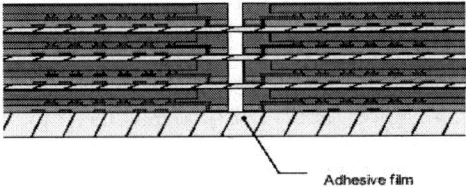

9 - Sawing of the rebuilt and stacked wafer.

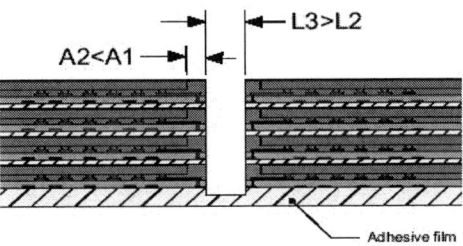

10 - Collective plating of the edges of
    kerf streets (electroless Ni+Au).

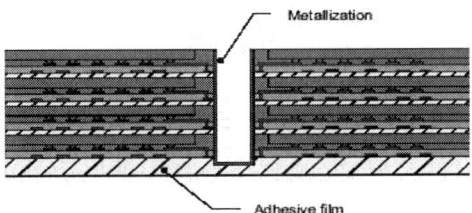

11 - Direct Laser patterning

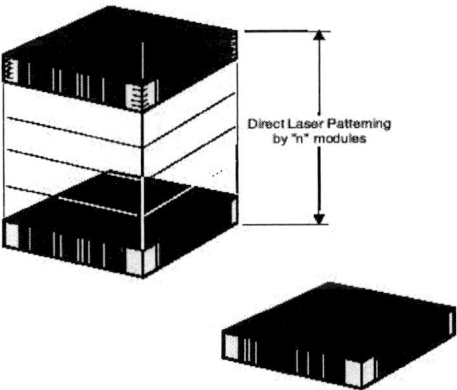

Figure 1 – Wireless Die-on-Die process

SQUEEGEE INFLUENCE ON BUMP METRICS
FOR STENCIL PRINTED WAFERS

Jeff Schake
DEK USA Inc.
Flemington, NJ, USA
jschake@dek.com

Guy Burgess
Flip Chip International, LLC
Phoenix, AZ, USA
guy.burgess@flipchip.com

ABSTRACT
Stencil printing fine powder solder paste on wafers and reflowing leverages the flexibility and productivity of standard SMT assembly equipment. As with any process, a vast number of variables need to be considered before launching successfully. Design of a proper stencil template is critical to print consistently formed individual solder paste deposits of correct dimensions. Stencil foil thickness and aperture size selection not only determines the capacity available for formation of the solder paste deposits, but also affects the ability of solder paste within the aperture to be conveyed to the wafer. Compliance to area ratio rules will result in high paste transfer efficiency and predictable bump size. Maintaining a sensible spacing between adjacent apertures will minimize defects. With the demand for finer pitch escalating, conventional design rules are being stretched to their limit. This study investigated the influence of five different squeegee designs on bump size and distribution using several aperture designs intended to challenge fineness of spacing. Metal trailing edge squeegees deposited the largest average paste volume. The experiment also produced excellent yield at tighter than recommended aperture spacing attributed to technology advancements in tools and materials.

Key words: solder paste, squeegee, stencil, wafer bumping

BACKGROUND
The technique of wafer bumping by stencil printing has been popularized by its speed and low cost of ownership. During the few seconds that it takes for the print head to circulate solder paste across a stencil, hundreds of thousands of wafer pads can be bumped in rapid succession. A standard automatic SMT stencil printing machine is quite suitable to be configured for the job, requiring vision alignment compatibility for small wafer fiducials, under stencil cleaning capability, and proper wafer support fixturing for a base level tool.

The rail system on a printer is capable of conveying wafers safely with the use of a properly designed chuck, which is typically a few inches larger than the wafer. The chuck is constructed of low mass material with a recessed cavity to a depth that matches the wafer thickness and whose boundaries fit the wafer outline. Thick wafers are secured by vacuum applied via machined channels routed within the nested area. Thin wafers may bend into vacuum channel openings, therefore porous stone may be required for these applications. Fully integrated programmable robot wafer handling systems offer a truly hands free process.

There are a wide variety of solder alloys used in stencil printing wafers consisting of basic eutectic tin-lead and including all types of lead-free alloys. In fact, the solder pastes for this process are available in a wider spectrum of alloy combinations than plating bumping techniques. The alloy powder combined with liquid flux ingredients at known ratios creates a solder paste rheologically suited for stencil printing. The flux is customarily a water-soluble based system, which is a more aggressive chemistry compared to no-clean formulas, and particularly well suited for protecting the fine alloy powder size pastes used to print wafers.

Use of water-soluble pastes also requires a flux residue removal procedure after wafers are heated in reflow. Removing residues is important for several reasons. These fluxes will continue to attack the surfaces it is covering even after reflow, which may jeopardize the integrity of the active IC if left standing. In subsequent flip-chip assembly, the assembly tool may utilize its vision system to recognize a clean bump pattern to reference for accurate die placement. Solder bumps will also have flux applied again to their surface in assembly and will bond much more successfully if new flux is applied to residue free surfaces. Furthermore, water-soluble flux residue removal will ensure the best opportunity of achieving complete underfill as the spaces between bumps shall be open to allow good wetting and flow of material.

Electroformed stencils are the recommended manufacturing technology for stencil printing wafers. The metal foil is typically nickel that can be comfortably plated up to any thickness specified between 25µm and 200µm. The additive manufacturing process creates apertures in parallel rather than by one at a time. The surface finish of aperture sidewalls is smooth and angle is non-tapered. Aperture size and position accuracy specifications are akin to the laser-cut process. Moreover, electroforming is capable of producing smaller holes in denser configurations. It is widely reported these stencils deliver enhanced transfer efficiency and print repeatability performance. [1]

The focus of this experiment is to evaluate the effect of squeegee type on performance of the stencil printing process, measured against bump size and distribution. However, the expectation of achieving good results with any squeegee design is unreasonable without understanding if this process is capable of producing the target bump size.

BUMP SIZE PREDICTION
The "truncated sphere model" has been a good tool to assist us in designing wafer bumping stencils. [2] We use the final reflowed bump height dimension that is required and work backward from that to determine the appropriate solder paste volume per pad. By using the equation shown in Figure 1, reflowed solder ball bump height and bond pad size are inserted to calculate solder bump volume per pad. Solder bump volume is translated to solder paste volume by multiplying the result by two, since solder paste consists of half flux and half metal by volume.

Bump Vol. "a" = $\pi h/6 \ (3r^2 + h^2)$

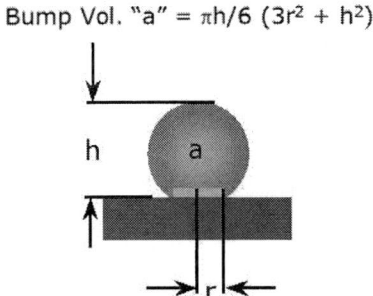

Figure 1: Truncated sphere formula

At this point the solder paste volume is known, but the aperture size still is not. The aperture size and stencil thickness dimensions need to be assigned in a combination that will transfer the correct solder paste volume in repeatable quantities while not bridging.

STENCIL DESIGN STRATEGY
The stencil design is perhaps the most critical design element to get right in order to achieve good bumping results with high yields. The thickness of the stencil metal foil is significant in determining the size of the aperture opening needed to achieve the required volume of paste to produce a target reflowed bump height. Thicker stencils are attractive because the openings can be reduced to ease aperture crowding and potentially lessen bridging concerns. However, paste transfer efficiency suffers with thicker stencils because paste tends to stick against the aperture walls instead of being flushed out to the wafer. Thicker stencils present more aperture wall surface area for the paste

to grip, which tends to hinder paste release. The level of solder paste debris left behind in the aperture after a print can be predicted by comparing the aperture opening and aperture wall areas, a term known as "area ratio". [4] This is the ratio of surface areas the solder paste contacts when residing in the aperture during the print (Figure 9 refers).

As the flux in a paste deposit occupies approximately 50% of its volume, the size of the aperture capacity should be twice the solder bump volume plus some additional capacity to compensate for expected paste transfer loss attributed to area ratio. Hence, the overprinting of pads is essential in stencil printing wafers to get enough paste volume for a distribution of sufficiently tall reflowed solder bumps, as shown in Figure 2.

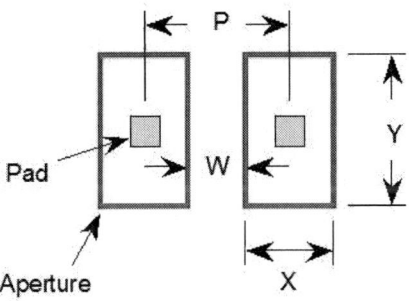

Figure 2: Aperture / pad diagram.

A suitable minimum spacing between apertures, "w", is a value equal to stencil thickness or a web to stencil ratio of 1. Expanding this spacing will discourage bridging defects. Achieving success on narrower spacing may be limited by stencil manufacture capability, aperture to wafer gasket compliancy, paste rheology, and printer machine stencil cleaner performance.

SETUP & TEST PROCEDURE
In our experiment we aimed to achieve a final reflowed bump height of 90μm per pad. Six different aperture sizes were designed, each with a unique width and all of equal lengths. Aperture geometry was oblong, resembling a rectangle with severely rounded ends, which has been documented to produce quite good results historically. [5] Four levels of spacing between adjacent apertures were tested, two of which are compliant with the design rule to match or exceed the stencil thickness dimension and two that are violating this. Figure 3 defines the apertures used.

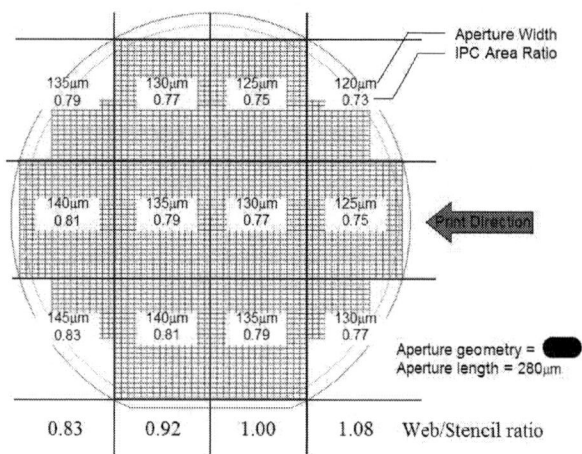

Figure 3: Aperture design.

Five different squeegee types were tested, each one allocated to a unique wafer. The difference between squeegee blades was material and shape, consisting of one metal blade, two hard polyurethane, and two soft polyurethane squeegee blades. One squeegee of each polyurethane hardness was a diamond shape mounted vertically to the holder, while all others were straight profile and mounted in trailing edge configuration. All squeegee blades measured the same length and blade tip angles measured 60° relative to the stencil.

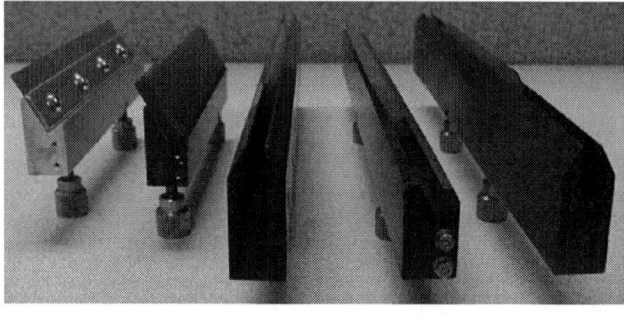

Figure 4: Squeegee types.

Setting up the automatic stencil printing machine consisted of test printing with each squeegee on blank wafers to determine the lowest pressure level applied that would result in wiping the stencil surface clean. The print speed was held constant for all squeegees. Table 1 lists the printing machine parameters used across all squeegee types.

Wafer ID	Length	Angle	Material	Shape	Hardness	Speed	Pressure
1	170 mm	60°	Metal	Trailing Edge	N/A	25 mm/sec	4.0 kg
2	170 mm	60°	Poly	Diamond	67 D	25 mm/sec	6.0 kg
3	170 mm	60°	Poly	Trailing Edge	70 D	25 mm/sec	3.6 kg
4	170 mm	60°	Poly	Trailing Edge	92 D	25 mm/sec	9.0 kg
5	170 mm	60°	Poly	Diamond	92 D	25 mm/sec	4.0 kg

Table 1: Squeegee and process parameter list.

In order to more practically simulate production conditions, two prints were performed on blank wafers before the proper patterned wafer was loaded and bumped. The stencil was not cleaned between any of the prints, however, it was in a known clean condition before running the first dummy print for each squeegee. The solder paste used was a commercially available Sn95.5/Ag4.0/Cu0.5 alloy and Type-6 powder size material. The prints were performed with the wafer placed in direct contact with the stencil bottom side (i.e. no snapoff or print gap). The printed wafer was then placed into an 8-zone forced hot air convection furnace to reflow the solder. Figure 5 contains details of the reflow profile. The furnace was equipped to supply the heated zones with nitrogen and oxygen levels were measured less than 50ppm. Flux residues were removed by dedicated cleaning equipment. Finally the bumps on the wafers were inspected and measured with a specialized optical metrology system, reporting yield and bump height.

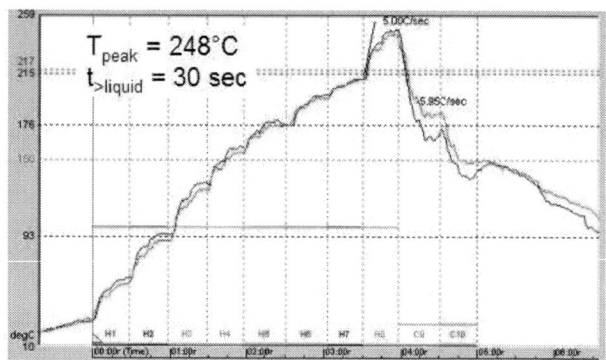

Figure 5: Pb-free reflow profile.

BUMPING RESULTS

All solder bump height measurement data was converted to solder paste transfer efficiency values in order to emphasize printing performance. Using the truncated sphere formula from Figure 1, the solder volume per bump is calculated. Doubling this volume will compute the quantity of solder paste deposited per pad through the stencil. Finally the solder paste deposit volume is divided into the aperture capacity to determine paste transfer efficiency. So transfer efficiency compares the print deposit solder volume to the volume within the boundaries of the aperture. 100% paste transfer efficiency means that the aperture is filled completely with solder paste during the print stroke and all of that material is completely transferred out of the aperture and onto the wafer bond pad when the wafer separates from the stencil.

There are two events that occur during printing that will diminish transfer efficiency, filling and release processes. Transfer efficiency loss can occur if the aperture is not completely filled with solder by the printing squeegee and will also result when paste sticks to the aperture walls during release. It is proposed that the aperture filling process is highly influenced by the printer hardware and process parameters, while the release process is dominated by stencil effects. Nevertheless, transfer efficiency does not identify the contribution from filling or release on performance, but represents a combined effect.

Figure 6 shows the difference between each wafer in average transfer efficiency scaled against the left axis. The metal trailing edge squeegee obtained the highest print transfer efficiency followed by the two soft polyurethane squeegees. The hard polyurethane squeegees showed poorest transfer efficiency trends. From previous experience, it is known that a sharper squeegee blade tip will achieve very good print results and this is the argument used to explain transfer efficiency differences between the different squeegee types.

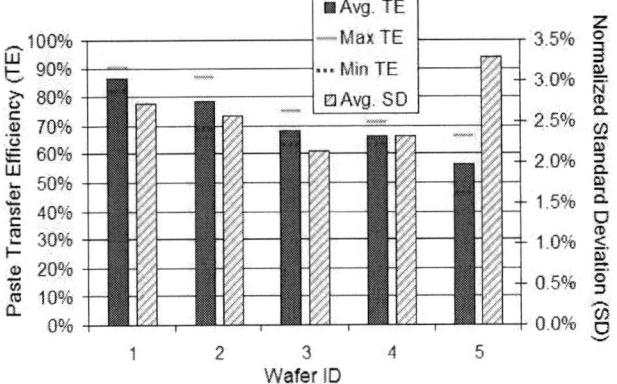

Figure 6: Print performance per wafer.

The difference in transfer efficiency performance between the diamond polyurethane squeegee types is unmistakable. A unique feature on both these squeegees the chamfer at the base of the angled tip. Figure 7 shows this feature on both and also models the predicted behavior of the tip during the print stroke. The hard polyurethane squeegee is quite rigid and maintains its original form during the print, while it is believed the soft polyurethane diamond tip is more sensitive to distortion under applied pressure. Such blade tip flexibility could serve as a benefit to generate a sharp line of uninterrupted contact against the stencil. Alternatively, the inflexible hard diamond tip squeegee maintains a larger surface area of contact with the stencil. This situation perhaps represents a dull blade scenario and leads to less effective distribution of solder paste into the apertures.

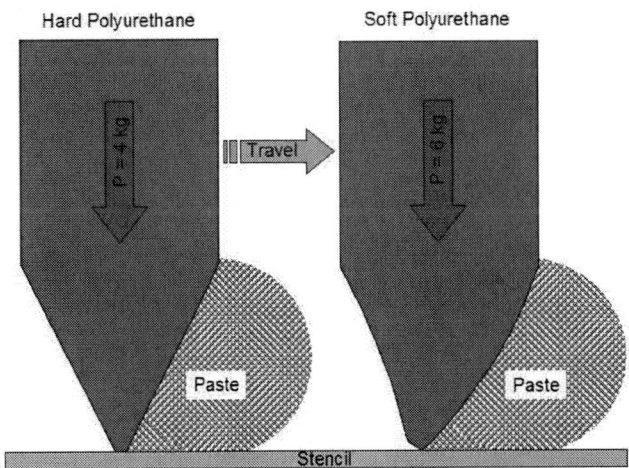

Figure 7: Diamond squeegees and tip flexibility.

Now addressing the difference between the two polyurethane trailing edge squeegees, the trend is consistent with the diamond squeegee observations as the softer blade gives higher transfer efficiency. However the difference in performance between the two polyurethane trailing edge blades is not nearly as significant as for the diamond squeegee comparison. It is interesting to note the unusually high level of print pressure applied to the hard polyurethane trailing edge blade used to print on wafer 4. At lower print pressure levels, the solder paste tended to escape under the blade leaving a thin but obvious film of residue coating the top of the stencil surface. Although feeling sharp to touch, this stiff blade tip may suffer a surface roughness condition that prevents homogeneous contact against the stencil under low pressure. High pressure is required to force the squeegee into intimate contact with the stencil to wipe the paste cleanly. Failure to wipe the paste clean will result in quite inconsistent transfer efficiency. High squeegee pressure is not considered an ideal circumstance to print fine pitch wafers, as this may cause material scavenging from the aperture to lessen paste transfer opportunity. It may also promote flux bleed under the stencil leading to increased occurrence of paste bridging. It also may be the case that the print speed used is just too fast for this squeegee blade type to cope with, and that print pressure could be relaxed at lower print speed.

It is not only desirable to use a squeegee blade that is capable of delivering high transfer efficiency, but also producing consistently well-formed deposits. Print uniformity is also important to achieve good bump coplanarity. Unfortunately this experiment did not show the wafer possessing highest transfer efficiency (i.e. wafer 1) to have lowest standard deviation. Referring to Figure 6 referencing the right side axis, the standard deviation of bump size is expressed as a percentage of its average. Wafer 3 is charted to have the most uniform deposits, which was printed with the soft polyurethane trailing edge blade. The pressure setting on this squeegee was the lowest of all wafers printed, suggesting the blade had very good clean contact with the stencil. Perhaps the reason this squeegee was not the best in transfer efficiency is due to its elasticity and tendency to scoop slightly into the apertures during the print and pilfer material away. However, the packing of solder alloy into the apertures is highly repeatable. The poorest performing squeegee for print uniformity is the hard polyurethane diamond type, matching its same low ranking for transfer efficiency. Figure 8 shows example photographs of high and low uniformity paste deposits.

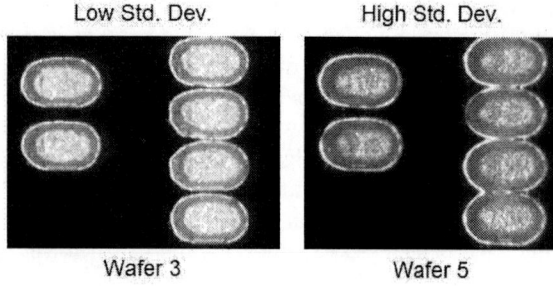

Low Std. Dev.   High Std. Dev.

Wafer 3   Wafer 5

Figure 8: Print deposit uniformity differences.

The stencil thickness and aperture sizes were designed on this stencil to exceed the IPC recommend minimum area ratio of 0.66 in order to obtain high solder paste transfer efficiency. [5] Studies have shown good correlation of transfer efficiency on aperture area ratio, with results of one such study shown as the "model" data in Figure 9. [3] It is interesting to see how much influence the squeegee and process have on paste transfer efficiency performance when the data is filtered by wafer and area ratio. Wafer 1 and several of Wafer 2 data points fit the model trend, while the remaining wafers tend to deviate as a result of influences from squeegee and process. Note that the "model" trend predicts fairly constant transfer efficiency across the range of area ratios considered for this test. Within this region of the area ratio curve solder paste tends to strongly favor adhesion to the wafer bond pad rather than the aperture side walls during its cycling through the aperture.

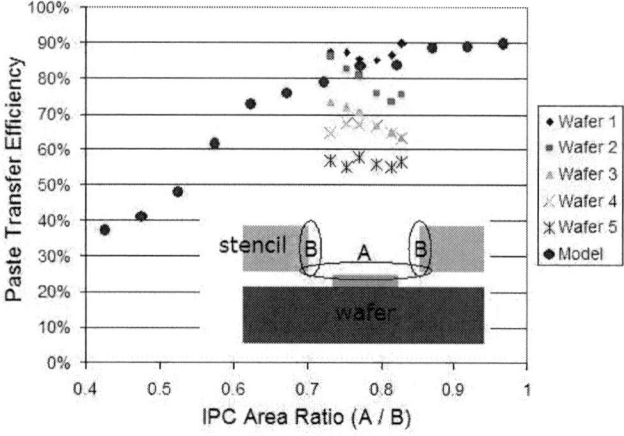

Figure 9: Aperture and process influence on bump size.

As stated earlier, a bump size target of 90μm was desired and several aperture sizes were considered to achieve this. Actual bump sizes produced from this work are plotted in Figure 10, sorted by wafer and aperture size. Each aperture size responded to the squeegee and process changes quite similarly (wafer 1 best, wafer 5 worst), in agreement with the transfer efficiency results in Figure 6. The two largest aperture sizes considered for wafer 1 accomplished the bump height goal.

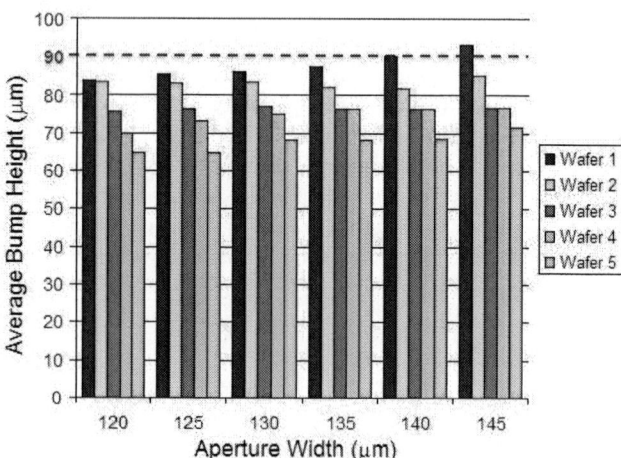

Figure 10: Average bump height data.

Finally, from a yield perspective, all wafers excluding one averaged above 99%. Wafer 2 produced a bump yield of 93%, attributed mainly to a slight print misregistration. Yield was determined by counting all die that did not show shorted or missing bumps and dividing that into the total number of die. The success of these results is significant considering that half of the wafer was printed with apertures separated at a distance less than the stencil thickness dimension. Such aggressive stencil designs are not generally recommended and good results are believed only achievable using superior quality stencils and solder pastes.

CONCLUSIONS
It is proposed the highest potential paste transfer efficiency should result for the aperture area ratio considered if:

- The squeegee blade tip is sharp during the print.
- Solder paste rolls in front of the squeegee.
- The stencil is wiped consistently clean.
- The preceding are achieved at low print pressure.

This study determined the metal squeegee to produce the best transfer efficiency in support of these claims, while also delivering satisfactory control of print uniformity. This blade is considered to be best at sustaining its original sharpness in use, and is not as sensitive to dulling compared to polyurethane types. The downside of the metal squeegee is vulnerability to damage in handling. Metal squeegees will not tolerate a drop test, nor will the stencil if it encounters a collision from a falling blade. It can also cut your hand if forcefully handled.

Area ratio is introduced as a useful tool to predict solder paste transfer efficiency. However, achieving expected performance is also contingent on selection of squeegee and process parameters. Almost certainly solder paste rheology will also contribute highly to process success.

The scope of this study was simplified to consider a common print speed and dial in the best print pressure for each squeegee blade type. However the ranking of these squeegees may also be skewed to favor those that perform particularly well at 25mm/sec. Future test strategy will consider using an equal pressure level across all squeegees. Print speed will be selected at the highest level that accomplishes wiping the solder bead cleanly across the stencil at an applied pressure that complies with 0.5-0.6 kg per inch of squeegee blade length.

AKNOWLEDGEMENTS
The authors wish to thank Ashok Viswanathan from Binghamton University for his process assistance in parts of this work.

REFERENCES
[1] Coleman, W.E. and Burgess, M.R., "Choosing a Stencil," Surface Mount Technology Magazine, 2006, 20, 7, pp 14-17.

[2] Schake, J.D. and Srihari, K., "Stencil Printing on 200μm & 250μm Pitch Pads in a 3-Row Perimeter Array," Technical Report - CSP/DCA Consortium, Universal Instruments Corporation, December 1998.

[3] Schake, J.D. and Srihari, K., "Solder Paste Release From Real and Scaled-Up Stencil Apertures," Technical Report - CSP/DCA Consortium, Universal Instruments Corporation, November 1998.

[4] Schake, J.D., "Investigation of Bump Distributions from the Stencil Printing Wafer Bumping Process," Master's Thesis, Binghamton University, Binghamton, N.Y., December 1998.

[5] IPC-7525 "Stencil Design Guidelines"

FABRICATION OF TAPERED THROUGH-VIAS ON (100) SILICON FOR WAFER-LEVEL PACKAGING

Huang Shuang Wu and Chia Yong Poo
Micron Semiconductor Asia Pte Ltd.
Singapore
markhuangsw@micron.com and chiayp@micron.com

ABSTRACT

Increasing demand for high-performance, high-density, low-power, smaller-sized portable electronic gadgets is driving the development of subsystem integration onto a single integrated circuit (IC) component. Methods of integration can be system-on-a-chip (SoC), system-in-a-package (SiP), and wafer-level packaging (WLP); currently, SiP is the most preferred method because only tested-good subsystems are integrated, thus producing a higher yield. Die- or wafer-level integration of these subsystems using through-silicon vias is one of the best methods to form a small-footprint, high-performance SiP.

This paper presents a silicon etching method to create through-wafer interconnections for die- or wafer-level three-dimensional (3D) integration and stackable IC packages. Focusing on the development of chemical wet-etch methods to fabricate tapered through-vias on (100) silicon wafers, the team evaluated IC surface protection during wet-etch processing. The affecting factors of through-via integrity, such as surface smoothness, contamination, and mask design and misalignment were also investigated. In addition, the team studied the effect of process temperature and etchant concentration on etch rates and characterized the profile and quality of through-vias. The as-formed through-vias are coated with dielectrics and metals, such as Ti/W, as well as Cu layers so as to provide an electrical path for further IC stacking.

Keywords: through-via, IC protection, anisotropic etch, wafer-level packaging

INTRODUCTION

Recent research and industrial pressure call for higher performance and density, lower power, and smaller size in portable electronic gadgets with very high imaging resolution and enhanced communication and computing functions, such as PDAs, MP3 players, and cellular phones. This demand has fueled the development of subsystem integrations onto a single IC component, such as SoC, 3D SiP, and WLP[1-3].

In the meantime, intensive marketing competition represents a major challenge for low-cost IC packages. It seems that SoC packaging has become less competitive due to IC yield concerns. Currently, SiP is the most preferred method because only tested-good subsystems are integrated, thus producing a higher yield. Die- or wafer-level integration of these subsystems using through-silicon vias is one of the best methods to form a small-footprint, high-performance SiP.

The main methods to form through-silicon vias can be categorized into dry etch and wet etch. The former can be used for fine-pitch and high–aspect-ratio via formation, while the latter is conducive to cost reduction through batch processing and easier to the subsequent metallization due to tapered via structure and smooth via surface ((111) crystalline plane), as seen in Figure 1.

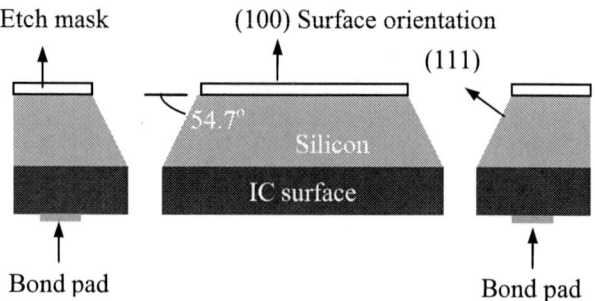

Figure 1. Illustration of pyramidal pits etched into (100) silicon wafer from backside using anisotropic potassium hydroxide (KOH) etchant bounded by (111) crystal orientation.

This paper presents a silicon etching method to create through-wafer interconnections for die- or wafer-level 3D integration and stackable IC packages. Focusing on the development of chemical wet-etch methods to fabricate tapered through-vias on (100) silicon wafers, the team evaluated IC surface protection during wet-etch processing. The affecting factors of through-via integrity, such as surface smoothness, contamination, and mask design and misalignment were also investigated. It should be noted that the etch mask was made using plasma-enhanced chemical vapor deposition (PECVD) technique instead of the conventional low-pressure CVD method. An extra metal mask was also applied so as to reduce pinholes on the silicon surface during the KOH etch process. In addition, the team studied the effect of process temperature and etchant concentration on etch rates and characterized the profile and quality of through-vias. The as-formed through-vias are coated with dielectrics and metals so as to provide an electrical path for further 3D SiP and WLP.

EXPERIMENT
Materials and Equipment
The main materials used include:
- 8-inch (100) silicon wafers of full thickness as the test vehicle
- Silicon oxide and silicon nitride by PECVD as wet-etch masks
- Ti/TiW/Cu thin metal coating by physical vapor deposition (PVD) as etch mask reinforcement
- Photoresist for pattern formation of etch masks and protection of active wafer during chemical mechanical planarization (CMP)
- KOH pellets for preparation of etching solution

The core equipment for this study comprises CMP for wafer backside polishing, PECVD for SiO_2/Si_3N_4 deposition, reflectometry system for SiO_2/Si_3N_4 thickness measurement, PVD system for Ti/TiW/Cu deposition, and aligner for double-sided patterning.

Process Overview

Figure 2 shows the main process flow of tapered silicon through-via fabrication and metallization. First, a wafer of full thickness was back-ground to a thickness of 250µm. The rough surface on the wafer was further polished through CMP with a precoated photoresist thickness of 10µm to protect the wafer's active side. The polished wafer was rinsed with DI water and dried for the subsequent PECVD deposition. The 1µm-thick dielectric films of SiO_2 and Si_3N_4 each shrouded the whole wafer at KOH-etch mask. An additional metal film of 500Å Ti, 1000Å TiW, and 2µm Cu was immediately coated onto both sides of the wafer to fortify the dielectric etch mask. The as-formed etch mask shown in Figure 2(b) was also used to protect the IC wafer from corrosion by strong alkaline KOH chemicals during the etch process at elevated temperature.

Next, the wafer backside was patterned to expose the silicon to be etched by running a standard photo process, as shown in Figure 2(c). The 10µm-thick positive photoresist was applied to dry-etch the etch mask. The front-to-back alignment procedure was carried out in a commercial double-sided aligner. Alignment marks were first created on the front side of a wafer in a manner similar to that used with a single-sided mask aligner. In creating alignment marks on the other side of the wafer, alignment marks on a plastic mask were stored electronically before loading the wafer under the mask. The alignment marks initially created on the wafer were viewed through a microscope situated under a wafer stage. The wafer was then aligned to the stored image. Once this was done, exposure was performed to complete the front-to-back alignment.

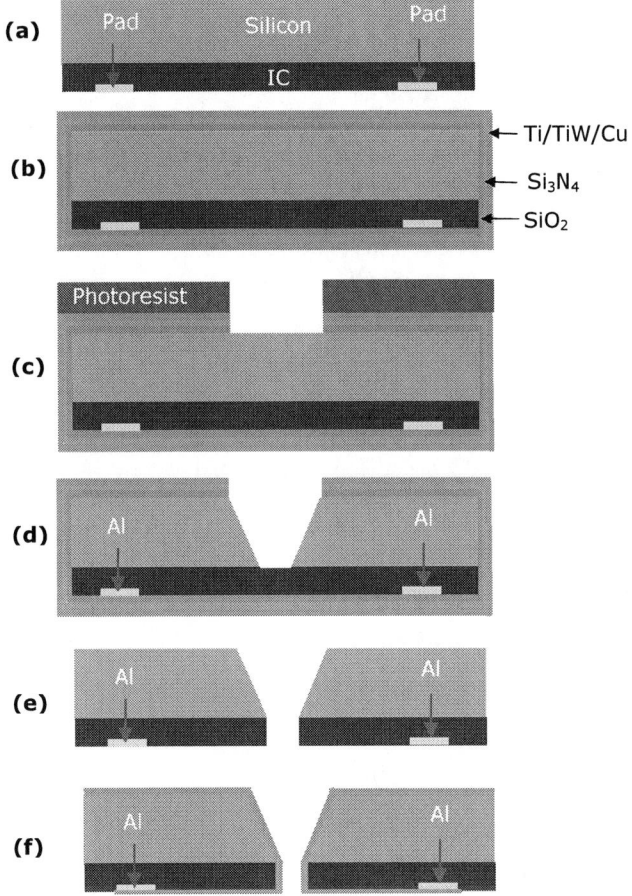

Figure 2. Tapered through-via fabrication and metallization sequence: (a) thinned and CMP-treated IC wafer (b) multi-layer etch mask deposition (c) etch mask opening (d) anisotropic KOH etch (e) reactive ion etch (f) passivation and metallization

The outside metal layers were then removed using concentrated H_2SO_4 and H_2O_2-containing acidic mixtures. The dielectric etch mask (SiO_2 and Si_3N_4) was plasma-etched using the mixed gases of CF_4, O_2, and Ar at a flow rate of 40:4:100 sccm, respectively. The plasma chamber pressure was controled at 80 mTorr and RF power set at 120W. About a 2-hour etching duration was required to remove the SiO_2 and Si_3N_4 etch mask. The photoresist was then stripped with acetone, and the wafer was cleaned with hot DI water. Next, the wafer was dipped into a KOH etch tank under different conditions (temperature, KOH concentration, and duration) for fabrication of the through-via on a silicon wafer.

After KOH etching, the wafer was treated with the afore-mentioned chemical to strip the external metal layer as etch mask reinforcement and further subjected to the reactive ion etch (RIE) to remove dual-layer etch mask (SiO_2 and Si_3N_4) and silicon oxide-based materials on the IC side, as shown in Figure 3. The as-formed through-vias were coated with silicon oxide as dielectric using PECVD and then coated with a metallic layer using PVD to realize the electrical interconnection for further 3D or WLP.

RESULTS AND DISCUSSION
Chemistry of KOH Etch

KOH was used to etch silicon anisotropically. The etching process is inherently a redox reaction involving oxidation of silicon and reduction of hydrogen in water[4-5]. Specifically, silicon atoms at the surface react with hydroxyl ions. The silicon loses four electrons to be oxidized.

$$Si + 2\,OH^- \rightarrow Si(OH)_2^{2+} + 4\,e$$

Simultaneously, the hydrogen in water gains the electrons from silicon and is reduced.

$$4\,H_2O + 4\,e \rightarrow 4\,OH^- + 2\,H_2\uparrow$$

The complex ion, $Si(OH)_2^{2+}$, further reacts with hydroxyl ions to form a soluble silicon complex and water.

$$Si(OH)_2^{2+} + 4\,OH^- \rightarrow SiO_2(OH)_2^{2-} + 2\,H_2O$$

Thus, the overall redox reaction is:

$$Si + 2\,OH^- + 2\,H_2O \rightarrow SiO_2(OH)_2^{2-} + 2\,H_2\uparrow$$

It is worth mentioning that the rate of H_2 generation is more dependent on etching temperature than KOH concentration, which was found to have some impact on the profile of silicon vias such as sidewall smoothness and undercut.

Protection of the IC Surface

It is a challenging task to protect the IC surface when the KOH etchant is applied to through-wafer fabrication. Without proper protection, the metal pads will be etched away during the silicon wet-etch process. First, polyimide films and dicing tapes, parylene coating formed by CVD, and spin-coated benzocyclobutene (BCB) coating were evaluated for their resistance to KOH. The use of these organic materials proved futile because of adhesion issues with the wafer. Shortly after being immersed into the heated, 30 percent KOH tank, the tapes and coatings started to lift off. We then focused on the inorganic and metallic protective coatings shown in Table 1.

Table 1: Evaluation of materials for IC surface protection in KOH solution

Material Types	Thickness (μm)	Deposition Methods	Resistance to KOH @ 75°C
SiO_2	1.0	PECVD	Pinhole
Si_3N_4	1.0	PECVD	Blistering
Cu	2.0	PVD	Peeling
Ti/TiW	0.5/1.0	PVD	Poor
SiO_2/Si_3N_4 /Ti/TiW /Cu	1.0/1.0 /0.5/1.0 /2.0	PECVD + PVD	Good

It can be seen from Table 1 that silicon oxide and silicon nitride deposited by PECVD caused either pinhole or blistering defects during the KOH-etch process. Due to its poor adhesion to silicon, copper peeling was observed in the KOH solution at 75°C for a very short time when it was used as the protective layer of the IC surface. Although a Ti/TiW layer is extremely adhesive to silicon, it was still poor in KOH resistance because it was found to be vulnerable to attack from KOH at elevated temperatures. However, the combination of all these layers as the IC protection layer was demostrated to be good enough to resist against KOH attack in an anisotropic silicon etch process. The combined layers can also be used as a silicon KOH etch mask.

Etch Mask Considerations

Theoretically, due to the crystal lattice structure of silicon, the final etch pattern will take up an oblong shape where the size is determined by the edge at maximum length aligning to the (100) and (111) atomic planes. In other words, the required through-via dimension can be achieved using different mask patterns, as shown in Figure 3.

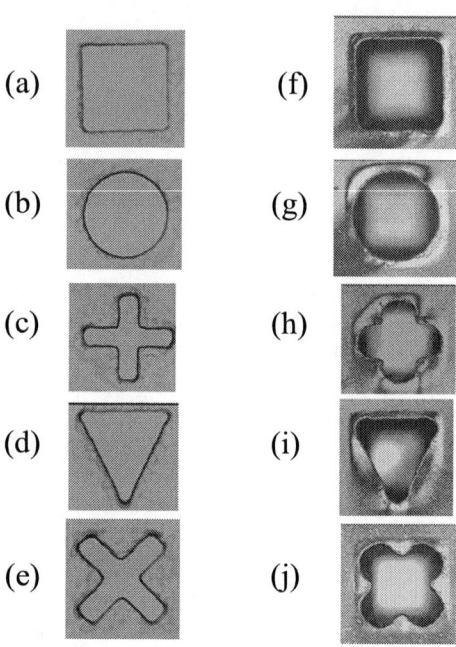

Figure 3. Different mask design for through-via formation showing the mask pattern before etch (a–e) and image of top via after etch (f–j). The etch mask used for all the patterns is SiO_2 and Si_3N_4.

However, the etch mask materials should not only have a high chemical resistance against the etchant used—the KOH etchant in our experiment, for example—but also possess good adhesion to bulk silicon in order to prevent the attack by the hydrogen generated during the KOH etch process. The H_2 gas may weaken the edge of the oxide mask, which gives rise to the delamination between SiO_2 and bulk silicon. In the worst case, this phenomenum can cause the etch mask to break off and make the through-via size out of control. It can be seen from Figure 3(f–h) that the undercut occurred to a certain degree in our experiemnt. Based on the overall redox equation, the rate of H_2 formation is much faster than that of silicon etching. Without proper control of silicon etch rate the aggregation of small H_2 bubbles can lead to the evolution of big H_2 bubbles, especially towards the bottom of tapered silicon vias. The attack from the entraped H_2 bubbles together with potential KOH undercut may further weaken the adhesion of SiO_2/Si_3N_4 etch mask to the silicon surface. The etch rate can be controlled very well by monitoring the process temperature and the concentration of KOH solution (to be discussed later).

In this experiment, we used the additional metal layers (Ti/TiW/Cu) as an etch mask reinforcement and thus efficiently control the through-via dimension. In the meantime, this additional metal layer tremendously improves the mask robustness in wafer handling to reduce surface defects such as scratches and dents.

Factors Affecting Etch Integration

KOH etchant also attacks SiO_2 at a relatively low etch rate, which will cause undercut and thus result in an enlarged through-via dimension. The obvious undercut occurred along the uneven silicon surface in Figure 4(a) due to poor adhesion of SiO_2 to bulk silicon. With CMP treatment, the undercut can be well restricted, as shown in Figure 4(b).

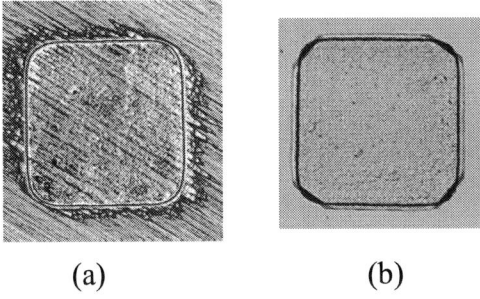

(a)   (b)

Figure 4. Comparison of top via profile on silicon showing different undercut from the samples of a) background silicon and b) CMP-treated silicon. The etch mask used is SiO_2 and Si_3N_4, and the pictures were taken after 30 min KOH (30 wt%) etch at 70°C.

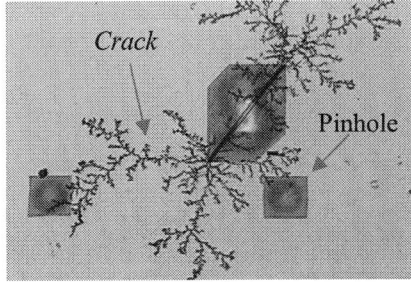

Figure 5. Photograph of surface defects on SiO_2/Si_3N_4 etch mask after KOH etch after through-via formation.

The experimental results showed that the additional metal layer (Ti/TiW/Cu) can be conducive to restricting the undercut. Another advantage of using this metal layer is to reduce surface defects. Figure 5 illustrates the defects, such as pinholes and micro-cracks, on an etch mask of SiO_2/Si_3N_4 after KOH etch at 80°C for 5 hours. The cracks initiated at room temperature from the corners of the

pinholes due to high stress of the silicon nitride mask. The pinholes are believed to be associated with the defects from PECVD silicon oxide and silicon nitride deposited at low temperature (280°C). These defects can be avoided by the incorporation of the above-mentioned metal layer.

Because KOH is a strong anisotropic etchant, etch-mask misalignment to the (100) silicon crystal lattice will give rise to larger through-via size than what is required. The extent to which the feature size grows depends on the amount of misalignment to the crystal lattice. Figure 6 shows the etching result of 45° mask misalignment to (100) crystal orientation. It can be seen from Figure 6(b) that two neighboring vias actually link together under transparent SiO_2/Si_3N_4 etch mask after 5.5 hours of KOH etch at 70°C. Therefore, the control of through-via dimensioning is very critical to some applications like fine-pitch interconnection. An enlarged through-via will cause an electrical short with its adjacent one after through-via metallization.

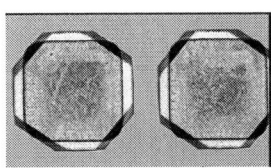 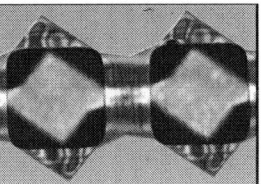

(a) Etched in 30 wt% KOH at 70°C for 1.5 hrs

(b) Etched in 30 wt% KOH at 70°C for 5.5 hrs

Figure 6. Photographs showing the enlarged via dimension due to etch-mask misalignment to (100) crystal orientation.

Control of Etch Rates

The silicon etch rate using KOH depends upon the process temperature and the concentration of the KOH solution. It can be seen from Figure 7 that the etch rate of (100) silicon increases as the process temperature rises and the concentration of KOH solution decreases. However, experimentation has found that the solutions of 20 percent and 25 percent KOH lead to the formation of rough via sidewalls, as shown in Figure 8. The rough etching may be due to the formation of insoluble precipitates. On the other hand, the high process temperature (90°C) also results in even worse rough etching than the result shown in Figure 8. Bubbling nitrogen during KOH etching improves the sidewall smoothness, possibly by minimizing the adherence of hydrogen to the sidewall.

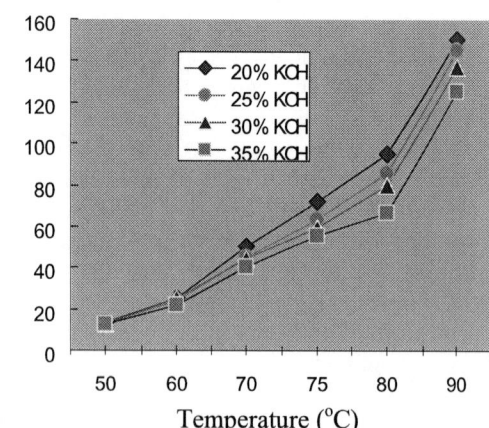

Figure 7. Effect of process temperature and the concentration of KOH solution on (100) silicon etch rates.

Figure 8. SEM micrograph showing the half-cut, tapered through-via with rough sidewall etched in 25% KOH solution at 80°C.

a) Patterns of through-via etched from wafer backside

b) Close-up of a tapered through-via

Figure 9. An example of an application using tapered through-via technology in 3D SiP and WLP.

Application of Tapered Through-vias

The as-formed, tapered through-via shown in Figure 8 has an outstanding advantage over a vertical through-via. The former is more suitable for dielectric deposition and metallization, while the latter is more feasible in fine-pitch or high-density applications such as WLP and 3D packaging. Figure 9 shows an example of a tapered through-via etched from wafer backside that can be used to provide the interconnection between the active surface and backside.

This approach is becoming more popular in 3D SiP and WLP[6-7].

CONCLUSION

The fabrication of through-vias on a (100) silicon wafer using KOH etchant was carried out for through-wafer interconnection with the objective of reducing the manufacturing cost through batch process. The factors that affect the through-via integrity include silicon surface roughness, surface contamination and etch mask misalignment. Rough silicon surfaces resulted in poor etch mask deposition. Surface contamination caused pinholes of the etched surface. Etch mask misalignment gave rise to undesired via shapes and dimensions The multi-layer etch mask formed at rather low temperatures was developed to protect the active wafer surface and to control the quality of silicon through-vias. The process temperature and etchant concentration not only affected silicon etch rates but also had an impact on the via-sidewall smoothness. This work provides an alternative approach for through-silicon wafer interconnects in 3D SiP and WLP.

ACKNOWLEDGMENT

The authors would like to thank Dr. B. K. Tan of SIMTECH in Singapore for his cooperation. Special thanks go to Mr. Vicknesh and Dr. Raman from IMRE for their contribution to the deposition of oxide and nitride mask.

REFERENCES

[1] Vaidyanathan Kripesh, Seung Wook Yoon, V.P. Ganesh, Navas Khan, Mihai D. Rotaru, Wang Fang and Mahadevan K.Iyer, "Three-Dimensional System-in-Package using Stacked Silicon Platform Technology," IEEE Transactions on Advanced Packaging, vol.28, No.3, August 2005

[2] Eugene M. Chow, Venkataraman Chandrasekaran, Aaron Patridge, Toshikazu Nishida, Mark Sheplak, Calvin F.Quate, and Thomas W.Kenny, "Process Compatible Polysilicon-based Electrical Through-wafer Interconncts in Silicon Substrates," Journal of Microelectromechanical Systems, Vol.11, No.6, December 2002

[3] Ronald Hon, S.W. Ricky Lee, Shawn X. Zhang, C.K. Wong, "Multi-Stack Flip Chip 3D Packaging with Copper Plated Through-Silicon Vertical Interconnection," Proceedings of 2005 Electronics Packaging Technology Conference, 7–9 Dec. Singapore, vol.2 pp. 384–389.

[4] H. Seidel, "The mechanism of anistotropic silicon etching and its relevance for micromachining," Proc. Transducers '87, Rec. 4th Int. Conf. Solid-State Sensors and Actuators, Tokyo, Japan, June 2–5, 1987, pp. 120–125.

[5] H. Seidel, L. Csepregi, A. Heuberger, and H. Baumgartel, "Anisotropic etching of crystalline silicon

in alkaline solutions I: Orientation dependence and behavior of passivation layers," J. Electrochem. Soc., vol.137, no.11, pp. 3612–3626, Nov.1990.

[6] S.W. Ricky Lee, Ronald Hon, Shawn X.D. Zhang, C.K. Wong, "3D Stacked Flip Chip 3D Packaging with Through-Silicon Vias and Copper Plating or Conductive adhesive Filling," Proceedings of 55th Electronic Components and Technology Conference, Florida, May 31–June 3, 2005 pp. 759–801.

[7] N.T. Nguyen, E. Boellaard, N.P. Pham, V.G. Kutchoukov, G. Cracium, and P.M. Sarro, "Through-wafer Copper Electroplating for three-dimensional Interconnects," Journal of Micromechanics and Microengineering, 2002 (12), pp. 395–399.

LITHOGRAPHY-GRADE CONTROLLED EXPANSION SUBSTRATES FOR WAFER LEVEL PACKAGING

Greg Rudd and Bob Cronk
SMI (Spectra-Mat, Inc.)
Watsonville, CA, USA

ABSTRACT

Wafer level packaging of opto and opto-electronic devices can result in significant cost savings. A substrate with good thermal conductivity is required where devices dissipate significant power. Additionally, using a substrate with a coefficient of thermal expansion (CTE) similar to the device can increase reliability when temperature-induced strain is a problem. Tungsten-copper metal-matrix composite (W/Cu) has long been used for package components and heat sinks. It has good thermal conductivity, but unlike most alloys, it can be made with any of a range of CTE's to match various other device and substrate materials. W/Cu has some ductility to relieve strain and is readily machined, plated, and brazed. Pure tungsten also has good conductivity and low CTE (~4.5). In our paper, we present data on W/Cu and pure W substrates up to 150mm diameter produced to specifications suitable for wafer level packaging. We describe the properties of the material and show the result of the measurement of flatness/bow, Total Thickness Variation (TTV), and surface finish. We also discuss how processes impact these parameters.

Key words: Wafer-level packaging, substrate, thermal expansion.

INTRODUCTION

Wafer level packaging of optical and opto-electronic devices can result in significant cost savings. A substrate with good thermal conductivity is required where devices dissipate significant power. Additionally, using a substrate with a coefficient of thermal expansion (CTE) similar to the device can increase reliability when temperature-induced strain is a problem. Tungsten-copper metal-matrix composite (W/Cu) has long been used for package components and heat sinks. It has good thermal conductivity (similar to that of aluminum), but unlike aluminum or most alloys, it can be made with any of a range of CTE's to match various other device and substrate materials. W/Cu has some ductility to relieve strain and is readily machined, plated, and brazed. Mo/Cu has similar properties, albeit at a slightly lower thermal conductivity.

In our paper, we present data on W/Cu wafers up to 150mm diameter produced to specifications suitable for wafer level packaging. We describe the properties of the material and show the result of the measurement of Bow, Warp, Total Thickness Variation (TTV), and surface finish. We also discuss how processes impact these parameters.

PROPERTIES OF TUNGSTEN-COPPER

The engineering properties of interest for a semiconductor substrate made of tungsten-copper are low, tailorable coefficient of thermal expansion combined with high thermal conductivity. The CTE of W/Cu can be varied within a range of values from 6.4 – 8.1 ppm/C° (Figure 1).

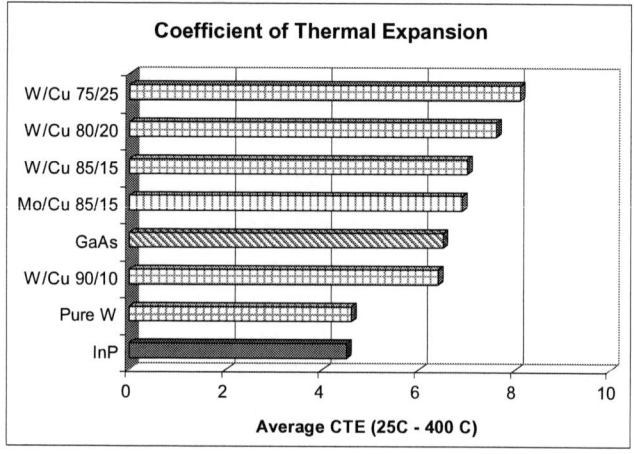

Figure 1. CTE of various substrates and semiconductors.

The thermal conductivity varies with composition also. It is better than typical aluminum alloys, but about half that of high purity copper (Figure 2).

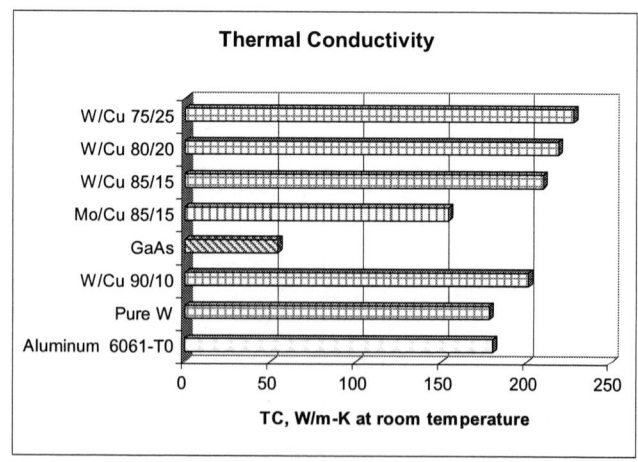

Figure 2. Thermal conductivity of various substrates compared to GaAs and aluminum as reference points.

WAFER LEVEL INTEGRATION OF TUNGSTEN COPPER

Miniaturization of electronic devices has led to the development of Chip Scale Packages (CSP) to meet the requirement of smaller circuit board area availability. Moving from a one-at-a-time chip packaging process to a wafer level packaging process reduces costs and can make delicate semiconductor wafer handling more robust resulting in improved yields.

A typical approach for packaging high power laser diodes that require CTE matched substrates is to solder attach a small thin submount made of tungsten-copper to the laser die. This is typically performed by metallizing the bottom of the die and the entire submount. A thin solder preform, on the order of 25 microns thick, is placed on top of the submount, the laser die is then placed on top of the solder preform and the assembly is heated to reflow the solder. The process can be quite costly due to the small size of all components involved, the exacting placement tolerances required and the slow nature of one-at-a-time processing.

If the packaging process can be performed at the wafer level, the level of assembly automation can potentially be increased and yields improved, reducing cost. In addition, functionality of the package can be improved by allowing more complex feedthrough wiring.[1]

CRITICAL DIMENSIONAL CHARACTERISTICS
Substrates used for wafer level packaging must meet the critical dimensional specifications required for semiconductor processing. Similar to semiconductor wafers, metal substrates must be extremely flat, have a consistent thickness across the entire substrate, and have a highly polished surface. Our process development has strived to meet the demands of lithography grade wafers. Lithography grade wafers are used for photolithography applications requiring very tightly controlled metrology. Photolithography yields are very closely related to focal plane variations in the semiconductor as well as well as surface defects and micro-topography.

The main shape characteristics of importance for our tungsten-copper substrates are:

- Bow

- Warp

- Total Thickness Variation (TTV)

- Surface Roughness

BOW. From *ASTM F534 3.1.2:* "The deviation of the center point of the *median surface* of a free, unclamped wafer from the median surface reference plane established by three points equally spaced on a circle with a diameter a specified amount less than the nominal diameter of the wafer." The median surface: is the locus of points in the wafer equidistant between the front and back surfaces.

Since bow is measured at the center point of the wafer only, a three (3) point reference plane about the edge of the wafer

is calculated. The value of bow is then calculated by measuring the location of the median surface at the center of the wafer and determining it's distance from the reference plane. Bow can be a positive or negative number. Positive denotes the center point of the median surface is above the three point reference plane. Negative denotes the center point of the median surface is below the three point reference plane.

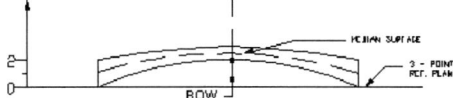

Figure 3. Bow Example.

WARP. From *ASTM F1390:* "The differences between the maximum and minimum distances of the median surface of a free, unclamped wafer from a reference place."

Like bow, warp is a measurement of the differentiation between the median surface of a wafer and a reference plane. Warp, however, uses the entire median surface of the wafer instead of just the position at the center point. By looking at the entire wafer, warp provides a more useful measurement of true wafer shape.

The location of the median surface is calculated exactly as it is for bow and shown in Figure 3. For warp determination, there are two choices for construction of the reference plane. One is the same three point plane around the edge of the wafer. The other is by performing a least squares fit calculation of median surface data acquired during the measurement scan. Warp is then calculated by finding the maximum deviation from the reference plane (RPD_{max}) and the minimum differentiation from the reference plane (RPD_{min}). RPD_{max} is defined as the largest distance above the reference plane and is a positive number. RPD_{min} is the largest distance below the reference plane and is a negative number.

$$\text{Warp} = RPD_{max} - RPD_{min}$$

Figure 4. Warp example.

Figure 4 is an illustration of the warp calculation. In this example RPD_{max} is 1.5 and is shown as the maximum distance of the median surface above the reference plane. RPD_{min} is − 1.5 and is shown as the maximum distance of the median surface below the reference plane. Note that warp is always a positive value. For the case in point:

$$\text{Warp} = 1.5 - (-1.5) = 3$$

The example shown in Figure 4 also illustrates the usefulness of taking both bow and warp readings. The median surface of the wafer shown intersects the reference plane at the wafer center; therefore, bow measurement would be zero. The calculated warp value is more useful in this case as it tells the user the wafer does have shape irregularities.

MANUFACTURING PROCESS

Tungsten's common raw form is a powder in the 1-5 micron size range. Powder metallurgy processing is required to produce useable shapes and components. Pure tungsten is very difficult to form to shape, but tungsten copper is readily machined. Tungsten copper is a true composite material meaning that virtually no alloying takes place between the tungsten particles and copper. The composite is composed of distinct phases of pure copper and pure tungsten.

There are several processes available to fabricate tungsten-copper into shapes. Spectra-Mat uses the following steps:

- Isostatic press 99.9+% W powder

Powder is packed into rubber molds and consolidated under high pressure using water as the pressing force. There is no need for binders, lubricants, or additives, just the pure tungsten powder is consolidated uniformly under constant pressure from all directions. The compacted powder is very porous and has the consistency of chalk, but can be handled and sawed.

- Sinter to desired density at 2000+ °C

The compacted tungsten is then heated in a hydrogen atmosphere furnace to over 2,000 C to solid state weld the tungsten particles together creating a strong porous "preform" with metallic appearance. The degree of sintering is controlled precisely to achieve the desired porosity.

- Infiltrate with OFHC copper

The pores of the porous tungsten preform are filled with pure molten copper by placing solid oxygen-free high purity copper strips in contact with the preform and heating in a hydrogen atmosphere to a temperature above the melting point of copper (1083° C) The molten copper infiltrates the pores by capillary action.

- Rough grind.

Wafers leaving material fabrication are in a crude form with respect to final thickness, final geometry, warp, and surface finish. Rough grinding of the wafers rapidly removes material and is the first step in wafer thinning. The process is designed for rapid material removal, and therefore thickness control and surface finish are crude compared to downstream processing. Typical surface finish achieved at the rough grind process is 64 micro-inch (1.6 μm) R_a. Approximate thickness dimensions are achieved by single sided grinding.

- Surface grind

The surface grinding step that follows rough grind is slower, and more precise with respect to Total Thickness Variation, surface finish, and thickness control. Typical surface finish achieved at this step is 32 micro-inch (0.8 μm) R_a.

- Turn O.D., machine the flat on the edge if desired.
- Anneal at 800-1000 °C

The mechanical work performed on the metal substrate during all of the machining steps introduces stress in the material which is relieved through a high temperature anneal process. This may be done with or without uniaxial pressure, depending on the application.

- Rough lap.

Rough lapping with relatively coarse abrasive is used to achieve flatness and remove machining marks. Double sided lapping is important to minimize residual stress resulting in bow and warp.

- Serialize the wafers.
- Repeat rough lap if required to remove serial number burrs.
- Polish lap

SMI uses fine (micron-sized) abrasive to meet surface finish requirements. Typically we use both profilometer and Zygo™ interferometer measurements of surface finish.

- Metalize.

The heterogeneous nature of tungsten-copper makes it desirable to metalize the surface for soldering or brazing. Slight variations on plating processes used for common alloys are also used for tungsten-copper, usually followed by a heat treatment process to create the adhesive bond between substrate and plated nickel layer. For example, electolytic nickel diffuses well into tungsten upon heat treatment, creating a strong bond between the initial metallization layer and both tungsten and copper. Electroless Ni also gives satisfactory results for some applications. Once an initial nickel adhesion layer is created on the wafer additional metallization processes can be performed. One example is vapor-deposited Au/Sn to eliminate the need for solder preforms.

At present, we make W/Cu substrates as thin as 0.010" (250 μm). Our goal is to make substrates to 0.006" (150 μm).

MEASUREMENTS

The surface finish obtained on a bare W/Cu wafer can range from better than 2 μinch R_a to 16 μinch R_a (0.05 μm to 0.4 μm) as desired, depending on the application. Metallized wafers can have better surface finish still. The tables below summarize typical form measurements of W/Cu wafers.

Thickness Data Summary table, recent lot

Sample	Wafer ID	Mean Thick	Min. Thick	Max. Thick	TTV
1	206039-26	268.99	266.63	270.81	4.18
2	206039-27	267.18	265.2	268.92	3.72
3	206039-28	269.13	266.95	271.33	4.38
4	206039-29	267.45	265.56	269	3.44
5	206039-30	268.04	265.92	269.68	3.76
6	206039-10	256.21	254.95	258.13	3.18
7	206039-09	256.12	254.29	258.31	4.02
8	206039-08	256.85	255.38	257.96	2.58
9	206039-06	256.31	254.85	257.62	2.77
10	206039-03	256.15	255.03	257.73	2.70
Mean					**3.64**

Units = micrometers (μm)

Bow and Warp Data Summary table

Sample	Wafer ID	Warp	Bow
1	206039-26	51.92	-21.67
2	206039-27	20.25	-1.46
3	206039-28	22.79	-1.27
4	206039-29	38.25	-2.3
5	206039-30	22.01	-7.56
6	206039-10	49.54	-18.63
7	206039-09	n.m.	24.53
8	206039-08	37.3	-15.34
9	206039-06	51.25	-20.77
10	206039-03	42.14	-22.44
mean		**37.27**	**-8.69**

Units = micrometers (μm)   Data provided by
Sigmatech, Inc , Tempe, AZ

CONCLUSION

SMI has developed processes to manufacture CTE-matching tungsten-copper substrates to be used in wafer-level packaging. They are capable of meeting form and finish specifications similar to those of GaAs and Si wafers. They may be useful in wafer-level packaging where it is desired to minimize thermally-induced stress on the chip. Similar processes can also been used to make Mo/Cu and pure W substrates.

---

[1] "An Expert Looks at the Issues: Dr. Tom Stefano on Wafer Level CSP's", Chip Scale Review, May-June 1999.

SU-8 BONDING FOR TRANSPARENT PACKAGING

C. Brubaker and T.Matthias
EV Group Inc.
Tempe, AZ 85284
c.brubaker@evgroup.com

M. Wimplinger
EV Group, E. Thallner GmbH
St. Florian, Austria

ABSTRACT

In recent years, a veritable explosion of optical micro-devices have entered the market place, from projection display systems such as LCD technology and Texas Instruments DLP™ systems, to optical sensing devices such as he CCD chips used in digital cameras and cell phones. The vast increase in the demand for these products (with device sales easily measured in the hundreds of millions) and highly competitive nature of the marketplace require the use of more cost effective means of creating transparent packages to protect these devices from the environment.

This paper will discus the use of SU-8 to perform a wafer lever bonding process to apply the transparent package to such devices. The use of SU-8 bonding medium for this process is ideal for a variety of reasons. First, it allows accurate definition of a precise separation gap between the device and the glass cover; which also prevents any relative tilt between the cover glass and the device (very important for proper optical processing). Second, the gasket pattern can be easily defined in the SU-8 via mask aligner. Third, the bonding process itself can take place at very low temperatures (less that 150ºC), which makes it easily compatible to packaging CMOS devices. Finally, when SU-8 is cured, it is impervious to any solvent (and virtually all chemistry that may encountered) creating effective protection for the optical device.

Keywords: SU-8 Bonding, Transparent Packing, Optical device

INTRODUCTION

In the last decade, there has been a significant increase, even a veritable explosion, of optical micro-devices entering the market.

This includes projection micro-display systems such a LCD, OLED, LCOS or Texas Instruments' DLP™ technology, which can be found in millions of multimedia projection systems in conference rooms and home theaters worldwide, as well as in a new generation of high definition, large screen television projection systems.

This also includes optical sensing devices such as CCD chips and CMOS image sensors. These devices have been in digital cameras for years, but have also recently seen an explosion with the introduction of cell phones with integrated photographic capability (at this point, it seems as if more models available include a camera than are without).

As is the case with all of the highly lucrative consumer products markets, competition is tight. Just as is the case for consumer semiconductors, it is critical that all aspects of device manufacturing become more cost effective. This is especially true for one of the most expensive steps – packaging.

Because of the nature of these devices, packaging needs to meet two primary requirements. First, because these devices are all intended to perform some sort of optical interaction with the environment (either to the environment, in the case of the projection (LCD, DMD, LCOS) systems,

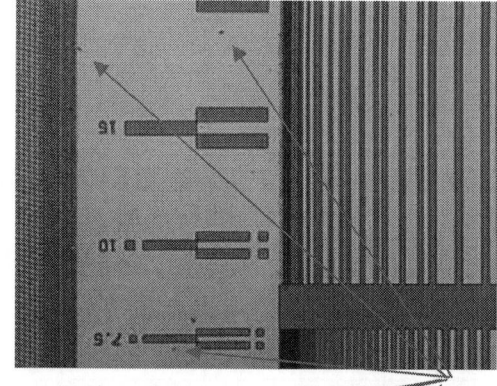

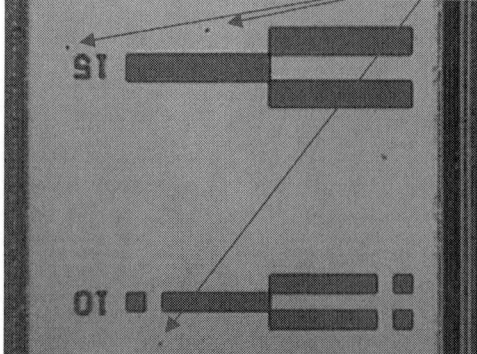

Particle defects on CCD

Figure 1 – Particle Defects on CCD – visible as same-location defects on images from different magnifications

or from the environment, in the case of the image sensors) some portion of the package must be transparent. Second, the package must be able to protect the device from negative interaction with the environment – most critically in the case of particles.

Particles will especially be a problem for these optical devices. In the first case, the system is actually a mechanical device operating on a micro scale – millions of tiny mirrors rotating back and forth to produce the digital image. Particles can interfere with the movement of these devices, causing lost or malfunctioning pixels. For an image sensor, particles can obscure one or more pixels, creating a black spot in the image (see Figure - 1).

Current technology has the packaging of these devices occurring at device level. Because each individual process has to occur with each and every device produced, the cost of packaging these devices is enormous. As has already been discovered in the semiconductor industry, the more packaging you can perform at wafer level, the more cost savings are gained.

In addition to the money savings based on economy of scale, other benefits occur from packaging at wafer level. First is consistency of packaging across a greater number of devices (less package variation). Second, especially critical to optical devices, is the improved parallelism between the device chip and the glass cover – very critical, since any tilt can create a distortion of image as light passes through the cover.

In order to perform this packaging at wafer level, some form of wafer-to-wafer bonding must be performed. Several techniques exist for this, borrowed from the MEMS industry. One way would be to etch the glass cover to create cavities, and then bond the glass so that these cavities are above the devices. However, this method has a few flaws.

First, it requires the performance of a deep etch process (sometimes hundreds of microns deep), which is both time consuming and uncommon in many semiconductor manufacturing facilities. Second and more importantly, is

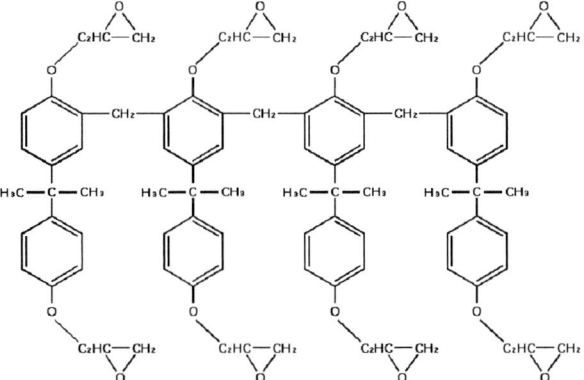

Figure 2 – SU-8 molecular structure

Figure 3 – High resolution features in 80μm thick SU-8 layer – photo courtesy Dalsa Semiconductor

that the etch may affect the optical properties of the glass, thus impairing the function of the device. Finally, there comes the difficulty of applying the bonding layer (adhesive, metal, or glass frit, for example) over the now topographically challenged substrate [3]

This paper presents and alternative method for creating this package – that of wafer bonding using SU-8 as an intermediate layer.

SU-8

SU-8 is a high-contrast negative tone photo-sensitive epoxy material that was originally developed for use in patterning of semiconductor devices. The base resin (the molecular structure is shown in Figure 2) was originally developed by Shell Corporation – the photo-sensitive version was developed and patented by IBM (US Patent No. 4882245 (1989) among others). It is currently primarily manufactured/distributed in several grades by MicroChem Corp (MCC) in Newton, MA.

The chemical qualities of SU-8 make it very well suited for use as a photolithographic material. The non-crosslinked resin is highly soluble in several solvents (gamma-butyrolactone – GBL - and cyclopentanone are the actual casting solvents used by MCC, but it is also soluble in MEK, PGMEA, Acetone…). This allows solutions to be manufactured with a very high solids content (>70%), allowing spin coating in excess of 500μm in a single coating.

Additionally, SU-8, while being sensitive to radiation below 400nm wavelength, is also highly transparent to radiation above 350nm. This brings two benefits – first, photo-patterning can be performed with near-UV (365nm) radiation in film thicknesses up to 2000μm (2 mm!!) thick. Second, because little radiation loss occurs through the film, the sidewalls of exposure very nearly approach 90°, allowing structures with an aspect ratio of 20:1 or more to be produced (Figure 3). Once exposed, the 8-part epoxy structure will form a dense network of crosslink, giving the

Property	Value
Density	1.2 g/cm^3
Young's Modulus	4-5 GPa
Tensile Strength	34 MPa
Max Stress	34 MPa
Poisson Ratio	.22
Stress	16-19 MPa
T_g - uncrosslinked	55°C
T_g - crosslinked	230°C
CTE	52ppm/°C
Thermal Conductivity	.2 W/mK
Thermal Stability	5% weight loss @ 364°C
Dielectric Constant (k)	4-4.5
Breakdown Voltage	>4x10^7 V/m

Table 1- Properties of SU-8 resin

material a large degree of thermal stability (in excess of 200°C).

SU-8 has one quality that makes it very ill-suited to be used as a photoresist – it is very difficult to remove, and is virtually immune to all common solvents. In act, the only (relatively) practical means to remove SU-8 involves as ashing above 600°C, extended baths in heated piranha, or removal in molten salts. Another method has been more recently developed making use of microwave based plasma, but even this is relatively slow at 1 μm/minute removal [4]. The problem is, since SU-8 is most valuable as a material in thicker films (>50μm), this means that the wafer would have to etch for >50 minutes to clear.

However, this quality, which makes it so ill-suited as a standard masking layer like other photoresists, makes it perfect for applications where the SU-8 will actually become a permanent part of the structure. As such, SU-8 found rapid acceptance into the MEMS industry

The physical properties of SU-8 make it just as valuable to use in MEMS devices as it chemical properties. Even the electrical properties show that it has some value as a dielectric as well, with a k value of 4.5 - similar to that of silicon dioxide. Table 1 shows a summary of some of SU-8's properties.

Because of these properties, SU-8 has been used for a staggering array of applications, including microfuidic devices, x-ray imaging systems, and mechanical gear or cantilever structures. The material has become greatly useful for developing elements of a device intended to remain as part of the structure. This also makes SU-8 an ideal gasket material when using wafer level bonding for a packaging process.

SU-8 FOR TRANSPARENT PACKAGES
SU-8 is an ideal candidate for use as a gasket material for creating a transparent package for optical devices. As mentioned above, once fully crosslinked, the physical properties of the material will serve to create a strong seal between the device wafer and the cover glass.

Because the material can be spin-coated onto the cover glass wafer, and due to the large number of formulations available for this material, it is a simple process to accurately define the spacing between the cover glass and the device in a range anywhere from ~ 5μm to hundreds of microns. The high degree of uniformity of the spin coat (typically less than ±3% of total film thickness across the full wafer) guarantees that the planarity of the glass cover will be highly parallel to the device (much smaller than the wafer, so the local thickness variation for a given device will likely be <<1%)

In addition, the fact that this material can be patterned like a photoresist also allows for very fine definition of the gaskets around the devices using a contact/proximity mask aligner. Also, the high aspect patternability of the material can allow the creation of narrower gaskets even for relatively large spacing between device and cover, allowing tighter pitch of devices on the wafers.

Finally, the bond process itself is low temperature, taking place at <150°C. This prevents the CMOS devices that are being protected from being at risk due to elevated thermal budget. Additionally, it reduces any thermal expansion mismatches between the silicon device wafer (typically ~ 3ppm/°C) and the glass (which is typically a formulation intended for high optical quality, and can have a thermal expansion anywhere in the range of <1 to 10 ppm/°C).

SU-8 BONDING PROCESS
As hinted above, the SU-8 bonding process breaks down into several steps:

Coating
This is the step that defines the height of the spacing layer for the package. To avoid fouling of any structures in the device, it is recommended that the carrier wafer only be coated for the bonding process. For thinner separation gaps (<20 μm), the lower formulations of SU-8 (-02 to -10) can

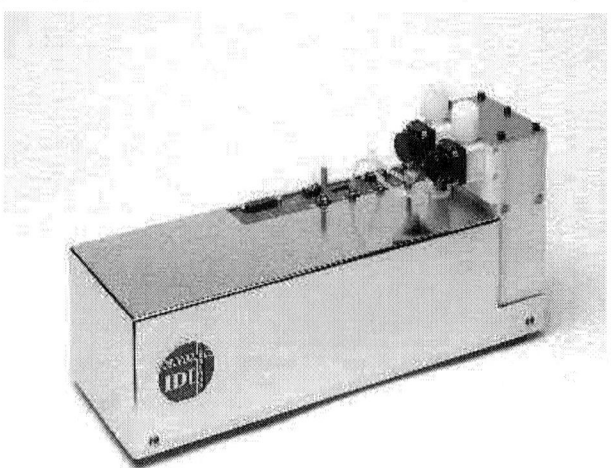

Figure 4 – Thick resist pump

be processed on a standard resist spin coater with little to no modification.

For thicker formulations and layers, several modifications to the resist system are recommended.

First, the materials at this point (especially for -25 and up) are becoming too viscous for standard photoresist pumps to handle – instead, a specialized thick resist pump (such as the IDI610HVP from IDI Cybor [5] – Figure 4) is required – this pump can process formulations as high as SU-8 100 (~50,000 cP).

Next, modification to the coater bowl is recommended as well. The first element is an integrated cover for the coating module (Figure 5), which can close after the dispense of the polymer to trap solvent in the bowl to prevent overly rapid drying of the solvent in the top layers of resist. This is especially critical with a material like SU-8, since a change in solids content of only 3% can cause a three-fold change in the material's viscosity. If the upper layers dry prematurely, they will begin to wrinkle as the lower layers continue to spread freely. In addition to the cover, a controlled exhaust flow is required to reduce the level of exhaust when the bowl is closed. Otherwise, closing the lid will effectively create a low-grade vacuum in the chamber, actually increasing solvent removal.

Soft Bake
Again, greater care needs to be taken with the thicker films. First, it is recommended that soft bake only be performed on a hot-plate, since convection oven systems effectively bake from the top layer down. This causes rapid solvent removal to occur initially from the outer surface of the polymer, creating a densified anti-diffusion layer which will hinder the evolution of solvent from lower regions

Figure 5 – Cover for coat module

However, when using a hotplate, care must be taken to avoid heating the substrate too fast – this can result in rapid vaporization of solvent in the region of the polymer closest to the substrate, which can result in the formation of bubbles.

In their data sheet [6], MCC recommends a two-step baking approach, with an initial bake performed at 65°C, and followed by an extended bake at 95 ° C. However, it is also possible to make use of a hot-plate module equipped with a wafer proximity system that can allow ramped approach of the substrate to the surface of the hot-plate, limiting the speed with which the substrate heats (and preventing vaporization).

Exposure
This is the first point where processing SU-8 for bonding differs from the standard processing of SU-8. As mentioned earlier, when processed in the standard fashion, SU-8 undergoes a dense cross-link process, resulting in a film that has thermal stability in excess of 200°C. Even above the nominal glass transition temperature (T_g) of 230°C, the viscosity change of the cured SU-8 film is so slight that, for bonding purposes, it is effectively still a solid (and thus, not bondable). In contrast, the T_g of the uncured film is 55°C. However, in the uncured state, the gaskets cannot be defined.

Instead, a compromise must be reached – the material must be processed in such a way that it reaches a *partially* cured state – sufficiently cured to survive develop and allow patterning of the gaskets, but not cured to the point where it will no longer reflow.

The exposure is the first step where this partial cure can be controlled. In order to slow down the cross-link process (which cannot be completed until the material experiences a post exposure bake, or PEB), the material should be exposed with a dose significantly less that the optimal dose for he film thickness. Bear in mind, the highly transparent nature of the material will ensure that the entire structure, top-to-bottom, will receive nearly the same exposure. Therefore, reducing the dose will primarily impact the amount of photo acid generated.

PEB
As mentioned above, this step is also used to control the degree of cure for the material. Technically, once exposed, SU-8 will crosslink, regardless of the application of a PEB. However, the PEB acts to increase the rate of reaction of the crosslink process. This rate is highly dependent on the bake that is performed, so by performing the PEB at a temperature lower than that recommended, the cure will occur more slowly, and thus in a more controllable manner.

Develop
For the develop, it is critical that the develop performed takes the minimum amount of time necessary to clear the unexposed features. Keep in mind that the film was

(deliberately) only partially cured – this means that, instead of the exposed region being totally immune to solvents, it is instead only relatively less soluble in the developer. It is also important to choose the proper solvent for develop. Even though materials such as acetone, MEK, GBL and cyclopentanone can be used for develop, they are powerful enough to begin attacking even the partially crosslinked regions. Instead, it is better to stick to the recommended developer (PGMEA) or similar materials (such as PGME, or a combination of the two).

Bonding

SU-8 has been discussed as a bonding medium in other circumstances [7, 8] In the bonding process, it becomes a race to see how much reflow can be accomplished vs. how quickly the material can complete its cure. As mentioned previously, the SU-8 bonding process can be completed with temperatures less than 150°C – however, the standard curing temperature for SU-8 is 95°C, and full cure will occur in <15 minutes at this temperature – 150°C would make it cure that much faster. And since it is impossible to determine the T_g of the partially cured film, there is no defined temperature "plateau" to hold at to allow time for the reflow – and thus the bonding – to occur.

The bond process will control both the overall final bond strength, and the final separation distance between the substrates. This is determined by an equilibrium state between the amount of force applied during the bond, and the resistance of the material at the bond interface (a function of the contact area of the material, and the viscosity of the material - which is in turn a function of the temperature ramp).

Please note that, if the temperature ramp is changed, then the equilibrium values will change as well. Since bonding of SU-8 is a race between reflow of SU-8 and cure of SU-8, it is difficult to predict ahead of time the result of a change in temperature profile.

Because of the sensitivity of the system to the amount of bonding force used and the temperature profile, it is very critical that a system is used that has a high degree of temperature and pressure uniformity.

RESULTS
Through application of the techniques above, successful bonding was achieved between the optical device wafer and the glass cover wafer using SU-8 2050.

Separation Gap
The separation gap between the two substrates could be estimated making use of a high-power microscope with a vernier gauge on the focus. By focusing first on the top surface of the SU-8 gasket (since the glass cover itself is too highly transparent to actually get a visual fix) then on the surface of the device wafer, the separation gap can be measured (see Figure 6). Based on these measurements, the separation gap was 110μm with a ±3μm variance across a 200mm wafer. Measuring across multiple bonded pairs, the variation of the mean separation gap was ±2.8μm.

Void-Free Bonding
In addition to the separation gap being consistent across the bonded pair, it is also critical the that the gaskets themselves be fully sealed for each device across the wafer. Any voids in the gasket can result in a loss of that device. Figure 7 shows an acoustic microscope image of a bonded pair, demonstrating very clear and void free gasket lines across the entire wafer.

Adhesion Strength
By this point, we have shown that we have been able to produce a uniform, void free bonding gasket using SU-8. However, neither of these is very important if the bond falls apart easily due to low adhesion strength. To test this, once

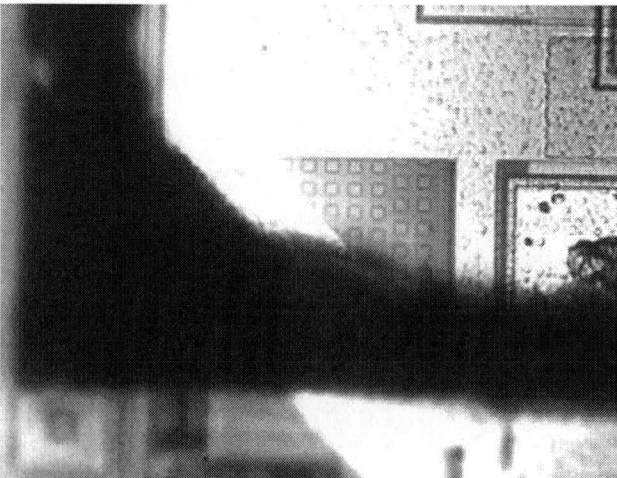

Figure 6 – Separation gap measurement – left – top of SU-8 gasket (159); right – device surface (48)

the bond was formed, the adhesion strength was tested by attempting to separate the wafers using a razor blade.

The bond was so strong that the glass material broke in the attempt to separate the two substrates. To try to further evaluate this effect, a pair of silicon wafers (one with 1000Å of oxide to mimic the glass surface) were bonded, with the same separation test performed (assuming the previously failure was due to the lower strength of the glass substrate).

Again, the wafers broke during the attempt to separate. However, in this case, in some areas, the gasket failed as well. Analysis of the gasket at the points of failure shows that the failure was in the bulk region of the SU-8 itself – the interface between the SU-8, the silicon, and the oxide did not fail (see Figure 8).

CONCLUSION
Wafer bonding through the use of SU-8 proves itself to be a valuable method to create stable, reliable packages to protect optical micro-devices from the environment while still allowing them to perform their function – optical interaction with the environment. The qualities of SU-8 lend themselves to be able to rapidly and easily define sealing gaskets to place a transparent cap over the deices at wafer scale.

By processing SU-8 using a resist processing system developed for thick resist processing, a contact/proximity mask aligner, and a wafer bonder capable of a high degree of temperature and pressure uniformity, it was possible to create highly reliable packages. The degree of planarity was high, with a ±3µm variation in a separation gap of 110µm across an entire 200mm wafer (giving an estimated <1µm

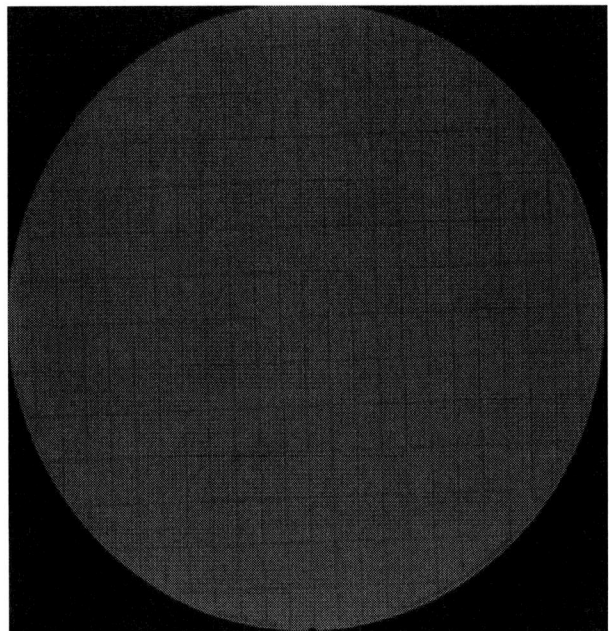

Figure 7- Acoustic Microgram of bonded 8" pair

variation for each device). The bond was also void free – guaranteeing the safety of the device. Finally, the bond itself was strong enough that the capping material itself is more likely to fail that the SU-8.

By making use of this process, the cost of packaging these devices will drop – passing on an overall savings for the production of these devices for use in the consumer market, ultimately making the end-use system (from projector to cell phone) less expensive as well, allowing commercialization of these products at an unprecedented level.

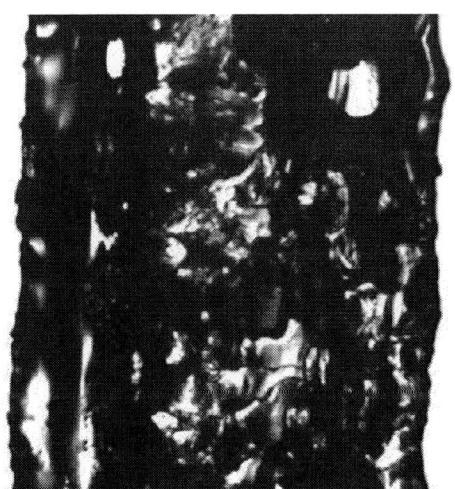 

Figure 8 – Separated surfaces of SU-8 bonding gasket – (left) – silicon wafer; (right) – Silicon w/ oxide

REFERENCES

1. Arensman, R., "TI Tries to Revive Stalled DLP Chip Sales," *Electronic Business Online* 6/27/2006 Retrieved August 4, 2006 from http://www.reed-electronics.com/eb-mag/article/CA6347021

2. Arensman, R., "Sensing a Sensor Shakeout?," *Electronic Business*, August 2006

3. Brubaker, C., Jennison, M., "Novel Concepts for BCB Processing," *Proceedings from the International Microelectronics and Packaging Society's 39th International Symposium on Microelectronics*, Oct. 8-12, 2006, San Diego, CA

4. K. Kim, D. Park, H. Lu, W. Che, K. Kim, J. Lee, C. Ahn, „A Tapered Hollow Metallic Microneedle Array Using Backside Exposure of SU-8"," *j. Micromech. Microeng.* 14 (2004) 597-603

5. Weider, B; "IDI/EVG Thick Resist Dispense System Evaluation Pump Model IDI M600 HVP," Retrieved August 4, 2006 from:
http://www.microchem.com/resources/IDI_Pump_with_SU-8.pdf

6. "NANO™ SU-8 2000 Negative Tone Photoresist Formulations 2035-2100" Feb, 2002; Retrieved August 4, 2006 from:
http://www.microchem.com/products/pdf/SU8_2035-2100.pdf

7. Sampath, S. K. ; St. Clair, L.; Wu, X.; Ivanov, D. V.; Wang, Q.; Ghosh, C.; Farmer, K. R.; "Rapid MEMS Prototyping Using SU-8 Wafer Bonding, and Deep Reactive Ion Etching," *Proceedingsof the 14th Biennial University/ Government/ Industry Microelectronics Symposium*, June 17-20, 2001

8. T. Glinsner, V. Dragoi, et al., "Wafer Bonding using BCB and SU-8 intermediate layers for MEMS Applications", Semicon Taiwan 2002 , Taipeh

METROLOGY FOR ULTRA-THIN WAFER AND DIE STRENGTH CHARACTERIZATION AND RELATED EDGE DAMAGE AND MODELING CHALLENGES

David Liu, Anwei Liu, Michael I. Current, Wojtek J. Walecki, and Ann Koo
Frontier Semiconductor
San Jose, CA, USA
www.frontiersemi.com

ABSTRACT

Thinned (~100 um) Si wafers with various surface damage removal treatments were tested to fracture in ball & ring and 3-point bend geometries. Non-linear displacements and Weibull-like fracture loading was observed in ball & ring tests while linear displacements and failure at much lower loads was observed in a 3-point bend stress geometry.

Keywords: three-point bend, fracture strength, dicing damage

INTRODUCTION

Packaging of thinned die, to dimensions of ~100 um, is a key challenge for the fabrication of "pocket and wallet" chip formats such as smart-cards and for multi-chip arrays [1]. Well established procedures for reliability testing of full thickness (~750 um) die are inadequate for evaluation of the properties and stress conditions for "ultra-thin" wafers and die.

EXPERIMENTAL

A series of thinned wafers, with a typical thickness of 100 to 105 um and with different post-thinning treatments of the ground surface, were measured with 3-point bend and "ball & ring" fracture tools. The wafer thickness was measured by IR light interferometry with the FSM 413 tool. Samples for the ball & ring testing were cut to 16x16 mm^2 and the 3-point bend samples were cut to 4x16mm^2.

The FSM Die Strength tool has a precision load column attached to either 3-point bend or ball & ring fixtures (Fig. 1). Load is applied by a stepping motor drive and is monitored by a piezoelectric force gauge.

The fundamental difference between these two geometries is that, for the 3-point bend, the dicing damage along the side of the sample is under stress during testing and contributes to the failure strength characteristics. For the ball & ring fixture, the dicing damage is outside of the support ring and does not contribute to the failure [2].

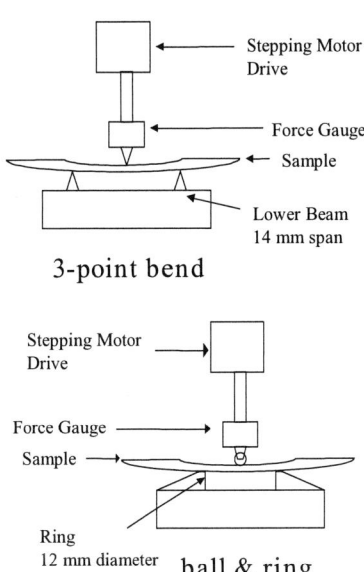

Figure 1. 3-point bend and ball & ring fracture test fixtures.

The 3-point bend samples were stressed over a span of 14 mm, with displacement stress applied at the middle of the span. The ball & ring samples were stressed by a central ball contact with a 2 mm radius of curvature across a 12 mm diameter ring support. The ball & ring fixture is considered to be a "simply supported" beam or ring, which allows the wafer to bend at the supporting points. The thinned wafers were cut into 14 to 30 fracture samples and tested in the FSM Die Strength tool. During the tests, the displacement Z motor drive was set with an initial moving speed of 6 um/sec and a final speed of 3 um/sec.

Table 1. Summary of results for 100 to 105 um thick samples for ball & ring testing.

Wafer	Die strength [N]			Weibull Characteristic Value	Weibull Slope	Number of samples tested
	Average	Stdev	Median			
1	3.49	4.87	1.82	3.35	1.13	20
2	3.31	2.56	3.12	3.66	1.42	20
3	4.80	5.34	3.89	4.88	1.42	30
4	5.04	5.36	4.20	5.23	1.19	20
5	6.60	7.01	4.15	6.48	1.02	30
6	6.00	4.47	5.09	6.69	1.07	14
7	6.67	5.96	4.71	7.11	1.28	20

RESULTS

Ball& Ring Testing

The general measurement results for the thinned wafers as measured in the ball and ring fixture are listed in Table 1. Each of the wafers was thinned to ~100 um and treated with various post-thinning procedures. The results are listed in order of fracture strength. The load-displacement behavior at fracture showed strongly non-linear behavior, with an approximately parabolic characteristic, as expected from the modeling of Chen et al. [3] (Fig. 2).

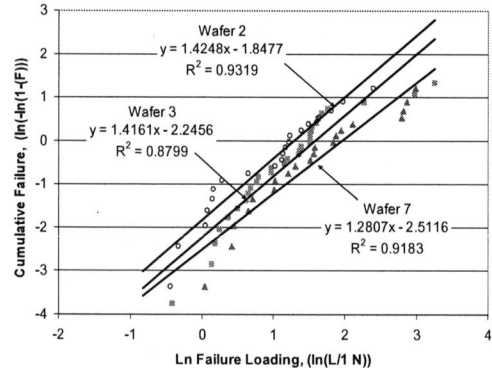

Figure 3: Weilbull plot of the cumulative failure distribution versus loading at failure for wafers with low (Wafer 2), moderate (Wafer 3) and higher (Wafer 7) fracture strength.

The Weibull Cumulative Failure Probability Function,

$$F = 1 - \exp(-(x/a)^b) \qquad (1)$$

with a "characteristic value" a, and a "slope" b, which leads to a log-linear behavior in the log-log Weibull function,

$$\ln(\ln(1/(1-F))) = b * \ln(x) - b * \ln(a) \qquad (2)$$

and is descriptive of the fracture observed in Fig. 3 with characteristics listed in Table 1.

3-point Bend Testing

Samples cut into 4×16 mm^2 pieces and stressed in a 3-point bend fixture experienced failure at much lower loading force and larger displacements. In the typical example shown in Fig. 4, a linear force-displacement characteristic was observed for the initial (faster) and final (slower) loading ramps up to the point of fracture.

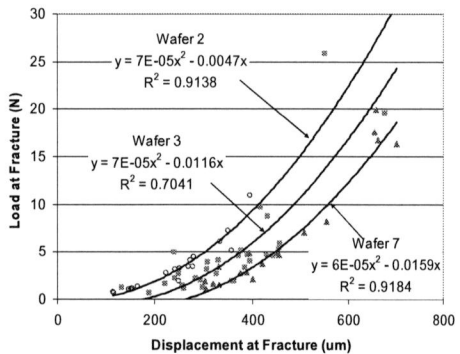

Figure 2. Results of load versus displacement measurements with ball & ring stress for wafers with low (Wafer 2), moderate (Wafer 3) and higher (Wafer 7) fracture strength.

Weibull distributions of the behavior of the cumulative failure distribution and the load force at the point of fracture show broadly linear behavior over much of the displacement range (Fig. 3).

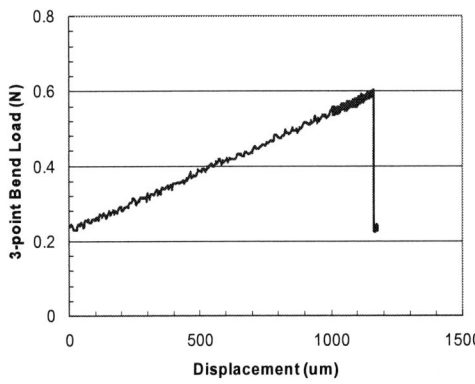

Figure 4. Load-displacement history for a die cut from Wafer 1 tested in a 3-point bend fixture.

The larger displacements at a given load, compared to the ball & ring tests, is partially due to the smaller cross-section of the 3-point bend samples and larger support span. For the 9 samples tested from Wafer 1, the failure load in the 3-point bend fixture was ~10 % of the fracture point observed in the ball & ring fixture (compare values in Tables 1 & 2).

Table 2. Fracture conditions for die cut from Wafer 1 tested with 3-point bend stress.

Sample #	Breaking Load (N)	Displacement at Fracture (um)
1	0.2787	1121
2	0.2874	1160
3	0.3498	1210
4	0.3822	1291
5	0.3306	1008
6	0.4457	1382
7	0.5304	1691
8	0.5304	1691
9	0.4008	1182

DISCUSSION
Ball & Ring Testing
The 16x16 mm^2 samples tested in the ball and ring fixture exhibited strongly non-linear displacements and much higher fracture loading than 4x16 mm^2 pieces tested in a 3-point bend fixture. The ball & ring samples showed broadly Weibull-like characteristics (Fig. 3) with a slope factor in a range from 1.1 to 1.4, close to the behavior (slope = 1) for a "random" failure mode distribution. However, attempts to fit the failure loading to the non-linear displacement model of Chen et al. [3] yield un-physically high Si fracture strengths. A likely factor in this story is the need to add the effects of sample sliding (or rolling) at the ring support boundary, particularly for the large displacements prior to fracture observed in thin (~100 um) samples, rather than the simpler assumption of a "simply supported" sample and zero-stretching at the support ring location.

3-point Bend and Ball & Ring Testing
The much lower failure loading observed for the 3-point bend samples compared to the ball & ring results for similar materials is likely due to the contribution of die cutting damage along the sides of the 3-point bend samples to the failure under stress. This suggests the uses of these two types of stress fixtures for testing different aspects of the reliability of thinned die in various packaging formats. While the ball & ring geometry isolates the failure from the effects of the die edge damage, 3-point bend testing more closely approximates die stress experienced by smart-card die in diverse "pocket and wallet" environments. The sensitivity of 3-point bend samples to die edge damage can be used to advantage in evaluation of various edge damage reduction strategies in a similar fashion to the ways ball & ring testing has been successfully used to evaluate reduction methods for residual stress from die thinning grinding and etch.

An additional factor to consider in the evaluation of the failure of thin chips is the need to check for cracking of hermetic seal films, which is likely to occur at lower stress than fracture failure of the die itself.

SUMMARY
Ultra-thin (~100 um) Si die exhibit significantly different characteristics when tested in 3-point bend and ball & ring geometries. Use of ball & ring fixtures can isolate the effects of post-thinning surface damage removal while 3-point bend geometries can be used to gauge the effects of die cutting damage and late processing steps.

REFERENCES
1. Marcos Karnezos, *Advanced Packaging* August, 2004.
2. C. Landersberger, et al, "New dicing and thinning concept improves mechanical reliability of ultra thin silicon", presented at "Int. Symp. on Advanced Packaging Materials" in Braselton, Georgia, March 2001.
3. L.D. Chen, M.J. Zhang, S. Zhang., J. Appl. Phys. 76(3), (1994) 1547-1551.

UTILIZATION OF DIE ATTACH ADHESIVES IN WAFER LEVEL ASSEMBLY OF CAVITY PACKAGES FOR IMAGE SENSORS

G. Humpston and M. Nystrom
Tessera Inc.
San Jose, CA, USA
ghumpston@tessera.com

S. Kanagavel, M. Previti, and M. Wilson
Cookson Electronics Inc.
Alphretta, GA, USA

ABSTRACT

High performance image sensors require encapsulation to provide protection from the environment. Preferably, the packaging is done at the wafer level in order to exploit the low cost and high yield of this method of assembly. Each device package must provide an air space between the front face of the semiconductor die and the underside of the package lid because the sensor array is covered with micro lenses, which are provided to maximise the low light sensitivity of the camera. One scheme for providing a stand-off distance between these components is to apply a picture frame of adhesive around each device on the wafer and then attach the lid wafer. This paper describes the materials attributes required of the adhesive and demonstrates how film and screen-print die attach adhesives can be adapted and used outside of their normal process bounds to achieve joints of satisfactory quality for this application.

Key words: Wafer level package, die attach adhesive

SOLID STATE OPTICAL SENSORS

Solid state optical sensors are finding application in an ever-widening variety of products. The largest markets by volume are camera modules for mobile phones and movement sensors for optical mice, followed by more camera modules for digital still cameras, video camcorders, web cameras and surveillance cameras (see Figure 1). However solid state optical sensors are also utilised in large quantities by industry in document copiers, bar code readers and positional control systems. Not only are all these markets large, but they are fast growing markets, many exhibiting compound annual growth rates in double digits [Prismark, 2006].

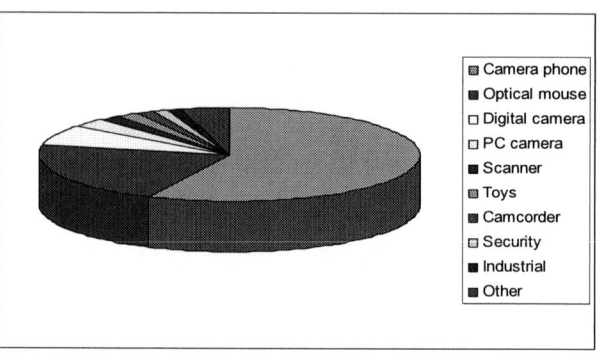

Figure 1 Predicted principal applications for solid state imagers in 2006 (Source: Prismark)

The majority of solid state image sensors are based on complementary metal-oxide semiconductor (CMOS) technology as this provides a more integrated solution than competing approaches. A CMOS image sensor comprises a 2D array of solar cells to provide the electro-optic conversion function, together with some additional electronics for picture and power management. Optical functions include things like automatic brightness and color adjustment. Some modern image sensors are able to provide an electronic version of an image in industry-standard file formats like "JPEG". Each light sensitive area on a chip with its associated electronics is referred to as a picture element, or "pixel" (see Figure 2).

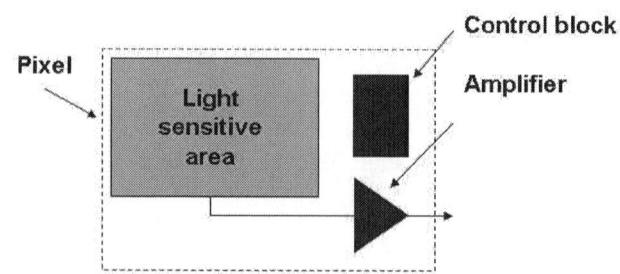

Figure 2 One pixel element of a solid state image sensor. The light sensitive area is smaller than the pixel dimensions because of the integrated electronics.

The electronics necessary to provide the functionality described above obviously takes real estate on the chip. Consequently, the dimensions of the solar cell in each pixel are reduced in proportion and this in turn adversely affects the ability of the image sensor to function in conditions where the light intensity is low. It is relatively easy to compensate for the light loss by placing a micro lens on each pixel. Micro lenses are pseudo-hemispherical in shape and serve to focus all of the light that impinges on each pixel on to the light sensitive area, as illustrated in Figure 3.

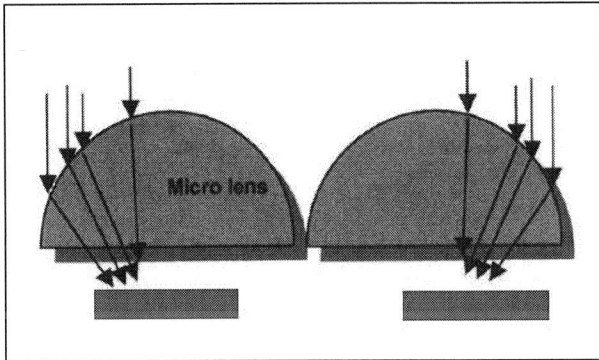

Figure 3 Micro lenses are used to focus the incident light on each pixel on to the light sensitive area, thereby increasing the low light sensitivity of the image sensor.

The micro lenses are typically only a few microns in height. To obtain the most benefit it is necessary to maximise the refractive index change at the external surface of the lens, which means there must be an air space above them. Observing this fundamental requirement means that solid state image sensors can only be housed in cavity packages.

WAFER LEVEL PACKAGING
Conventional semiconductor die are fabricated in parallel at the wafer level, then diced and singularly housed in discrete packages. This means that the costs of assembly are incremental for each die packaged. Wafer level packaging (WLP) is an alternative approach where the die are packaged while still in wafer form and the wafer is then singulated to free the packaged die. WLP has the advantage that the costs of packaging are shared among the good die on the wafer, greatly reducing packaging costs per die. However, WLP has proved to be technically extremely challenging and it is only recently that the materials and expertise to make it possible have become commercially available.

Conceptually, a wafer level cavity package for a solid state image sensor is very straightforward to achieve. What is required is to place a lid wafer over the device wafer, where the lid wafer is glass and spaced from the device wafer by means of a picture frame of a sealing material that surrounds each die. This arrangement is depicted in Figure 4. Practical realization is more complex, although products

based on this approach are manufactured commercially. The subject of this paper is the choice of material used for the picture frame seal, its method of introduction in to the assembly and the process use to affect the bonds to the silicon and glass wafers.

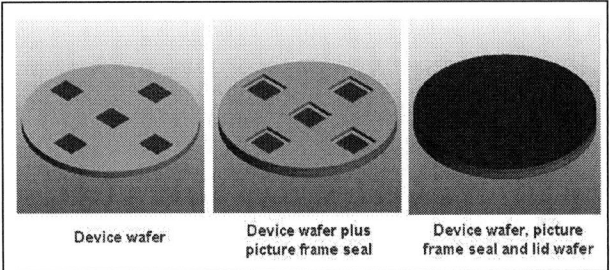

Figure 4 Cavity package formed at the wafer level by placing a picture frame of adhesive around each die and attaching a lid wafer. The assembly is then singulated to free individually packaged die.

PICTURE FRAME SEAL
The picture frame seal used to create cavity packages at the wafer level must possess a number of attributes as follows:

- Bond to borosilicate glass. The close match in thermal expansivity of certain grades of borosilicate glass makes them an ideal choice for a lid wafer on a silicon wafer.
- Bond to the wide variety of materials used as the capping layer on CMOS image sensor wafers. Examples include, silica, silicon nitride and benzocyclobutene (BCB).
- Hermeticity. A package for a solid state image sensor has to meet the conflicting requirements of being hermetic, to prevent the ingress of water vapor, which can corrode the semiconductor and fog the glass lid, while having a vent to the environment so the thin glass lid does not unduly distort when there is a change in ambient pressure. The compromise is usually a hermeticity in the region of 10^{-5} to $5 \times 10^{-8} \mathrm{cm}^3$.atm/sec of helium.
- Controlled flow. The adhesive material is required to have sufficient flow to accommodate wafer topography, warp and bow, but without extruding to the extent that it contaminates bond pads on the chip or interferes with the optical path.
- Form a joint with thickness and planarity controlled to exacting specifications across the wafer area. The key metrics are: Thickness 40μm +/-5μm, tilt <5μm across 200mm wafer.
- Low modulus after cure because of the low fracture toughness of glass and silicon. A modulus of less than 5GPa is usually appropriate for this application.
- High optical opacity. CMOS circuit function can be adversely affected if light falls on it so that by utilizing an opaque picture frame material to cover

43

the active circuitry, lower noise and more reliable performance can be achieved.

- Chemical stability. Must exhibit low weight loss on exposure to lead-free soldering cycles, low liquid and water uptake and low permeability to water vapor.
- Low materials and process cost.

Most commonly, the picture frame seal is created using photolithographic processes and materials like photoimageable BCB, or a non-photoimageable seal material in conjunction with photoimageable resists. This approach generally meets all of the attributes required above with the exception of materials and process cost. Photoimageable processes require a relatively expensive infrastructure and equipment set, a controlled light environment and a photomask that is custom for each wafer. Photoimageable chemicals are also relatively expensive and not particularly friendly towards the environment so there are often hidden process costs that must be accounted for, associated with the disposal of process waste and unused material.

Two alternative seal formation processes that appear attractive on the basis of materials and process cost are screen printing and punching from sheet of standard die attach adhesives. The challenge is to see whether a materials and process can be developed for each so that the seal exhibits the other attributes required.

PUNCHED ADHESIVE FILM

Adhesive film is manufactured in large quantities for die attach applications. It is typically produced in rolls that measure tens of meters long by up to 1m wide. One standard thickness is 40um and thickness control over the roll area meets the tilt specification in the as-received condition. The materials are typically translucent, have low moisture adsorption and permeability, and bond well to a wide variety of materials.

After a number of initial trials, the materials selected for detailed study were MC 500 and MC 700 series, both sourced from Mitsui Chemicals Inc. Although the exact chemical formulation is not disclosed, these materials are essentially polyimide modified to incorporate a proportion of epoxy with a silica filler. The difference between the two formulations is reflected in the process temperatures, the MC700 series requiring lesser initial thermal activation but a higher temperature cure schedule. Being polyimide-based die attach materials these film materials have quite low modulus, around 2GPa. They are designed to be used in a three stage process, that is, four heating cycles, as illustrated graphically in Figure 5. During the first stage the adhesive film is attached to the back face of the chip, followed by a bake to express residual solvents. In the second, slightly higher temperature step, the exposed face of the film is attached to a substrate. Finally, the bonded assembly is given an extended bake that converts the adhesive in to fully cured form. In the case MC 795 material the bake itself is a

two-step process, intended to replicate the wire bond and molding stages of manufacture of a plastic device package.

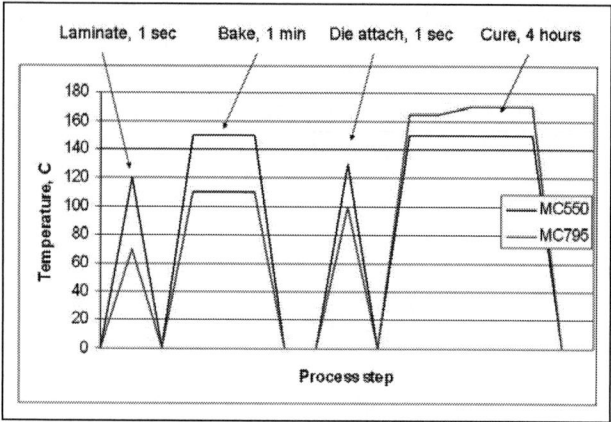

Figure 5 Recommended process conditions for using two die attach films from Mitsui Chemicals Inc. They differ principally in the thermal activation required.

Shaping of sheet adhesives has been readily accomplishable since the advent of numerically controlled punches. These equipments use a matching punch and die set to remove a piece of material, with shape defined by the tool set, from a location that can be specified to tolerance of a few microns over a 300mm wafer. Thus by repeated punching operations the surplus material can be removed from the sheet, leaving a lattice work of connected picture frame seals. An example is given in Figure 6.

Figure 6 A 200mm diameter wafer of adhesive film with rectangular windows formed in it using a numerically controlled punch

As their name suggests, die attach materials are designed to attach semiconductor die to substrates, with the next process step usually being one of wire bonding. Although wafer-to-wafer bonding is essentially the same, a key difference is the joint area. A 200mm wafer being in the region of 1,000-10,000 times larger than an individual die. One of the first problems encountered was of a high incidence of voids in the wafer level joints. The origin of these was traced to adsorbed water vapor in the die attach film and the problem eliminated by the introduction of strict controls on the humidity level in storage environments and other pre-usage protocols.

An important parameter in using film adhesives is the minimum width of web that must retained between each aperture for the latticework of material not to distort during punching or handling. Although the adhesive itself is only 40um thick, it can be supplied attached to a polyester backing sheet of considerably greater thickness and rigidity, typically 75μm. The adhesive film and backing sheet are punched together in a single operation and the backing film is removed after the adhesive has been aligned and placed on the wafer. The backing sheet greatly improved the mechanical rigidity of the adhesive lattice. Although the minimum web width that could be punched without distortion was determined to be 100μm, for 200mm wafers the web widths need to be 200μm or greater to avoid undue problems during handling. Figure 7 shows a detail of 300μm, 200μm and 100μm wide webs.

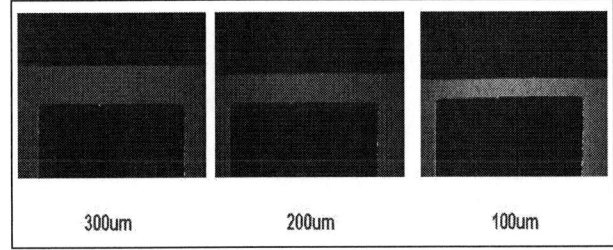

Figure 7 Webs of adhesive 300μm ,200μm and 100μm wide. When supported by a backing sheet, adhesive with a 200μm wide web this was about the minimum that could be handled at 200mm diameter without the weight of the sheet causing stretching or deformation during handling.

As mentioned above and illustrated in Figure 5 the die attach adhesive is designed to be subject to three thermal cycles to affect a join. Whilst attempting to mimic the specified process conditions on 200mm diameter wafers a problem was encountered whereby the relatively massive chucks of the wafer-to wafer bonding equipment could not change temperature as rapidly as recommended in the processing conditions for die attach. The problem was most severe on the cool-down part of the cycle, which was more than an order of magnitude slower than desired. The additional thermal energy imparted during the first heating cycle resulted in excessive cure of the adhesive and unreliability of the second bond. Also, from a simple economic perspective, a necessity to subject the wafers to

four heating cycles would have curtailed throughput on a very capital intensive machine.

The solution adopted was to develop a two step bonding process, which is depicted graphically in Figure 8. In the first stage all three parts are placed in contact by application of a compressive force and the temperature is then ramped rapidly to a set point. Following a brief period, typically two minutes, at that temperature the assembly was allowed to cool to room temperature. At some later time the bonded assembly was placed in an oven for four hours to complete the cure. Because these process conditions represent a significant departure from the recommended usage of the die attach material a statistical design of experiments was conducted to establish the sensitivity of each variable and its range limits, to enable a reliable process to be devised.

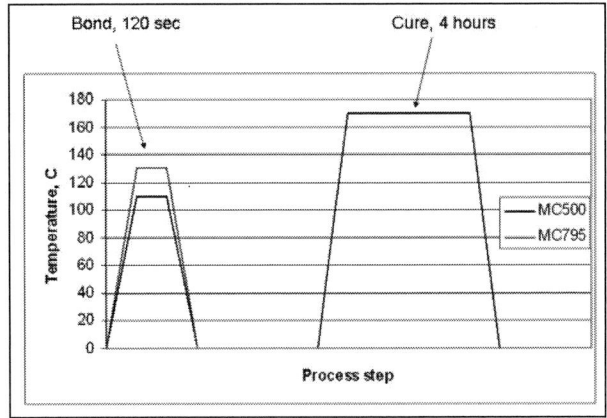

Figure 8 Single step bonding process developed to permit die attach adhesives to be used for wafer level packaging.

SCREEN PRINTED ADHESIVE
The development of a printable paste adhesive to facilitate WLP is seen as a lower cost, higher throughput process for manufacturing final assemblies versus the standard die attach film or lithography process. The challenge is to develop an adhesive that can meet all design criteria.

Major design criteria for WLP process include:
- Printable material keeping dimensional stability.
- Maintain 40μm +/-5μm spacing between the wafer substrates.
- Adhesion to various glass and wafer passivation layers.
- Curing/bonding process that enables High Volume Manufacturing (HVM).
- Low outgassing material during cure.
- Low outgassing material after cure in reliability aging testing.
- Low permissibility to moisture and vapor.
- High optical opacity

The first requirement is a system that can be screen printed. This is not necessarily as easy as first expected. To facilitate HVM, the printed material must not slump after initial print, maintain window frame dimensions, not wet out the bond pad area and attain its width to maintain a seal of the CM OS. The printing must be on a glass or silicon wafer substrate. The two different wafers must then be aligned at higher temperature >120°C to maintain integrity of the CMOS device. Initial bonding at elevated helps eliminate outgassing and fog formation inside the window frame area, but presents a material handling concern as most liquids will reduce viscosity and increase "slump" under higher temperature, thus presenting a technical challenge to the design of the WLP paste material. The viscosity change of the printed adhesive versus temperature conforms to a numerical model developed by William, Landel, and Ferry [VanKrevelen 1997] and is expressed as:

(1) $\quad \log \eta(T) = 13 - \dfrac{17.44 * (T - Tg)}{51.6 + (T - Tg)}$

Where:

η = viscosity
T = temperature
Tg = the glass transition temperature of the base polymer in the uncured state

Utilizing expression (1), the temperature dependent viscosity behavior of polymers can be described. Figure 9 is a plot of a model system and a developmental candidate formulation measured via dynamic temperature ramp at two different stress levels, 100 and 300 Pa. At 100 Pa, the shear stress is not sufficient to cause the material to yield, thereby maintaining a relatively stable viscosity across the temperature ramp. At 300 Pa, the weak forces are broken down causing the material to shear thin with increasing temperature and behavior then resembles a model system.

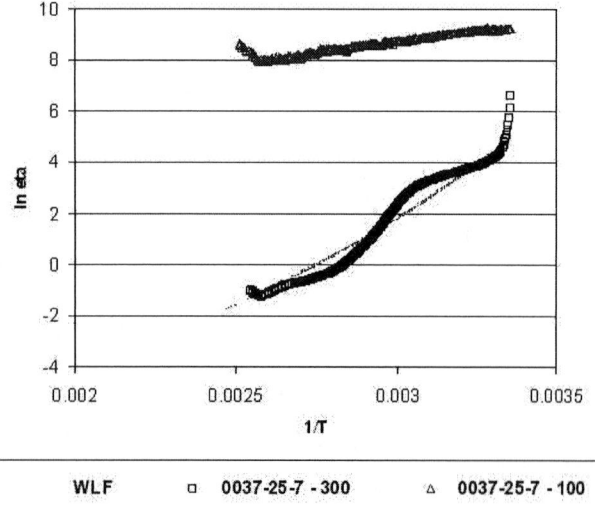

Figure 9 Ln η vs 1/T

A difficulty encountered during the printed adhesive development is balancing the ability of the material to be printed vs. the spread or slump that is seen during heating and attach. The adhesive must have a low enough viscosity to be printed, but maintain its shape during heating and bonding. This requires deviation from the numerical model across a wide temperature range.

One way to reduce slumping was to develop a "hybrid" polymer system. The hybrid adhesive system incorporates a portion of polymer that is cross-links under exposure to ultra violet (UV) light at 265 nanometers. The UV exposure provides dimensional stability of the printed adhesive polymer system and initial adhesion. The adhesion is formed by a small percentage of cross-link density of the UV polymeric portion of the adhesive. This ensures rigidity for the deposited window frame to maintain shape. After placement of the two wafers and exposure to UV for a few seconds, the joined wafer assembly can then be moved to a final cure heating stage allowing the placement table free to begin processing of the next wafer. This process flow is illustrated schematically in Figure 10.

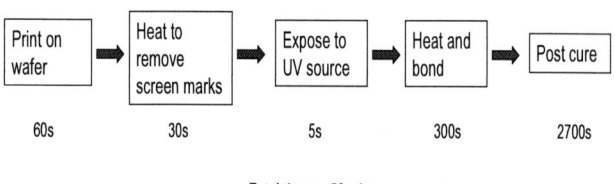

Figure 10 Proposed dual cure process map

Maintaining the 40µm +/-5µm thickness of the adhesive is done using spacer bead technology which is commercially available. This addition of silica is also employed which decreases the coefficient of thermal expansion (CTE) of the resultant material. The additional silica filler also assists in lowering total outgassing of the adhesive. The silica filler is a commercially available and a well studied phenomenon.

The remaining portion of the adhesive consists of low outgassing epoxy which is cured for one hour at 165°C. The epoxy material is a space grade polymer system and is extremely low outgassing upon cure. This is required as the window frame adhesive is sandwiched between both wafers and any outgassed material cannot escape the cavity. Figure 11 is a Thermo Gravimetric Analyses (TGA) of the system upon final cure after initial UV exposure. Resulting outgassing species by weight was 0.481%.

High optical opacity is achieved by addition of a carbon black pigment to the polymer system. A special grade and morphology was chosen to reduce outgassing and allow UV energy penetration through the printed window frame pattern.

46

The resulting silica-filled hybrid polymer system is extremely low outgassing, both during and after thermal processing as well as in reliability testing. The unique hybrid system allows a low cost WLP assembly that can be achieved with minimal capital compared to film or lithographic processing.

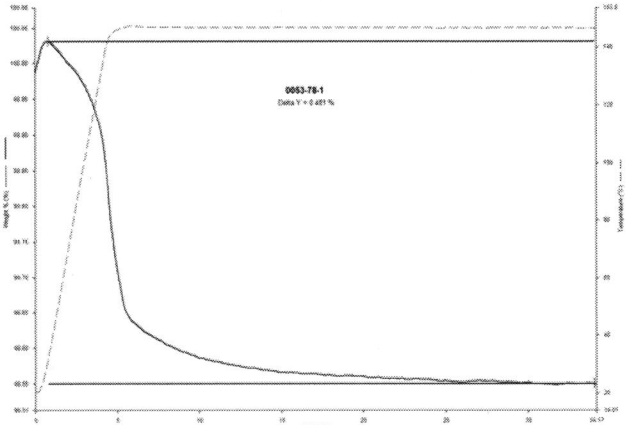

Figure 11 TGA Weight Loss During Cure after prepolymerization

PRODUCT RELIABILITY
Packaged solid state image sensors must pass a number of environmental tests to be deemed suitable for use in camera modules for cell phones. These tests and the associated pre-conditioning processes, where relevant, are given in Table 1. This Table also lists the results obtained for the film and reveals adequate performance. The screen print adhesive is still undergoing full assessment, the results of which will be reported at conference. Nevertheless, leading trials on a statistically small batch of components yielded encouraging results.

Test	Condition	Film Adhesive	Printed Adhesive
Wafer thinning	700 to 200um	Pass	TBD
Wafer dicing	Glass and silicon	Pass	TBD
MSL2	125°C, 24hrs 60°C/90%RH, 168hrs 3x Pb-free reflows	Pass	TBD
HTS	150°C, 1000hrs	Pass	TBD
TMCL	-40°C / +125°C, 1000 cycles	Pass	TBD

Table 1 Typical reliability tests used to determine the fitness for purpose in camera phones of solid state imagers packaged at the wafer level. The film when used with the newly devised process conditions, meet the required standard. Results for the screen printed adhesive will be reported at conference, but leading indicators are favorable.

CONCLUSIONS
Die attach adhesives possess many of the materials attributes required for wafer level assembly and are readily available at low cost. However die attach is a unitary operation involving small parts whereas wafer level assembly is a ganged process performed on relatively massive components. It has been shown that through modification of the process conditions, both film and screen printed die attach adhesives can be used to fabricate cavity packages at the wafer level.

REFERENCES
Prismark, 2006. Prismark semiconductor and packaging report summary, July 2006.
VanKrevelen, D.W., 1997. Properties of Polymers, Eslevier Science B.V., p. 406.

BCB WAFER BONDING WITH ELECTRICAL INTERCONNECTS

Praveen Pandojirao-S, Rachita Dewan, Dan O. Popa, and J.-C. Chiao
Automation & Robotics Research Institute
The University of Texas at Arlington
popa@arri.uta.edu

ABSTRACT

Wafer bonding is a critical step in micropackaging, guaranteeing that the active MEMS/IC devices are protected from the operating environment, and that devices achieve vertical integration. We have recently proposed a 3D wafer-level packaging technique for MEMS devices that is based on stacking of silicon and glass wafers manufactured using different micromachining technologies. The proposed wafer stacks can incorporate IC electronics, MEMS, microfluidics and microoptics, and are bonded together using patternable polymers such as Benzocyclobutene (BCB).

In this paper we report progress in simultaneous wafer level bonding and interconnect formation without the need for conventional via drilling and filling. We provide fabrication process parameters, preliminary interconnect DC characterization, and design guidelines for selecting interconnect diameters in typical MEMS applications.

1. INTRODUCTION

Wafer bonding is a mature technology in Level 0 packaging of MEMS, and in fabrication of MEMS substrates, such as SOI. Wafer bonding processes are dependent on the type of substrates (e.g. Silicon, glass, etc), and the type of bonding interface material (e.g. polymer, direct – fusion, anodic, eutectic, etc) [11R]. Wafer bonding using dielectric polymers, such as PI 2610, S1818, and, more, recently, BCB (benzocyclobutene) has been proposed for stacking IC wafers in three-dimensional (3D) electronics for a long time [5R]. The properties of BCB - excellent mechanical strength, very low out gassing, less sensitivity to surface preparation, make it a very attractive polymer for wafer bonding. Previous work proposes the use of BCB, and in particular Dry-Etch BCB, as an adhesive layer for accomplishing 3D integration of microfluidics, microelectronics, and MEMS [1R, 2R, 5R, 6R].

Fabrication of 3D electrical interconnects is an essential step in the context of wafer-level vertical integration. At the wafer level, a popular way of forming vertical electrical interconnects between wafers involves bonding the wafers, followed by wafer thinning, deep reactive ion etching (DRIE) through the stack, and deposition or plating of interconnect metal [5R]. At the die level, 3D interconnects are typically achieved by formation of solder bumps before bonding the wafers, as commonly done in flip-chip bonding [3R].

In our earlier work [1R-2R] we demonstrated the formation of microfluidic vias by using fully cured Dry-Etch Cyclotene™ 3000 series BCB. Additional studies showed that partially cured BCB does not reflow into micromachined cavities during the bonding step. In this paper, we repeat this work using a different grade BCB, namely the Photosensitive Cyclotene™ 4022-35, and, in addition, we form vertical electrical interconnects between bonded wafers. Thus, we combine the mechanical strength of BCB bonding between wafers with the formation of electrical interconnects via solder bumps. We use a thin layer of photosensitive BCB as the wafer bonding interface and spacer material, and a 3 μm height of the solder bumps.

In [7R] we provided preliminary evidence demonstrating the feasibility of forming vertical interconnects in this manner. In this paper we present additional electrical measurements of interconnects, as well as design guidelines for how to choose the interconnect diameter in two MEMS packaging applications: low-frequency, high current powering of thermal MEMS, and high-frequency, low current powering of RF-MEMS.

The results presented in this paper are part of a larger on-going 3D integration effort currently taking place at the ARRI's Texas Microfactory™ at the University of Texas at Arlington, and at the Bennington Microtechnology Center in Vermont - a one-of-a-kind MEMS pilot packaging facility in the US. The paper is organized as follows: in Section 2 we describe the fabrication, bonding and reflow process for the wafer stacks; in Section 3 we provide experimental evidence through resistance measurements that the patterned wafer bonds and interconnects can be formed simultaneously; in Section 4 we provide design guidelines for sizing the microbumps in the two specific application scenarios; finally, Section 5 concludes the paper.

2. FABRICATION, SOLDER REFLOW AND WAFER BONDING

Benzocyclobutene (BCB), Cyclotene 4022-35, is used to bond a Si wafer to another Si/Pyrex glass wafer. This grade of BCB being photosensitive in nature can be patterned using standard lithographic techniques. It is designed for a film thickness range of 2.7 – 6.9μm. However, for this research a 3μm thick layer of BCB on each wafer is used to bond them. Vertical electrical interconnections between the bonded wafers are formed through solder bump reflow. The solder reflow temperature is bound by the curing (180-250°C) and glass transition temperatures (>350°C) of BCB. Two solder compositions have been used - 95Sn/5Au with a near eutectic melting temperature of 217°C and 80Au/20Sn with a eutectic melting temperature of 280°C. The reflow temperature for 95Sn/5Au and 80Au/20Sn are 250°C and 320°C respectively. Hence, solder reflow can be performed simultaneous to BCB curing for 95Sn/5Au solder composition. However for 80Au/20Sn composition, solder reflow is performed after bonding wafers using BCB. Solder metal is deposited either in alloy form or as alternating layers of Au and Sn to reach the desired composition [3R]. The Au–Sn alloy exhibits inter-diffusion to form an Au-Sn compound, even at room temperature. This is evident from the grayish color of the solder after deposition, even though the top layer is of gold. The sequentially deposited multiple layers get homogenized after reflow [4R]. The fabrication process requires a number of established steps such as photolithography, lift-off, e-beam evaporator, Sputter deposition, solder reflow, alignment and bonding in an EVG 501 wafer bonder.

A silicon (Si) wafer is first cleaned using RCA clean. The wafer is patterned using a 1μm thick layer of positive photo resist, Shipley S3612, for base metal deposition. It consists of sequentially deposited three layers of Titanium, Platinum and Gold, each 0.1μm thick deposited through an E-beam evaporator. Base metal defines the area for solder reflow and also isolates the solder bump from the contact pad. After evaporation, lift-off is performed. AP3000 adhesion promoter and Photo BCB are then spin-coated to 3μm thickness on the resulting Si wafer. BCB is exposed, patterned and developed to define the area for bonding. Partial curing or hard curing of BCB is done in an inert gas environment according to its curing profile which is a function of time and temperature. The wafer is again patterned with a 1μm thick layer of S3612. The solder metal pattern is aligned on top of the already deposited base metal pad. This is followed by a multilayer Sn-Au deposition. For 95%Sn 5%Au bumps, a sequential deposition of Sn (600nm) and Au (20nm) is done several times to achieve the desired solder thickness. A second liftoff in acetone for 5-10 minutes leaves behind the multilayer solder only in the required regions. The fabrication process is illustrated in Figures 1 and 2.

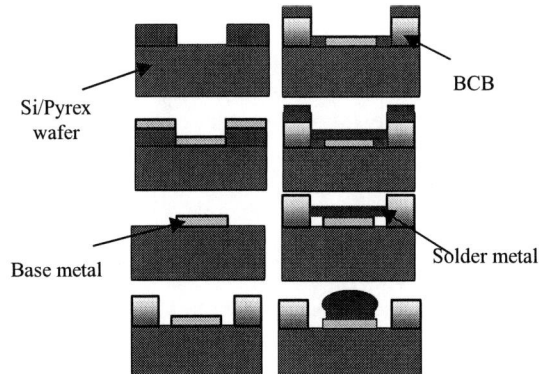

Figure 1: Process flow for formation of solder bumps in BCB channels

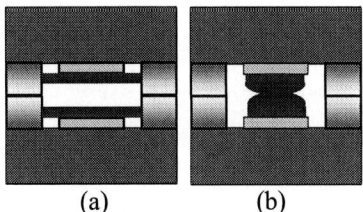

(a)   (b)

Figure 2: Forming 3D wafer stacks (a) Aligning two wafers and (b) Simultaneous wafer bonding and electrical interconnect formation

Silicon and Pyrex glass wafers were bonded using photosensitive BCB with different curing percentage from 35% - 95% cured BCB. The curing was done in a nitrogen environment using a Blue-M oven. Wafers with the cured BCB layer are aligned and bonded using the EVG 501/620 bonder and aligner at 3340N pressure, 250°C and for 30 min. No reflow was observed in the patterned channels for wafers bonded with 55% or more cured BCB layer in a similar manner to bonding results obtained for Dry-Etch BCB **[5R]**. Void-free and solvent-free bonding results were obtained. However, residual solvent was found trapped between wafers bonded with BCB cured at less than 55%. A razor blade was inserted between the silicon and Pyrex wafer to check for the bond strength **[6R]**. It was observed that either silicon or Pyrex wafer cracked and broke or BCB was film residue remained on one wafer only. This indicates that the bond strength at the BCB to BCB interface is higher than that at the BCB to substrate interface **[7R]**.

The solder reflow profile for Au-Sn alloy is a bell shaped profile with equal heating and cooling rates. The reflow temperature is around 30-40°C higher than the melting temperature of the alloy and the total time above the melting temperature should not be more that 4-5 min. Solder reflow experiments were performed with two different solder compositions of 95Sn/5Au and 20Sn/80Au. A 4 inch wafer was diced into quarter inch pieces for reflow testing. Each quarter of the wafer has an array of 40x40 solder bumps. 10

such samples were reflowed in a nitrogen environment to avoid the oxidation of the solder metal. The height of solder metal and base metal before reflow was recorded to be 2.5µm. However, height of the solder bump after reflow was more than 4µm. The basic microbump reflow profile results have been published in our earlier work [7R].

3. PRELIMINARY CHARACTERIZATION OF 3D INTERCONNECTS

4 inch Silicon and Pyrex wafers are bonded using the process described in the previous section. A total of 6400 solder metal depositions each on Silicon and Pyrex wafers are reflowed to form the electrical interconnect between them.

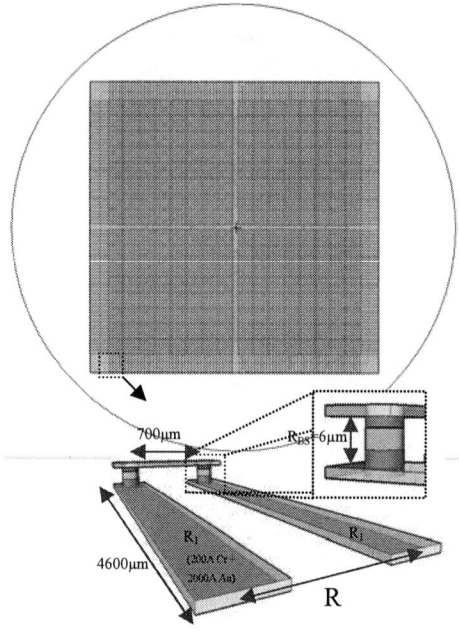

Figure 4: Schematic shows bonded 4-inch wafers with 1600 interconnects per quadrant. Close-up diagram shows metal contact pads exposed for electrical testing.

Our previous work [7R], has shown evidence of simultaneous wafer bonding and electrical interconnect formation. Further resistance measurements were carried out on similar samples. The average resistance measured between consecutive pads located on the same wafer was 121.8 ohms, with a standard deviation of 0.3 ohms. A constant current from 5mA to 15mA insteps of 1mA was driven through the resistance network and the resulting voltage drop across the contact pad was measured. The experiments are conducted at the Texas Microfactory ™ at UTA (Figure 5). Figure 6 shows plots of the voltage across the contact-pads versus the current. It can be seen that the resistance (indicated by the slope of the curve) gradually

increases as we move from pads 1:2 to pads 1:40. Resistance calculation through V-I measurements are tabulated in Table 1. The measurements were made on two different sets of bonded wafers and results were repeatable. While these results do not directly measure the resistance across individual interconnects, they do show that interconnects form after wafer bonding and reflow. Further measurements are necessary for full AC and DC interconnect characterization.

Table 1. Experimental results showing the resistance measurement between contact-pads 1:2 to 1:40

	Resistance (ohms)		Resistance (ohms)
Pads 1 and 2	120.0719091	Pads 1 and 22	373.2527273
Pads 1 and 3	133.4985455	Pads 1 and 23	394.225
Pads 1 and 4	145.3549091	Pads 1 and 24	400.6790909
Pads 1 and 5	155.4507273	Pads 1 and 25	421.8159091
Pads 1 and 6	148.5662727	Pads 1 and 26	412.7295455
Pads 1 and 7	148.5492727	Pads 1 and 27	424.0287273
Pads 1 and 8	182.516	Pads 1 and 28	436.4250909
Pads 1 and 9	190.1855455	Pads 1 and 29	446.5361818
Pads 1 and 10	198.2881818	Pads 1 and 30	451.5304545
Pads 1 and 11	208.9227273	Pads 1 and 31	483.2581818
Pads 1 and 12	240.6146273	Pads 1 and 32	492.9495455
Pads 1 and 13	262.1563636	Pads 1 and 33	502.1600909
Pads 1 and 14	234.4849091	Pads 1 and 34	496.5934545
Pads 1 and 15	244.6220909	Pads 1 and 35	518.2903636
Pads 1 and 16	292.3858182	Pads 1 and 36	565.9435455
Pads 1 and 17	298.8749091	Pads 1 and 37	565.4174545
Pads 1 and 18	288.3048182	Pads 1 and 38	569.0449091
Pads 1 and 19	296.1806364	Pads 1 and 39	556.2554545
Pads 1 and 20	341.7954545	Pads 1 and 40	553.1125455
Pads 1 and 21	359.7949091		

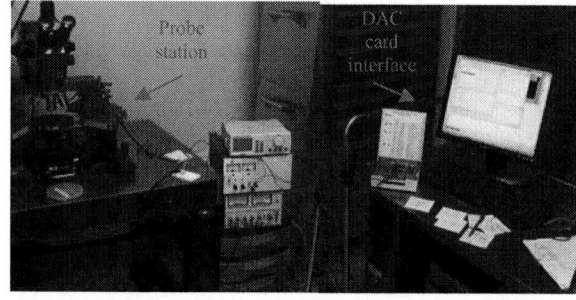

Figure 5: Electrical testing and characterization setup at the Texas Microfactory™ cleanroom.

4. MODELING OF 3D INTERCONNECTS

Modeling of 3D interconnects is done to select the correct dimensions for interconnects in terms of their diameter and pitch. These parameters determine the interconnect densities, and affect electrical properties such as frequency-dependent inductance, resistance and capacitance. Small length of interconnects is desired for faster propagation and low losses at higher frequencies (>30 GHz), as parasitics increase the RC delays and reduce the speed of the operation [8R]. A thermal analysis also needs to performed

50

to determine the temperature distribution across interconnects, and verify that they do not fail during operation. Guidelines on selecting the interconnect dimensions based on analytical and numerical modeling are therefore very useful prior to fabrication.

Two different analyses have been carried out namely: S1) low frequency analysis which includes a high current, low density and low frequency, and S2) high frequency analysis which includes a low current, high density and high frequency. The first analysis is useful in determining the maximum current carrying capacity of interconnects in a typical electro-thermal MEMS packaging application. The second analysis is useful in understanding the skin effect at higher frequencies, resulting in an increase in the effective dissipation through interconnects in electrostatic or RF-MEMS packaging applications.

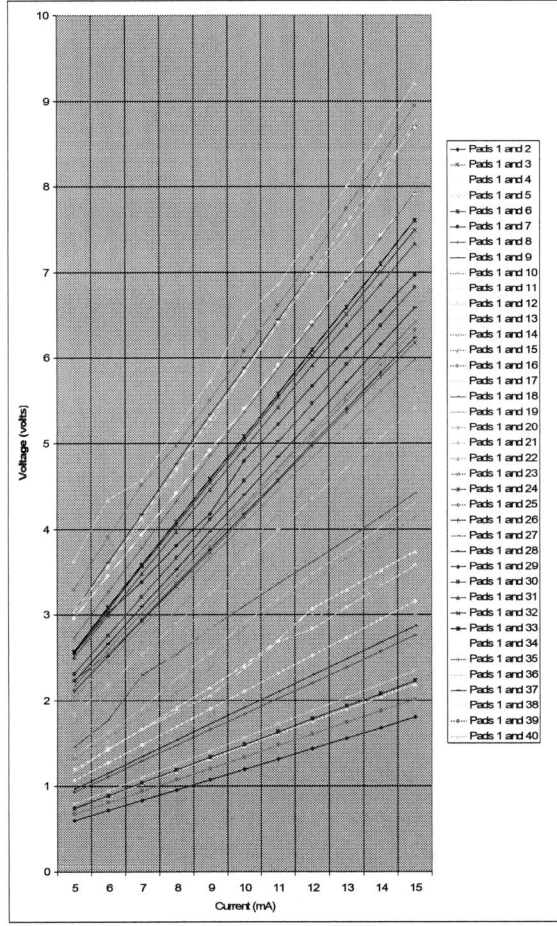

Figure 6: Equivalent resistance between contact pads of bonded wafer pairs, pads 1:2 to 1:40.

For S1 simulations, we used ANSYS® to numerically predict the steady state temperature distribution across interconnect due to joule heating. As the interconnect diameter changes the resistance of the interconnect increases as the square of that diameter. The purpose of S1 is to determine minimum values for the interconnect diameter at which the temperature across it is equal to higher than the re-melting temperature of the solder. The 3D model consists of interconnect, includes base metal and solder metal deposition, and a Silicon and Pyrex glass substrate, as shown in Figure 7.

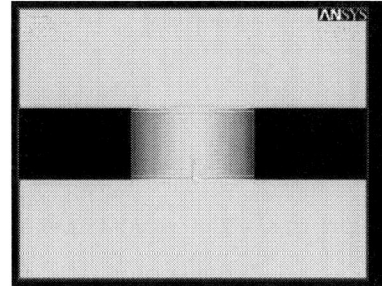

Figure 7: 3D model of interconnect in ANSYS.

Symmetrical boundary conditions were applied as a single block across the interconnect. Simulation results were evaluated for two solder compositions used to fabricate the interconnects – 95Sn/5Au and 80Au/20Sn. The electrical resistivity for these two solder compositions is estimated from known values of pure metal composition. A constant current of 100mA was applied across interconnects to assess their maximum current carrying capability at different interconnect diameters. This value of current is typical in the operation of electro-thermal MEMS devices. From the simulation results, it was observed that interconnect diameter smaller than 10µm for 95Sn/5Au and 5µm for 80Au/20Sn cause slight increase in interconnect temperature, though not enough to melt the interconnect. The steady state temperature values obtained are tabulated in Table 2. Of course, the values obtained from the simulation results depend on the modeling assumptions and the material properties, and should only be used as qualitative design guidelines.

Table 2: Maximum temperature across interconnect at different diameters

Interconnect Dia (µm)	Temperature for 95Sn/5Au (°C)	Temperature for 80Au/20Sn (°C)
1	1263	76.2
2	149.1	25.7
3	49.1	22.02
4	30.7	21.3
5	25.1	21.3
10	21.3	21.0
20	21.0	21.0

For S2 simulations, we took into account the skin effect that plays a significant role at high operating frequencies. This effect is the tendency of alternating current at higher frequencies to distribute itself within a conductor such that the current density at the surface of the conductor is greater than that at its core. This limits the cross-sectional area of the conductor available for carrying the current, thus increasing the resistance of the conductor to the flow of current through it. The effective resistance is higher than that at DC or low AC frequencies [9R]. A simple formula for the depth of penetration of current in the conductor (skin depth) in vacuum is given as [10R]:

$$\delta = \frac{1}{\sqrt{\pi f \mu \sigma}},$$

where δ the skin depth, f is is the frequency of operation, μ is the permeability of vacuum = 1.26×10^{-12} H/μm, σ is the electrical conductivity of the material. The value for skin depth is calculated at frequencies varying from 100MHz – 50GHz. The electrical conductivity of 80Au/20Sn is approximated to that of pure Au and the electrical conductivity of 95Sn/5Au was approximated to that of pure tin. Figure 8 shows the plot of skin depth vs. frequency for both solder compositions. At higher frequencies the depth of penetration decreases i.e. the current is more concentrated at the outer surface of the interconnect.

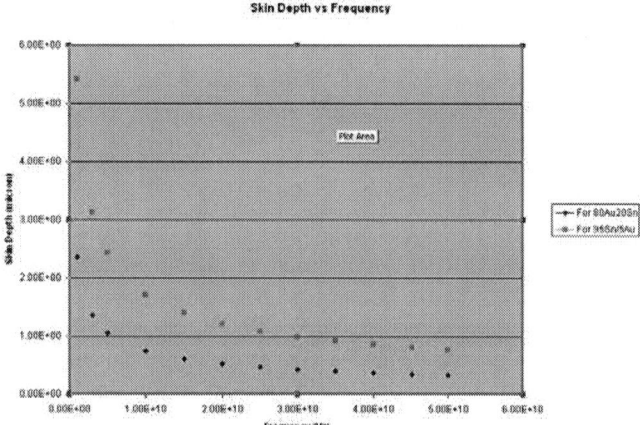

Figure 8: Plot of skin depth vs. frequency for 80Au/20Sn and 95Sn/5Au

The available area of interconnect for current flow decreases which in turn increases the effective resistance of interconnect. The area of interconnect is calculated as:

$$A = \pi[r^2 - (r-\delta)^2] = \pi(2r\delta - \delta^2),$$

where, r is the radius of the interconnect and δ is the skin depth. If the interconnect height is l, the effective resistance becomes:

$$R = \rho \frac{l}{\pi(2r\delta - \delta^2)}$$

This implies that the resistance of interconnect for smaller diameters is high at high frequency of operation. Thus the power loss across interconnects due to heat dissipation (i.e.

Joule heating) would be higher at high frequencies and lower interconnect diameters. Figures 9-10 plot the AC power dissipation for 100mA currents at different operating frequencies and varying values of interconnect diameters.

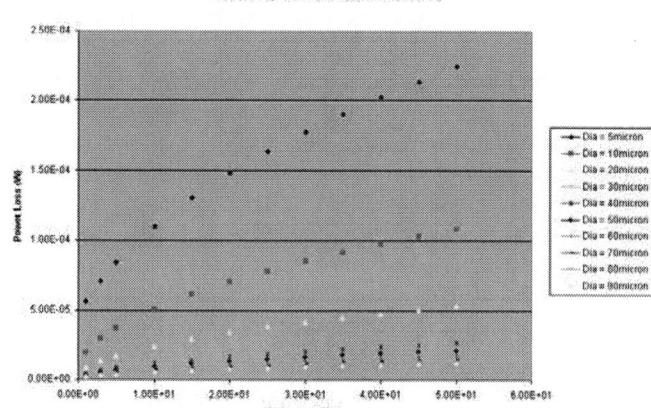

Figure 9: Power losses vs. frequency for different interconnect diameters for 80Au/20Sn

We notice that at an operating frequency of 1GHz, the penetration depth is 2.36μm for 80Au/20Sn and 5.42μm for 95Sn/5Au. Hence good guidelines for selecting the minimum interconnect diameter for 80Au/20Sn should be 5μm, and for 95Sn/5Au it should be 15μm. Again, the calculated skin depth and power loss values for S2 analysis depend on the material property assumptions and should be viewed as qualitative design guidelines.

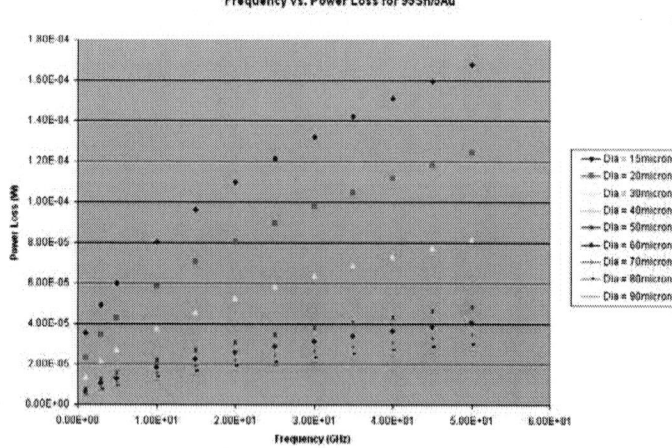

Figure 10: Power losses vs. frequency for different interconnect diameters for 95Sn/5Au

5. CONCLUSIONS AND FUTURE WORK

This paper presents preliminary simulation and experimental results for forming 3D interconnects between bonder wafer pairs using BCB. Preliminary experimental results are presented validating the process of simultaneous wafer bonding and interconnect formation. Preliminary modeling results provide guidelines for selecting minimum interconnect diameters for 80Au/20Sn and 95Sn/5Au in low-frequency/high frequency 3D packaging applications.

While small interconnect diameters (<10μm) help in achieving a high interconnect density, the power losses due to heat dissipation and skin effects are undesirable. Furthermore, it is also more difficult to achieve precise wafer alignment with small interconnect diameters. Given that typical MEMS devices are powered via pads designed for wire-bonding with dimensions between 50 to 200 μm, the proposed wafer-level stacking process is suitable as an alternative.

Future work will include direct AC and DC interconnect characterization, characterization of parasitic capacitance to determine interconnect pitch guidelines, and packaging of actual thermal, electrostatic, and RF-MEMS devices using these interconnects.

ACKNOWLEDGEMENTS

This work was conducted in collaboration with the Bennington Microtechnology Center in Bennington, Vermont. It was supported in part by the Infotonics Technology Center in Canandaigua, New York, and in part by the Office of Naval Research. The authors would like to acknowledge Stanford Nanofabrication Facility at Stanford University and Nanofab Research & Teaching Facility at UT Arlington for the use of their fabrication facility during this research.

REFERENCES:

[1R] Dan O. Popa, Harry E. Stephanou, T. Huang, J.-Q. Lu, B.H. Kang, "BCB Wafer Bonding Compatible with Bulk Micro Machining", in International Electronic Packaging Technical Conference (InterPack2003), Hawaii, July 2003.
[2R] Dan O. Popa, Jeongsik Sin, Woo Ho Lee, Harry E. Stephanou, T. Hwang, Y. Oh, Rensselaer Polytechnic Institute; E. M. Leonard, Eastman Kodak Co. "BCB wafer bonding for microfluidics", in Proceedings-SPIE The International Society for Optical Engineering, 2004
[3R] Takao Ishii, Shinji Aoyama, Masami Tokumitsu "Fabrication of 0.95Sn-0.05Au Solder Micro-Bumps for Flip-Chip Bonding", in J. of Electronic Materials, Vol. 30, No.6, 2001.
[4R] Kim J., Lee C.C., "Fluxless wafer bonding with Sn rich Sn-Au dual-layer structure", Materials Science and Engineering A, v 417, n 1-2, Feb 15, 2006, p 143-148.
[5R] F. Niklaus, R. J. Kumar, J. J. McMahon, J. Yu, J.-Q. Lu, T. S. Cale, and R. J. Gutmann, "Adhesive Wafer Bonding Using Partially Cured Benzocyclobutene for Three-Dimensional Integration", in J. Electrochem. Soc., Vol. 153, Issue 4, pp. G291-G295, 2006.
[6R] Yongchai Kwon and Jongwon Seok, "An Evaluation Process of Polymeric Adhesive Wafer Bonding for Vertical

System Integration", in Japanese Journal of Applied Physics, Vol. 44, No. 6A, pp. 3893–3902, 2005.
[7R] Praveen Pandojirao-S, Rachita Dewan, Ping Zhang, Dan O. Popa, J.-C. Chiao, Harry E. Stephanou, "Electrical interconnects for 3D wafer stacks", to appear in Proceedings of IMAPS Conference, San Diego, 2006.
[8R] Aggarwal A.O., Raj P.M., Sundaram V., Ravi D., Koh S., Mullapudi R., Tummala R.R., "50 micron pitch wafer level packaging testbed with reworkable IC-package nano interconnects", Proceedings - Electronic Components and Technology Conference, v 2, 2005 Proceedings - 55th Electronic Components and Technology Conference, ECTC, 2005, p 1139-1146
[9R] Cheng D., Electromagnetic Waves, Prentice Hall, 1993.
[10R] Bai S.J., Fasching R.J., Prinz F.B., "High density multi-layer connection technology for MEMS and CMOS applications", Proceedings of SPIE - The International Society for Optical Engineering, v 5116 II, 2003, p 536-542.
[11R] Tong, Q., Y., Gosele, U., "Semiconductor wafer bonding: science and technology", John Wiley & Sons, New York, 1999.

WAFER-TO-WAFER AND CHIP-TO-WAFER INTEGRATION SCHEMES FOR SYSTEMS-IN-A-PACKAGE AND 3D INTERCONNECTS

Thorsten Matthias
EV Group Inc.
Tempe, AZ, USA
T.Matthias@EVGroup.com

Stefan Pargfrieder, Herwig Kirchberger, Markus Wimplinger and Paul Lindner
EV Group
St. Florian/Inn, Austria

ABSTRACT

The technical feasibility and the unique technical potential of systems-in-a-package (SiP) and 3D interconnects have been analysed in detail during the last years and are widely acknowledged in the industry. Today the focus lies on innovative manufacturing and integration schemes, which meet both, the economic and technical demands. Even though the fundamental principles of SiPs and 3D interconnects are very similar, the wide range of applications requires a variety of different manufacturing processes.

Layer transfer on wafer level using aligned wafer bonding has the advantages of higher throughput, enhanced cleanliness, and the flexibility that standard fab equipment can be used. 3D stacking using chip-to-wafer bonding focuses on the yield, the "good known die" issue. Dies of different size can be integrated, e.g. several small dies on one big base die, enabling unchallenged small form factors. Furthermore the dies can be produced on different substrate materials, on different wafer sizes, and in different fabs by different producers. This flexibility results in very short time-to-market. The modular concept allows integrating standard components, which significantly reduces the development costs.

Very often the thinned dies have to be processed on both sides prior to bonding. A new ultra-thin wafer handling concept based on temporary bonding enables to process the back-thinned wafers using standard equipment. Precise alignment of the temporarily bonded wafers and subsequent permanent wafer bonding are key for achieving high performance devices. In this paper the recent advances in the integration and handling schemes are discussed with a higher emphasis on the manufacturing requirements.

Key Words: System-in-a-package, 3D interconnect, wafer bonding, 3D integration schemes

INTRODUCTION

The International Technology Roadmap for Semiconductors predicts continuing development towards smaller geometries, higher frequencies and larger chip sizes. The growing number of transistors per unit area results in an increasing number of interconnects. However it is pointed out that the traditional path of downscaling together with material innovation is no long-term solutions for the performance requirements [1].

3D Interconnect using chip stacking offers an approach to vertically connect two integrated circuits for the purpose of shortening the wiring distance between them. This technology enables increasing levels of integration, which increases the performance and the functionality while reducing cost, size, weight and power consumption [2]. Furthermore for next generation applications reduction of signal time delay and parasitic losses, increase of number of neighboring devices and extension of bandwidth are of crucial importance [3,4]. For on-chip clock distribution 3D exhibits a better performance because of the smaller die size [5]. Optical interconnects are an interesting approach for wafer-level heterogeneous hyper-integration [6].

Aligned wafer bonding is a wafer-to-wafer 3D interconnect technology where the wafers are aligned, bonded and interconnected face-to-face or back-to-face, and then thinned-back prior to additional stacking processes or prior to dicing. Wafer bonding and wafer-to-wafer alignment are well-established technologies, which are originating from the MEMS manufacturing. Heterogeneous subsystems like ASIC and MEMS devices might require strongly different, sometimes excluding process steps. Aligned wafer bonding enables heterogeneous functional stacks like CMOS with memory, mixed signal or bipolar (RF). Also heterogeneous material combinations like silicon with compound semiconductors (III-V,..) are possible. Wafer-level packaging using aligned wafer bonding allows combining the cap element with additional functional features thereby enabling functional packages and increasing the functional density.

Figure 1 shows the process flow of aligned wafer bonding. The two substrates are first aligned to each other in the bond aligner. Then the wafer stack is clamped on the bond chuck and transferred to the wafer bonder. This process separation is necessary due to technical as well as economical reasons. Depending on the specific bonding method, wafer bonding involves high forces, elevated temperatures, reducing gas atmospheres or vacuum. These conditions are not compatible with high precision alignment stages for volume production. The economic advantage of this process separation is that typically one bond aligner can support several wafer bonders due to the different cycle times.

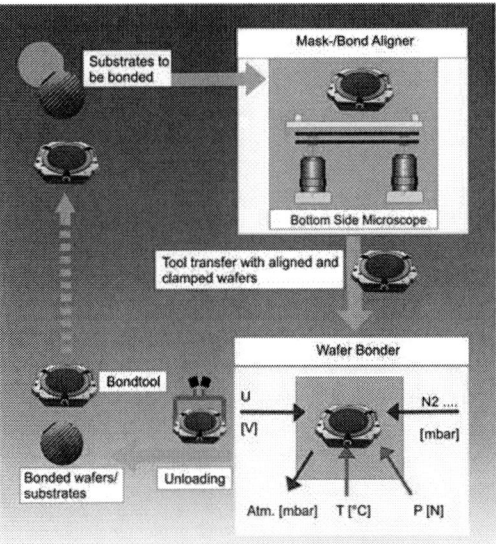

Figure 1: Process flow of aligned wafer bonding

The main alternative is chip-to-wafer (C2W) integration. As single dies are mounted on the base wafer, this approach enables mounting of multiple dies next to each other on one big base die. As via formation after bonding is difficult for C2W devices, via first approaches are the preferred integration scheme for C2W. A metal-metal thermo-compression bond transforms the bond pads into electrical interconnects. For the most promising methods, Cu-Cu bonding and the Cu-Sn solid-liquid-interdiffusion (SOLID) process, metal ion diffusion is the main bonding mechanism [7].

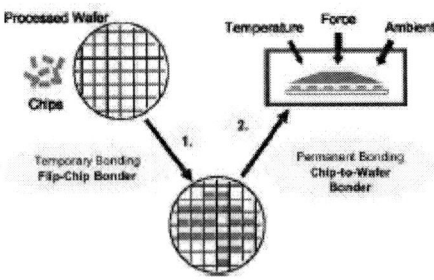

Figure 2: Advanced chip-to-wafer (AC2W) bonding: Schematic process flow

A new integration scheme, the advanced chip-to-wafer (AC2W) bonding targets these applications. As the diffusion rate is proportional to temperature, pressure and time it is not possible to perform the bonding as single-die process. The basic idea behind AC2W is to split the bonding process into two sub processes. The alignment and temporary pre-bonding is performed on a pick-and-place machine, whereas the permanent bonding of the dies is performed as a batch process in a dedicated bond chamber (Figure 2).

The AC2W bond chamber enables controlled process gas pressure or vacuum and programmable heating and cooling ramps. The permanent bonding process requires a controlled, homogeneous force to be applied on every single chip. One outstanding feature of the chip-to-wafer bonder focuses on the "known good die" problem (Figure 3), which arises if non-working die on the bottom wafer are not bonded with any top device. The result is a possible shift in the center of gravity away from the center of the bottom substrate. This results in the requirement to also shift the center of gravity of the bonding process. To address this, the center of force can be shifted within a 150mm diameter. In addition to variable center of force, different process technologies require different absolute force and variable coverage of the bottom wafer with top chips due to different-sized top chips, a bottom wafer with a different grid size, or variable yield. To handle these varied force requirements, the chip-to-wafer bonder enables programmable bond force from 150N up to 40kN.

Figure 3: Shifted contact force due to "Known Good Die" (KGD)

Figure 4 shows a schematic of 3D chip stacking. Two processed wafers are aligned and bonded face-to-face. The top wafer is then thinned down to a few microns thickness, and then high aspect ratio via holes are etched through the backside of the thinned wafer to provide vertical electrical connections between the two wafers. In this manner, electrical connections that may otherwise have been several centimeters long are now only a few microns instead. This is dramatically enhancing the performance of such circuits [8,9]. After thinning, the vias become accessible from the topside allowing the process to be repeated, so that a stack of multiple wafers can be achieved. A final thinning step can then be performed on the backside of the resulting wafer stack.

There are several fundamental differences between W2W and C2W approaches in terms of technology, economical considerations and also manufacturability [10]. Yield and throughput are considered the most important factors. Other factors are the feasibility and capability for post-bond processes, the interconnect density and the number of stacking levels. Depending on the application each of the integration schemes may have significant advantages.

Wafer-level processes have the inherent advantage that the cycle time is independent of the number of dies on the wafer. This is also true for all other processes down the line prior to dicing, so the more subsequent process steps the bigger is the throughput gain due to wafer-level processing. The throughput for single-die processes scales with the number of dies. As the cycle time for wafer-level processes is nearly independent on the wafer size, W2W shows even bigger throughput advantages for 300mm wafers, whereas for 150mm wafers (e.g. MEMS or compound semiconductor devices) C2W can be competitive for medium to large dies. For pick-and-place tools there is a trade-off between speed and accuracy.

Figure 5 shows a throughput comparison for W2W and C2W alignment (incl. pre-bonding clamping on bond fixture). In case of low chip count of only a couple of hundred dies the high-speed C2W machines outperform the W2W approach for low alignment accuracy requirements of 10 micron or above. However, high alignment accuracy on C2W comes always at the cost of throughput due to the necessity for vibration damping and alignment control feedback loops.

Wafer Bonding for 3D Interconnect

The research and development during the last decades resulted in avariety of well-established bonding methods. But the specific requirements of CMOS processing limit the potential bonding methods. Only low temperature methods can be used in order that the devices are not destroyed. The best-investigated wafer bonding techniques for 3D interconnects applications are Cu-Cu diffusion bonding [11-13] and adhesive bonding using Benzocyclobutene (BCB) [2,14-18]. Recently Silicon direct bonding gained more interest due to the development of wafer surface plasma activation, which enables to reduce the annealing temperature to below 400°C [19].

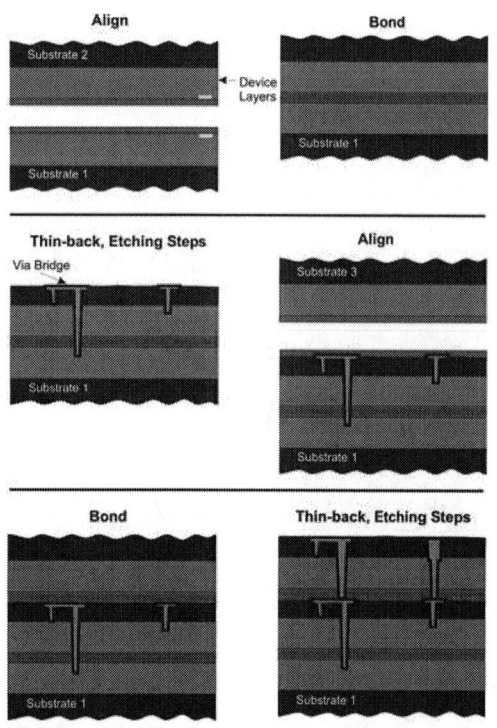

Figure 4: 3D Interconnects through aligned wafer bonding

Kwon et al. investigated the correlation between BCB layer thickness and bond strength. The experiments revealed that the critical adhesion energy increases linearly with BCB layer thickness. But even thin BCB layers starting from 0,4µm thickness achieved sufficiently high bond strength enabling reasonable aspect ratios for the inter-wafer via interconnects [15]. The bottom wafer of the wafer stack gives mechanical support during the back thinning of the top wafer. Therefore no handling substrates are required, which keeps the process flow

simple. A three step thinning process consisting of backside grinding, chemical mechanical polishing (CMP) and wet etching enables remaining wafer thicknesses of below 1µm [17]. Lu et al. evaluated the impacts of bonding and thinning on the electrical properties of state-of-the-art 130 nm technology CMOS SOI test structures and reported that the bonding and thinning process flow does not significantly affect 130 nm technology CMOS device parameters [18]. The packaging compatibility has been analyzed by autoclave and liquid-to-liquid thermal shock (LLTS) testing as well as wafer sawing. Neither degradation of bond strength nor delamination of the BCB layer has been observed [17,18].

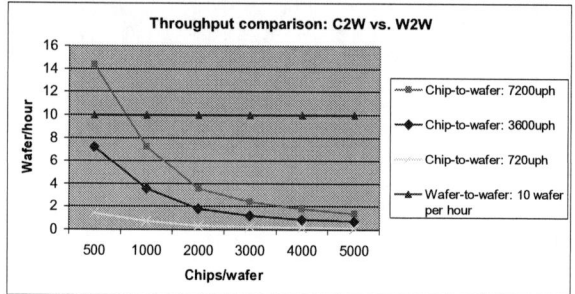

Figure 5: Throughput comparison: W2W vs. C2W alignment [10]

The process of Cu-Cu wafer bonding as well as the correlation between the bonding process, the Cu microstructure and the resulting bond strength have been investigated in depth [11,12]. Recently Morrow et al. reported that they have achieved a wafer-level Cu-Cu 3D integration approach capable of delivering yields required for high-volume production [13].

The advantage of the Cu bonding technique is that the bond pads directly connect the two devices electrically and thermally. Thereby the process steps of high aspect ratio via etching and filling are avoided. The advantage of adhesive bonding is that the required specifications for wafer surface properties and cleanliness are relaxed as any particulates are embedded in the adhesive layer.

Typically after bonding one of the wafers has to be back thinned. Any un-bonded area (void) might result in wafer breakage during back grinding. Therefore a void in the bond interface is not acceptable and the wafers have to be rejected from further processing.

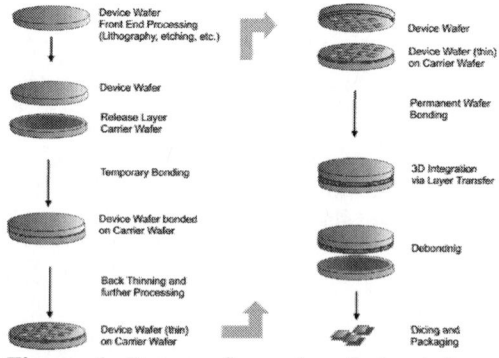

Figure 6: Process flow of wafer-level layer transfer applying temporary and permanent wafer bonding.

56

Wafer-level Layer Transfer

The alternative approach to bonding bulk wafers and subsequent thinning and processing is to perform the wafer thinning and final processing prior to bonding the device wafers. A widely accepted possibility in the semiconductor manufacturing industry to prevent thin wafers damages during processing is to temporarily bond the device wafer to a suitable rigid carrier substrate prior to the back thinning process.

Figure 6 shows the process flow for wafer-level layer transfer using temporary and permanent wafer bonding. The originally thick device wafer is bonded with its active surface to a carrier wafer using an adhesive intermediate layer. After backside processing, including the thinning process and further process steps (lithography, etching, via filling, CMP,...), the thin device wafer, supported by the rigid carrier substrate has to be permanently bonded to the second device wafer. The possible alignment methods are the same as for the bulk wafer approach. Very often transparent alignment can be used as Silicon gets transparent in the low-micron thickness range. Finally the rigid carrier substrate has to be released either in parallel to the permanent bonding of the device wafers or during a separate process step.

The adhesives for temporary bonding allow the release of the device wafer by using different approaches: UV release adhesives de-bond after exposure in UV light, thermal release adhesives have to be heated above a characteristic release temperature and solvent release adhesives have to be dissolved in a chemical solvent for de-bonding of the device wafer from the carrier. Adhesive layers for temporary bonding can also be differentiated regarding their physical condition: Liquids (waxes and thick resists) and dry-film laminates.

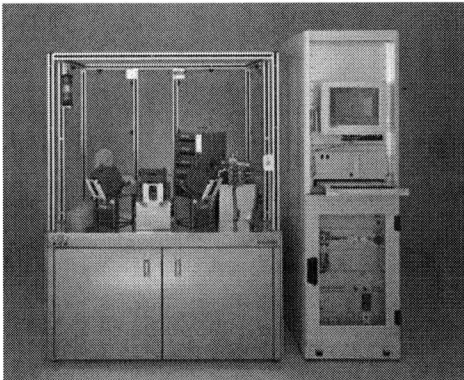

Figure 7: EVG®850 Temporary bonder for wax bonding

Subsequent advantages of using a rigid carrier are the protection of the active surface of the device wafer during grinding and polishing procedures and the flattening of probably strongly warped wafer material. The growing demand on these methods is a result of the increased importance of applications based on compound semiconductors or ultra-thin wafers.

The temporarily bonded stack, consisting of device wafer, intermediate layer and carrier wafer, is generally further processed by using several different techniques (e.g. lithography, etching, etc.). Therefore several different aspects have to be taken into account before selecting the intermediate layer for the targeted application.

Maximum Temperature Capability

Today's available waxes or dry-film laminates enable maximum process temperatures up to 200°C. High temperature capable resist can withstand process temperatures up to 300°C, before showing degradation or delamination effects.

As matter of fact, high temperature resist is usually released and de-bonded by using a solvent bath. Chemical resistivity of this resist against solvents in processes used after bonding, as well as fully automation and integration into a production line of this approach might be challenging.

Chemical resistance

The chemical resistance of the intermediate material to the chemicals used during the processing of the stacked wafers (back-grinding, coating and developing, etching. etc.) is a further key issue. A chemical attack, especially in the border areas of the intermediate layer, can affect the properties of the intermediate material and therefore might influence the further backside processing steps or substantially change the de-bonding behavior. In case of unexpected release of the intermediate layer due to poor chemical resistivity, the protected front-side (via the intermediate layer) of the device wafer might also be influenced.

Vacuum capability

In various processes in the semiconductor manufacturing line (e.g. etching) the handling of the bonded stack in a vacuum environment is a need. Trapped air in the interface between the bonded wafers as well as in the intermediate material itself tend to discharge if processed in a vacuum chamber especially through the thin and brittle device wafer.

Therefore optimized process conditions have to consider various aspects like device and carrier wafer flatness, good coating uniformity in case of wax/resist usage (mainly focusing on the edge exclusion zone). In case of tape usage, the thickness variation of the tape (TTV, total thickness variation) and void free lamination are reflecting main issues. For all applications reliable and well-adjusted bonding conditions (e.g. good pressure uniformity) are a key factor to enabling high yield processes.

Ease of processing

This criterion involves mainly the ease of material application (spin coating in case of wax/resists and lamination in case of dry-film tape), the ease of the de-bonding process (heat release, UV-release or solvent release) and its degree of achievable automation as well as the ease of cleaning of the device wafer after de-bonding.

On wax applications the wax remains on the device as well as on the carrier wafer after the de-bonding process. Therefore, basically an intensive usage of solvents is required for removing and cleaning of the device wafer as well as carrier wafer. This might be done in a wet bench or a cleaning chamber alternatively. Resist applications will also require a final cleaning step of both surfaces of device and carrier wafer respectively, after the de-bonding in a solvent bath.

With usage of dry-film laminates there is basically no remaining material left on the wafers after de-bonding. The tape can be removed with peel-off technology. Eventually remaining residuals on the bonded surface of the wafers might be removed in a cleaning chamber. To overcome the need of solvent usage, a protective coating applied on the wafers, which can be resolved with water can be proposed as an alternative.

Thickness Variation
As the remaining thickness of the device wafer is focusing on ever-lower values, the thickness variation across the wafer is getting more important, even in a volume-manufacturing environment.

In order to keep the device wafer within the requested thickness uniformity after back thinning, it is imperative that the intermediate layer as well as the carrier wafer have comparable or even lower thickness variation. Waxes and resist applications are usually applied with spin coating, where the achieved uniformity is well below <1%. Suppliers of dry-film laminates are working on this issue since many years and have shown TTV (Total Thickness Variation) specification <2µm, which has been proven to be sufficient e.g. for the mechanical back-grinding process.

Others
Other important criteria, which have to be considered before committing to the type of intermediate adhesive, comprise sensitivity to UV light, adhesive bond strength, thermal expansion and environmental considerations.

Temporary Bonding and De-bonding with Waxes
Reversible wafer bonding, using low or high temperature waxes (up to 170°C), requires a wax coating step in liquid phase. The highest level of uniformity for the spin-coated wax is essential as mentioned above. Solid wax sheets (e.g. Wafergrip from Dynatex) are also commercial available, but the achievable degree of automation is lower than with liquid waxes.

The process of wax coating and bonding as well as the de-bonding process developed on fully automated systems will be reviewed in more detail.

The process for wax coating and reversible bonding was developed on an EVG®850 temporary wafer bonding system. The EVG®850 is a fully automated temporary

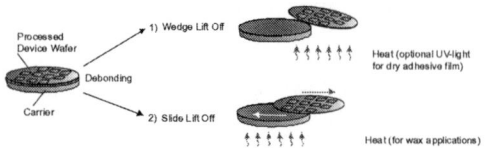

Figure 8: De-bonding process flow

bonding system up to 200mm substrates configured for wax bonding with a wax coating station, hot and chill plates and a bond chamber. The EVG®850 can also be utilized for thick resist bonding. The device wafers and carrier substrates are stored in cassettes for automated loading (Figure7).

A robot arm performs the wafer handling during the whole process. After optical pre-alignment the device wafers are spin coated in a coating module. The wax is dispensed at elevated temperatures onto the active device wafer surface or the carrier wafer. The temperature of the coating chuck can be adjusted (heating capability) to ensure a uniform layer thickness of the spin coated wax layer across the wafer. The thickness of the spin coated wax layer ranges

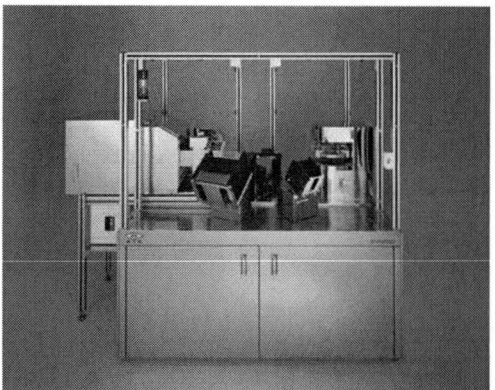

Figure 9: EVG®850 De-Bonder for thermal release de-bonding

up to some tens of µm (e.g. 20-50µm). The temperature uniformity of the heated spinner chuck is better than ±0.5%. After coating the substrates get transferred to a bake module in order to remove the solvent prior to bonding. The bond occurs in the bonding chamber under controlled atmosphere. To get void-free bonds this step is performed under high vacuum (5.10^{-2} mbar) and controlled temperatures with temperature uniformity down to ±1% up to 250°C. A uniform force in the low kN range is applied to achieve best bonding results. After the bonding step the wafer pair is loaded to a chill module and then transferred to the receive cassette.

The bonding strength for wax intermediate layer bonding is about 2 J/m2, a value large enough to withstand even harsh mechanical processes like grinding and polishing.

For the release of the device wafer from the carrier substrate two methods can be used. The wax can be dissolved in a solvent (solvent release) or it is fluidized by heating up (thermal release) in a de-bonding module. The de-bonding module enables a so called "Slide Lift-off" approach (Figure 8). Figure 16 shows a fully automated EVG®850 for thermal release de-bonding. The carrier wafer can be reused for further processes after de-bonding and cleaning.

The EVG®850 De-bonder in Figure 9 is configured for film frame de-bonding. The processed wafer stack and a standard metal film frame are transferred from the send

58

cassettes to the film frame mounting module which applies a special polymer carrier tape (e.g. dicing tape) onto the metal frame and the wafer stack at once. The mounted wafer stack (device wafer is facing the film frame tape) is transferred with a special robot end-effector (see Figure 10) into the de-bonding module. In a controlled manner (temperature controlled top and bottom chuck, wedge compensation, dwell time, wedge-off or slide-off de-bonding) the thin device wafer is detached from the carrier substrate and remains on the film frame. Finally the device wafer and the carrier substrate are cleaned from the intermediate layer residues on a cleaning module, thus the carrier wafer can be immediately reused and the device wafer is ready for further back-end processing (testing, dicing, packaging).

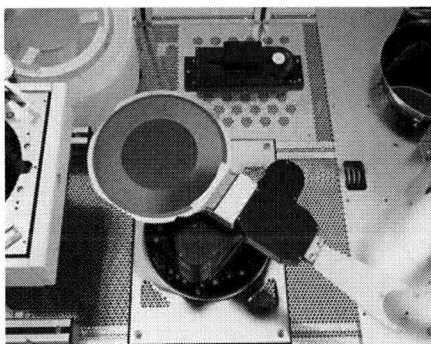

Figure 10: Special End-effector for Film Frame de-bonding

The EVG®850 De-Bonder can also be easily configured for temperature or UV-release of dry-film intermediate layers. As an alternative, the device wafer can be directly de-bonded from the carrier without usage of the film-frame and dicing tape. Fully automated handling and unloading of the thin device wafer can then be performed e.g. with Bernoulli-End-effectors.

Optional de-bonding on customized carrier substrates (e.g. wafer transfer on vacuum release trays or electrostatic chucks, ESC) are further possibilities on the EVG®850 De-Bonder.

Temporary Bonding with Dry-Film Laminates
The recently increasing popularity of dry-film adhesive tapes (e.g. Sekisui, Revalpha, Rexpan) especially for thermal-release bonding in the thin wafer industry can be mainly attributed to the ease of the application, the enhanced thermal release temperature and the improved TTV.

The EVG®850 Temporary Bonder with dry-film lamination is the first integrated fully automated tool which addresses these industry trends for the high volume manufacturing. A new method for cutting the dry-film laminate is implemented in the EVG®850 and shows several advantages compared to laser or blade cutting technologies. The dry-film adhesive tape, supplied to the machine on a reel-to-reel basis, gets punched out according to the requested dimension and is laminated onto the carrier substrate with optical tape-to-carrier

alignment within +/-15µm after automatically removing both protective films. Figure 12 is illustrating the process flow in the dry-film lamination module.

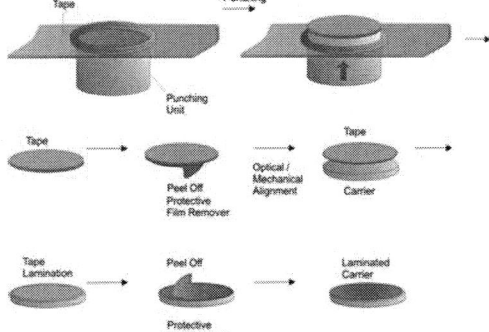

Figure 12: Schematic process flow of the dry-film lamination module on the EVG®850 fully automated Temporary Bonder for dry-film adhesives

The laminated carrier gets optically inspected (tape failures, voids, particles) before it is transferred to the bonding module (right side in Figure 11). This module is configured similarly to the EVG®850 Temporary Bonder for waxes. The device wafer can be coated optionally with a protective coating and will be bonded to the laminated carrier substrate in a bond chamber under controlled process parameters (vacuum, temperature, force, pre-bond separation distance, mechanical alignment).

Figure 11: The fully automated EVG®850 Temporary Bonder with dry-film Lamination

Advantages of the punching technology compared to the cutting technologies (laser and blade cutting) are better edge quality for minimum interference with the bonding and back-grinding process, no carrier edge degradation through cutting blades, more flexibility in tape dimension and shapes (tape diameter smaller than carrier substrate diameter possible, thus enabling pyramid structured assemblies for minimum wafer edge breaking rates during mechanical back-grinding) and almost no wear of the punching module components (punch and die-plate).

CONCLUSIONS
In this paper innovative manufacturing processes for vertical integration are reviewed. Chip-to-wafer (C2W) and wafer-to-wafer (W2W) are both viable approaches. C2W offers a higher flexibility and shorter product

development cycles. W2W has several limitations, but has high potential for high device performance.

A 2-step C2W bonding process is presented combining the flexibility of a pick-and-place tool with the interconnect performance of a metal-metal bond. A variety of well-established wafer bonding techniques exists. Depending on the requirements of the process and of the final application the appropriate bonding method has to be carefully chosen. Cu-Cu diffusion bonding and BCB bonding have been proofed to be fully CMOS compatible.

Thin wafer handling based on temporary bonding allows backside processing using standard fab equipment. Fully automated temporary bonding equipment (EVG®850 Temporary Bonder) is capable to mount the device wafer onto a carrier substrate with its active side facing the intermediate adhesive layer, thus preparing the device for secure and reliable back-grinding and backside processing steps. The intermediate adhesive layer can either be applied in liquid form (wax or thick resists) via spin coating or in rigid form (dry film adhesive tape) via fully automated lamination.

REFERENCES

[1] International Technology Roadap for Semiconductors 2003

[2] J. Lu, Y. Kwon et al., "Processing Technology for High Density Multifunctional Integration (HDMI) using Wafer Bonding and Monolithic Inter-Wafer Interconnection", (invited), in 19th International VLSI Multilevel Interconnection (VMIC) Conference, pp.445-454, Singapore, November 18 - 20, 2002.

[3] Conference proceedings of "3D architecture for semiconductor integration and packaging", April 13-15 2004, San Franscisco, CA

[4] Symposium proceedings of "3D Technology, Modelling and Process Symposium", April 13 2004, San Franscisco, CA

[5] Kuan-Neng Chen, Mauro Kobrinsky, Brandon Barnett and Rafael Reif, "Comparisons of Conventional, 3D, Optical and RF Interconnect for Clock Distribution," *IEEE Trans. on Electron Devices,* 51(2), pp 233-239, 2004.

[6] P. D. Persans, M. Ojha, R. J. Gutmann, J.-Q. Lu, A. Filin, J. Plawsky, "Optical Interconnect Components for Wafer Level Heterogeneous Hyper-Integration", Mat. Res. Soc. Symp. Proc. Vol. 812, F6.11.1

[7] Holger Hübner, et al., "Face-to-face chip integration with full metal interface", Advanced Metallization Conference, San Diego, 2002.

[8] A. Rahman, A. Fan, J. Chung and R. Reif, "Wire-Length Distribution of Three-Dimensional Integrated Circuits", IEEE International Interconnect Technology Conference, May 24-26 1999, San Francisco, CA, pp.233-235, 1999.

[9] S. J. Souri and K. C. Saraswat, "Interconnect Performance Modeling for 3D Integrated Circuits with Multiple Si Layers", IEEE International Interconnect Technology Conference, May 24-26 1999, San Francisco, CA, pp.24-26

[10] T. Matthias, S. Pargfrieder, M. Wimplinger, P. Lindner, „SiP and 3-D interconnects", Advanced Packaging Magazine Aug./Sept. 2006, pp32

[11] R. Reif, C. S. Tan, A. Fan, K.-N. Chen, S. Das, N. Checka, „3-D Interconnects using Cu Wafer Bonding: Technology and Applications", Advanced Metallization Conference, San Diego, Oct. 1-3, 2002

[12] K. N. Chen, C. S. Tan, A. Fan and R. Reif, "Morphology and bond strength of copper wafer bonding", *Electrochemical and Solid-State Letters,* 7 (1), pp G14-G16, 2004.

[13] P. Morrow, M. Kobrinsky, S. Ramanathan, C. Park, M. Harmes, V. Ramachandrarao, H. Park, G. Kloster, S. List, S. Kim, proceedings of Advanced Metallization Conference 2004

[14] Y. Kwon, J-Q. Lu, R.J. Gutmann, P.P. Kraft, J.F. McDonald and T.S. Cale, "Wafer bonding Using Low-K Dielectrics as Bonding Glue in Three-Dimensional Integration", The Electrochemical Society, 2001 Joint Meeting, San Francisco, Ca, September 2-7, 2001

[15] Y. Kwon, J. Yu, J.J. McMahon, J.-Q. Lu, T.S. Cale, R.J. Gutmann, "Evaluation of Thin Dielectric-Glue Wafer-Bonding for Three Dimensional Integrated Circuit-Applications", Mat. Res. Soc. Symp. Proc. Vol. 812, F6.16.1

[16] Frank Niklaus, „Adhesive wafer bonding for microelectronic and microelectromechanical systems", PhD Thesis, Stockholm, 2002

[17] S. Pozder, J.-Q. Lu, Y. Kwon, S. Zollner, J. Yu, J.J. McMahon, T.S. Cale, K. Yu, and R.J. Gutmann, "Back-End Compatibility of Bonding and Thinning Processes for a Wafer-Level 3D Interconnect Technology Platform", IITC 2004

[18] J.-Q. Lu, T.S. Cale, R.J. Gutmann, "Dielectric adhesive wafer bonding for back-end wafer-level 3D hyper-integration", ECS Meeting, Honolulu, Oct. 3-8, 2004

[19] Viorel Dragoi et al., "Low temperature MEMS manufacturing processes: plasma activated wafer bonding", Mater. Res. Soc. Symp. Proc. Vol. 872 (2005), J7.1.1

PATTERN EFFECTS ON ELECTROPLATED COPPER PILLARS

Arthur Keigler, Bill Wu, Jim Zhang and Zhenqiu Liu
NEXX Systems, Inc.
Billerica, MA, USA

ABSTRACT
Electrodeposited copper pillar structures provide advantages for some packaging applications. Economical manufacturing requires maximizing the deposition rate while achieving the coplanarity, flatness, and morphology required by the overall packaging process. Copper pillars are typically 50 to 80 microns tall with diameters from 40 to 100 microns. Adjacent heat sink pad structures may be formed during the same plating operation. Deposition control is strongly mediated by the organic additive package and depends sensitively on boundary layer thickness. This paper describes the effects of deposition rate on the process operating window for various pattern structures.

Key words: flip chip, wafer bumping, electroplating, copper, WLCSP

INTRODUCTION
Copper pillars provide several advantages in comparison to other flip chip attachment processes:

1) Copper has one forth the electrical resistance of PbSn solder bumps, reducing power consumption and heat generation within the package.

2) Copper has higher current density capability then PbSn bumps allowing for a smaller bump diameters and/or fewer I/O per device since fewer parallel power and ground connections may be required.

3) Better thermal conductivity of copper provides >3 times the thermal conductivity of solder bumps; in some applications therefore additional heat transfer pads may be fabricated during the same operation.

4) Copper pillars do not collapse, allowing for very fine pitch without compromising stand-off height, enabling perimeter attachment to lead-frame or die-center array without redistribution.

5) Copper is not an alpha-particle emitter, providing a low-alpha solution that doesn't require expensive materials.

6) Copper plating for package attachment allows straightforward integration with copper inductor and air-bridge type passive circuitry.

A disadvantage of copper pillar structures for packaging is the higher elastic modulus of copper relative to PbSn which results in higher built-in stress after soldering due to thermal expansion coefficient mis-match between the silicon and the package. This effect is partially mitigated since the Cu-pillar moves the weak Cu-Sn intermetallic interface away from the die surface and puts it at the package solder interface where it may be embedded in solder; at the die interface is a less fragile copper to copper joint. A cross section view of copper pillar with solder cap is shown in figure 1.

Figure 1. Cross section of Copper pillar with SnAg solder cap.

Functionality of copper pillars depends on several structural factors including co-planarity of pillars and pads within a die, flatness on top of pillars or pads, and roughness of deposit. In the case of thin wafers low as deposited stress and post copper annealing stress are important to avoid bending the wafer after thinning prior to singulation.

EXPERIMENTAL
Several different photoresist test pattern structures were deposited in a commercial plating system wherein the boundary layer thickness at the wafer surface can be controllably varied. Resist pattern variables that are typical for various flip-chip copper pillar applications were used to investigate their effect on process operating window. These were pillar diameter, variation of feature size within die, and feature pitch. The other main contributing factor to copper pillar processing, chemistry type and operating point, was investigated by comparing materials from several different vendors.

Measured variables were determined using a combination of optical and SEM microscopy and thickness measurement using a Tencor P12 contact type scanning profilometer.

The typical composition of Cu chemistry is listed in the table 1.

Table 1 Typical composition of copper chemistry

Cu (g/L)	H2SO4 (g/L)	Cl PPM	Additives
22	190	50	Commercial package

RESULTS AND DISCUSSION

Primary effects studied were the deposition rate, the chemical additive system, and the local resist pattern diversity. These are presented and discussed below.

EFFECT OF DEPOSITION RATE

A die pattern with several different pillar diameters was used to investigate the morphology dependence on deposition rate for two different boundary layer thicknesses. Precise estimation of boundary layer thickness was accomplished using limiting current measurements and is described in detail by Wu[1]. Boundary layer thickness was altered by changing the speed of the fluid agitation mechanism in the plating cell. For most results a thickness of approximately 70 microns, comparable with a fountain cell type of plating system, was compared to a thickness of 15 microns in a vertical Shear Plate process cell.

Through mask feature sizes were 80, 100, and 150 micron diameter circular openings in 100 micron thick dry film photoresist.

Several commercial copper chemistries, typically with three additive components, suppressor, leveler, and accelerator, were used for these experiments. Additives are adjusted to be on target to vendor specifications using CVS measurements.

Figure 2a, 2b, and 2c shows two effects as the deposition rate is increased. One, as the rate is increased the height difference between the three size openings increases. Two, the morphology changes, first the pillar surface becomes dome shaped and secondly at higher rate the deposit shows a cauliflower shape indicative of dendritic growth.

Figure 2a. Pillar morphology at 0.5 um/min deposition rate with 15 micron b.l. thickness.

Figure 2b. Pillar morphology at 1.0 um/min deposition rate with 15 micron B.L. thickness.

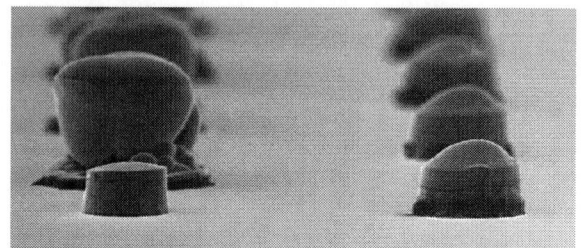

Figure 2c. Pillar morphology at 2.0 um/min deposition rate with 15 micron B.L. thickness.

A combination of suppressor, accelerator, and leveler organic molecules are used to slow down the copper deposition reaction and thereby reduce the influence of primary current and inhibit its inherent tendency to deposit thicker at edges of features (i.e. form dished deposits) and thicker on isolated features or small features. Although the overall set of reactions is complicated and of much scientific interest which is beyond the scope of this paper, there are several main effects the understanding of which is important for controlling production baths.

First, the suppressor and accelerator molecules form complexes with chloride and cuprous ions. These compete for absorption sites on the metal copper surface and differently influence the reduction reaction of copper ions. Although both suppressor and accelerator impede the reaction, with more accelerator adsorption, the reaction rate will be increased at local deposition site. Second, the concentration of additives in the plating bath is very dilute, in PPM level. The mass transfer of additive molecules from bulk solution to the interface of copper and electrolyte therefore has a significant effect on the growth interface reaction rates. Third, in general, for a higher current density or deposition rate more additive transport to the interface is needed to maintain the same level of kinetic control.

Photoresist pattern effect on bump shape and deposition rate can be understood with respect to Figure 3. A copper pillar resist pattern is shown in cross section with approximately a 1:1 aspect ratio in 70 micron thick resist along with a depiction of a thin boundary layer and the associated species which must diffuse through said boundary layer.

Organic additives and copper ions diffuse through the boundary layer to the growing metal copper surface. Copper ions are consumed by the deposit. Organic molecules may be either incorporated into the deposit or have their activity altered by the deposition reaction. For example, the accelerator combines with Cu+ to form accelerant at the interface between the copper and electrolyte. Meanwhile, the spent accelerant is transported away from the interface and consumed at bulk solution. Supplying sufficient accelerator to the growth interface is very critical to ensure relative high deposition rate at the deposition site [2].

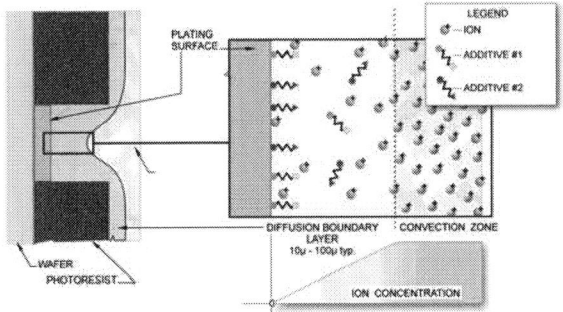

Figure 3. Boundary layer activity involved in through mask electrodeposition of copper

At a relatively low deposition rate, there are enough additives and cupric ions supplied to the growth interface due to low demand from the reaction. The diffusion of additives through the diffusion boundary layer is not mass transfer limited. Large features and small features receive similar surface concentrations of additives, so that there is less variation of deposition rate between large and small feature, as shown in the Figure 2(a).

At higher deposition rates the mass transfer of additives to the interface reaction starts to play a role. The large feature receives more accelerator than small feature due to having a lower aspect ratio. The variation of reaction rate between large and small feature becomes larger as shown in Figure 2(b).

At high the deposition rate, for example 4 microns per minute, both copper ions and organic additives are mass transfer limited and dendritic growth occurs, as shown in Figure 2(c).

EFFECT OF BOUNDARYLAYER THICKNESS
Deposition with a thicker boundary layer shows that the deposit degradation occurs at lower current density. Figure 4 shows that pillars deposited at 0.5 microns per minute with a thick boundary layer are comparable to those deposited at 1.0 microns per minute with a thin boundary layer. At 1 micron per minute the thick boundary layer is starting to limit the supply of additives and Cu ions to the Cu/electrolyte interface.

Figure 4a. Pillar morphology at 0.5 um/min deposition rate with 70 micron B.L. thickness.

Figure 4b. Pillar morphology at 1.0 um/min deposition rate with 70 micron B.L. thickness.

A thin diffusion boundary layer allows for more economical processing by allowing for a higher deposition rate for a given feature size and aspect ratio, and a uniform boundary layer ensures wafer level uniformity of pillar bumps.

INFLUENCE OF CHEMISTRY
Some applications require both copper pillars and thermal spreader or conductor bus type of features. A test pattern was used with 80 micron diameter circular bumps and 400x100 micron rectangular bars in a mixed array of approximately 160 micron pitch.

Figure 5 shows the normalized height difference between small and large features, where each point is the average delta for 5 die on the wafer, and normalization is done with respect to the average of 5 sites measured within each die.

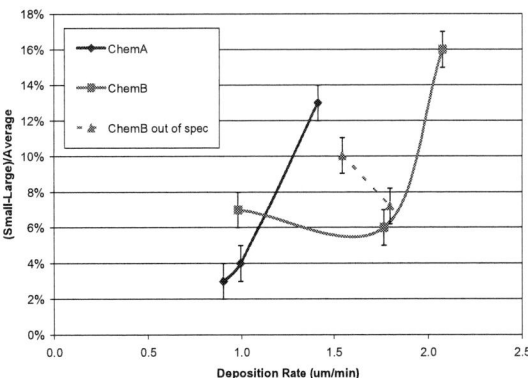

Figure 5. Deposition rate dependence of height difference between small and large feature for 60 micron deposit in 85 micron resist. Small feature is 80 micron diameter; large feature is 400x100 microns.

For the both chemistries higher deposition rate causes the feature size difference within the die to get worse. Also it is important to note that different chemistries have strengths and weaknesses in this regard.

Chemistry A has a lower absolute difference in thickness, but the feature size uniformity degrades strongly at deposition rates above 1 micron per minute. Chemistry B does not provide as low of a feature height difference at low deposition rates, but is able to sustain that difference up to almost 2 microns per minute.

Since the additive system is the only difference between the chemistry A and B, the additives in the Chemistry A are shown to have a stronger leveling effect than chemistry B. However, it seems like the concentration of additives in the Chemistry B is higher than the Chemistry A, so it is able to maintain the leveling function at a relative wide range of deposition rates.

Chemistry B also shows an anomalous effect of worsening feature size height difference at lower deposition rate. This unexpected behavior was checked several times, and shown to be aggravated as the additive package was further from the optimum operating condition.

SUMMARY

Patterned through mask deposition for copper pillar application can be an economical process for flip chip packaging. Diffusion of organic species to the growth interface and competing reactions at the interface determine the deposit shape and uniformity. For a given feature size and aspect ratio the deposition rate must be matched to the boundary layer thickness to achieve repeatable high quality copper pillar and pad wafer.

REFERENCES

[1] B. Wu et. al, "Boundary Layer Control in Industrial Plating Cell", J. Electrochemical. Soc., V152, C272, 2005.
[2] D. Barkey, et. al, "Studies on High-Aspect-Ratio Via-Filling for Wafer-Level 3D Integrated-Circuit Packaging", Peaks Conf. 2006

C4NP - DATA FOR FINE PITCH TO CSP FLIP CHIP SOLDER BUMPING

Eric Laine and Klaus Ruhmer
SUSS MicroTec, Inc.
Waterbury Center, VT, USA

Luc Belanger and Michel Turgeon
IBM Canada Ltd.
Bromont, Quebec, Canada

Eric Perfecto and Hai Longworth
IBM Microelectronics
Hopewell Junction, NY, USA

David Hawken
IBM Microelectronics
Endicott, NY, USA

ABSTRACT

High-end microelectronic packaging is increasingly moving from wire bonds to solder bumps as the method of interconnection. Flip chip in Package (FCiP) requires many small bumps on tight pitch whereas Wafer Level Chip Scale Packaging (WLCSP) typically requires much larger solder bumps on a greater pitch. There are various solder bumping technologies used in volume production. These include electroplating, solder paste printing and the direct attach of preformed solder spheres. Each of these technologies has important limitations in scaling from fine pitch FCiP to WLCSP. Electroplating is better suited to fine pitch, whereas solder paste printing and solder sphere attachment work well for coarser pitches. C4NP (Controlled Collapse Chip Connection New Process) has proven to be suitable for this entire range of solder bump pitch.

C4NP is a novel solder bumping technology developed by IBM and commercialized by Suss MicroTec. C4NP addresses the limitations of existing bumping technologies by enabling low-cost, fine pitch bumping using a variety of lead-free solder alloys. C4NP is a solder transfer technology where molten solder is injected into pre-fabricated and reusable glass templates (molds). The filled mold is inspected prior to solder transfer to the wafer to ensure high final yields. Filled mold and wafer are brought into close proximity/soft contact and solder bumps are transferred onto the entire 300mm (or smaller) wafer in a single process step without the complexities associated with liquid flux. C4NP technology is capable of fine pitch bumping while offering the same alloy selection flexibility as solder paste printing. The simplicity of the C4NP process makes it a low cost, high yield and fast cycle time solution for both, fine-pitch FCiP as well as WLCSP bumping applications.

This paper provides a summary of the most recent manufacturing and reliability results of C4NP bumped, 300mm wafer, high-end logic device packaging. This includes reliability data for C4NP lead free solder bumped devices attached to organic chip carriers. Mold fill data for CSP type dimensions is also included. Relevant process equipment technology and the novel requirements to run a HVM (high volume manufacturing) C4NP process are reviewed. The paper also summarizes the C4NP manufacturing cost model and elaborates on the cost comparison to alternative bumping techniques. The data in this paper is provided by IBM's packaging operations at the Hudson Valley Research Park in East Fishkill, NY and Bromont, Quebec.

Key Words: Flip chip, wafer bumping, lead free, CSP

INTRODUCTION

Many new electronic packaging applications are pushing the limits for weight, size, reliability, cost and high speed performance. At the same time, environmental considerations are driving new material requirements. These factors are driving a migration from wire bond to flip chip as the preferred method for connection from the semiconductor chip to the chip carrier or printed circuit board, and from leaded to lead free packages. Wafer bumping is becoming more pervasive, and several bumping processes have been established, each with different strengths. There is a need for one cost effective bumping technology that can address all requirements.

Earlier work has shown that Injection Molded Solder techniques have the potential to effectively address the challenges of wafer bumping. Realizing these benefits in production requires an integrated toolset that can support increasingly demanding manufacturing requirements. This paper reports on the latest improvements in the development of the C4NP production technology.

C4NP Process Flow

The C4NP process starts with a glass mold in which the bump pattern for an entire wafer is replicated as a mirror image of cavities in the glass mold. These cavities are filled with solder as the mold is scanned below a fill head. The fill head contains a reservoir of molten solder and a slot through which the solder is injected into the mold cavities. The cavity depth and diameter determine the volume of the solder bumps that will be subsequently formed on the wafer. The filled mold is inspected automatically and then aligned below a wafer with exposed UBM pads facing the mold. Mold and wafer are heated above the solder melting point and then brought into contact. The solder forms spherical balls which transfer from the mold to the UBM regions on the wafer, where they preferentially wet and solidify. Wafer and mold are separated, and the mold is cleaned for reuse. Figure 1 describes this process flow.

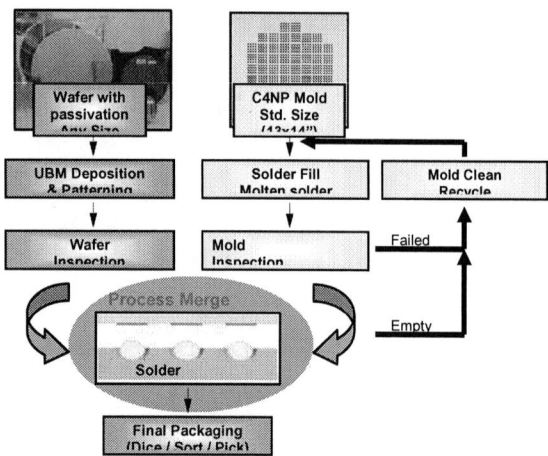

Figure 1: C4NP Process Flow

Mold Processing

C4NP molds are formed using borofloat glass plates, which have a coefficient of thermal expansion (CTE) close to silicon wafers. Photolithography is used to pattern and etch cavities whose diameter and depth precisely determine the volume of the solder bump, as well as defining the bump pitch and location. The molds are scanned beneath a solder injection head which fills the cavities with liquid solder precisely to the surface of the mold. Therefore, the solder volume transferred to the wafer is directly a function of the glass cavity volume.

Solder Transfer from Mold to Wafer

The solders used for wafer bumping do not wet to the glass mold, so upon heating, the solder alloys form spherical balls in the cavities, as described in Figure 2. The reflowed balls protrude above the surface of the mold by 10 – 20 um depending on ball size and cavity. Note from Figure 2 that the balls are not uniformly formed at the center of the mold cavity. The alignment of the mold cavities to the

corresponding UBM pads is sufficient to assure that the solder wets to the correct UBM pad.

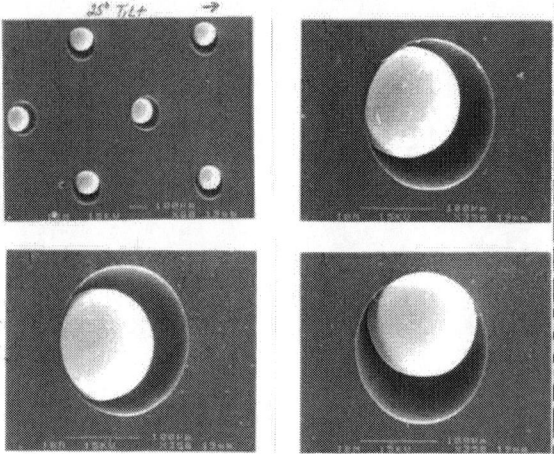

Figure 2: Reflowed solder spheres in glass mold cavities prior to transfer to wafer.

The filled molds are aligned with the wafer as shown in Figure 3. After alignment, the mold and wafer are heated and are brought into close proximity/contact, allowing the molten solder balls to wet the appropriate UBM pads where they preferentially remain when the wafer and mold are separated.

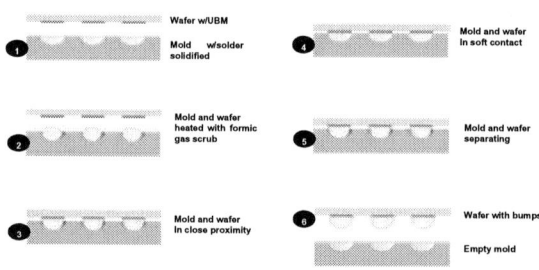

Figure 3: Solder transfer process sequence.

The solder transfer process takes place in a reducing gas environment which assures clean, oxide free, solder and UBM surfaces and avoids the need for liquid flux and subsequent cleaning. After solder transfer, the wafers are ready for test, dicing and subsequent packaging.

Technology Applications

The C4NP process allows great flexibility in the selection of solder bump alloys. Earlier work has shown the compatibility of the process with materials such as those noted in Table 1, with the exception of Pb-3Sn, which has a higher melting point. Other low melting point solder alloys are possible.

	Pb–3Sn	Sn–37Pb	Sn–0.7Cu	Sn–3.5Ag
Melting temp. (°C)	315–321	183	227	221
Elastic modulus (GPa)	9.5	29	—	37
Yield strength (MPa)	5.8	16.1	21.4	33.9
Tensile strength (MPa)	9.8	31–46	31	55

Table 1: solder bump alloy properties.

Changing solder alloys is accomplished by changing the fill head in the mold fill tool, which is done in less than an hour. The process temperatures of the solder reservoir and the mold can be adjusted to accommodate a particular material's characteristics. This flexibility allows C4NP to be backward compatible with existing solder alloys and UBM stacks as well as to enable the use of new solder materials and UBM stacks, including Pb Free variations. In addition, "dopants" can be added to the solder to reduce tin whiskers, improve electromigration performance, etc.

Production and Cost Considerations

For production, the mold processing and solder transfer shown in Figure 1 are accomplished by specialized tools designed to fill the molds, inspect the filled molds, and transfer the solder from the filled molds to semiconductor wafers as shown in Figure 3.

For volume production, the C4NP toolset includes a solder transfer cluster tool and mold stocker to support a production rate of 300 wafers per day (WPD) of 300mm or smaller wafers. Production rates for C4NP are independent of wafer size.

The per wafer cost for production wafer bumping is a function of the following cost determining factors:

A: Personnel cost
B: Consumable and material cost
C: Equipment maintenance and support
D: Equipment depreciation
E: Building overhead (footprint, cleanroom)
F: Wafer yield
G: NRE cost per part number/bump pattern
H: Chemistry supply and waste treatment
I: IP cost/IP wafer toll

As part of this work, a sophisticated cost model has been developed to compare the cost of C4NP wafer bumping with alternative technologies such as electroplating or screen printing of solder. By modeling a variety of cases, C4NP has emerged as the lowest cost fine pitch flip chip bumping technology.

One of the most critical differences between C4NP and alternative bumping technologies is the use of molds. A minimum number of molds are required depending on the number of wafers per day with a particular bump pattern. The cost of molds directly impacts the per wafer bumping cost. It is therefore critical for C4NP equipment technology to minimize the number of molds needed. Also, the number of reuses of a given mold is critical in determining bumping cost.

It is reasonable to assume that molds can be used several hundred times. It is beyond the scope of this paper to provide actual per wafer costs. The numbers depend on the individual company information which is often considered proprietary. However, the various cases which have been investigated show a per wafer cost reduction by using C4NP instead of electroplating. The "per wafer cost" reduction accomplished by C4NP ranges from approximately 10% to 30%.

C4NP for Lead Free Solder Bumping – Manufacturing Data

Lead free solder bumping has been one of the most important drivers for new bumping technologies such as C4NP. Lead free bumping is also impacting the choice of the UBM stackup. Since lead free solders typically have a high Sn content, such as Sn2.0Ag and Sn0.7Cu, they are consequently highly reactive with Cu. There are several ways of addressing this issue, one of which is by using electroplated Ni as a barrier layer. C4NP is compatible with any solder wettable surface, including Cu, Ni and Au. The UBM construction is the primary influence on the package reliability. The solder deposition method is a secondary factor.

Manufacturing data for 300mm wafers bumped using C4NP with lead free solders and a Ni plated UBM has been collected. Figure 4 depicts the chip test vehicle used to collect 200 um pitch manufacturing and reliability data. Features of this test vehicle are described later.

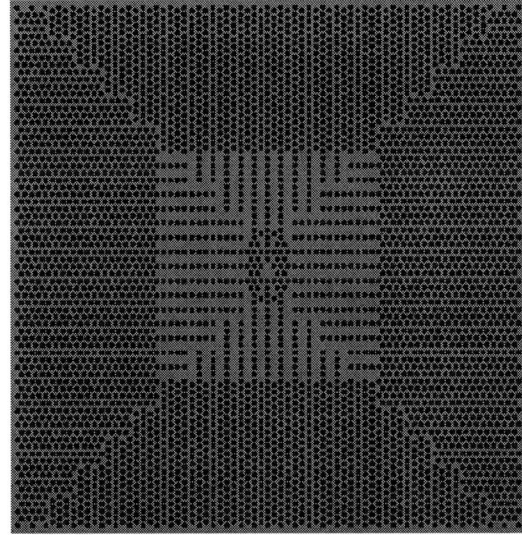

Figure 4: 200 um pitch test vehicle C4 pattern.

Figure 5 describes metrology data for ten 300mm wafers bumped with SnCu and SnAg solders using C4NP. For each wafer, the solder height mean and standard deviation, along with the solder diameter mean and standard deviation, are charted. This data represents a measurement of all 1.27 million bumps on each of the ten wafers.

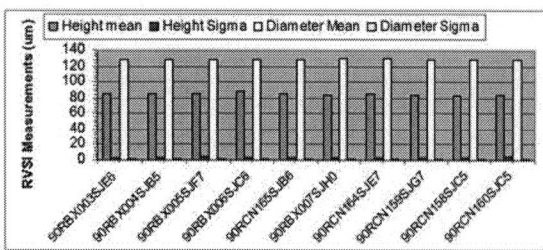

Figure 5: 200um pitch metrology data

Figure 6 depicts the C4 footprint for the 150 um pitch chip test vehicle used to collect manufacturing and reliability data.

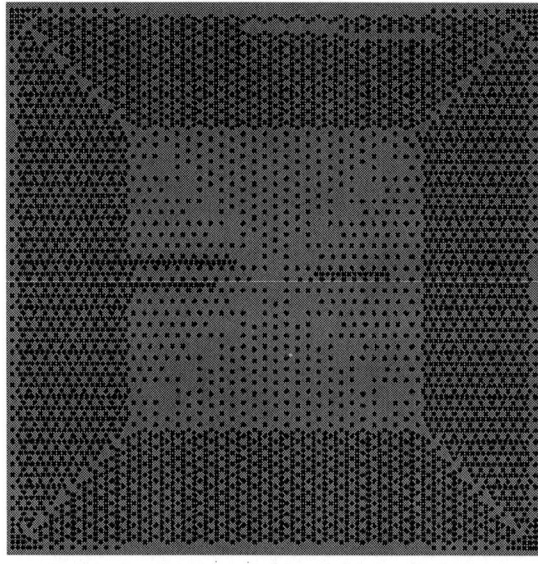

Figure 6: 150 um pitch test vehicle C4 pattern.

Figure 7 describes the C4 bump solder metrology data for this 150 um pitch test vehicle, bumped with SnCu and SnAg solders using C4NP. The data is presented in the same format as Figure 5.

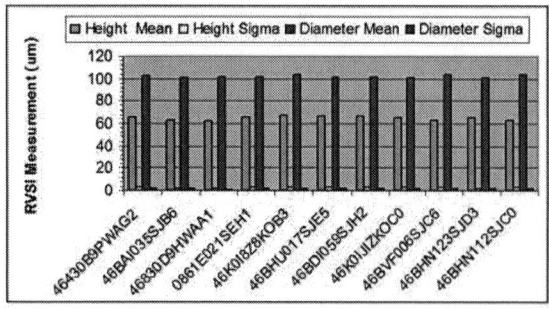

Figure 7: 150 um pitch metrology data.

Solder volume (and resulting bump height) was purposely varied from wafer to wafer in order to test the process window in subsequent reliability testing. This was done by selecting molds with cavity volume distributions at the limits of the specification. The distributions of this

metrology data compare favorably with electroplating. In addition, the standard deviations are expected to improve with higher quality molds from external suppliers.

To evaluate the solder joint strength of C4NP transferred bumps compared to plated bumps, a series of bump shear tests have been performed. The tests were run on SnCu and SnAgCu Pb-free solders. The solder bumps were transferred onto thick plated Cu UBM pads. As shown in Figure 8, the shear strengths for C4NP bumps were equivalent to those for electroplated bumps. The higher shear strengths of the SnAgCu bumps are due primarily to the higher yield strength and hardness of this alloy compared with those of the Sn0.7Cu alloy.

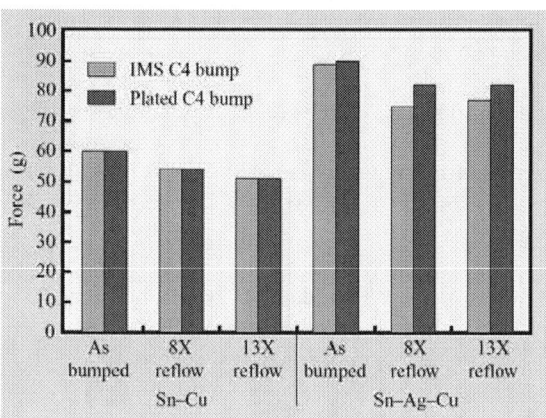

Figure 8: Shear strength comparison: C4NP bumps compared to electroplated bumps.

One of the concerns of any solder deposition technology is the formation of voids at fabrication which may expand and grow during subsequent reflows. During the C4NP fabrication, wafers are routinely inspected for voids by X-Ray in nine locations, approximately 50 C4 bumps per location. To date no voids of any size have been found. This is to be expected, since C4NP uses solid solder alloys that are melted and subsequently transferred to wafers. There is no conversion of the solder to a paste containing flux or to a plating chemistry, so there is no evolution of organics or trapped gasses during reflow. A sample of qualification wafer data is found in Table 2.

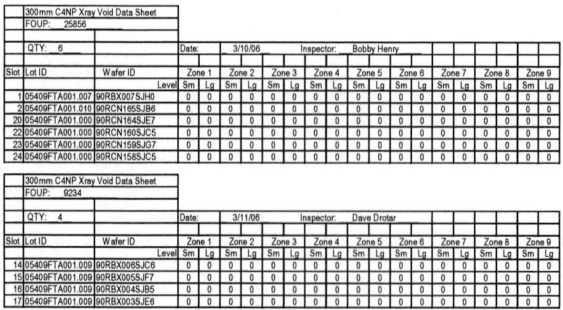

Table 2: x ray void inspection data.

Reliability Data

In order to test reliability of solder bumps manufactured using C4NP, a test vehicle with the following attributes was used:

Test Vehicle Description

- Chip size: 14.67 x 14.76 mm
- Pitch: 200um
- Wafer size: 300mm
- # of Chips in wafers: 271
- # of C4's in chip: 4,699
- Total C4s in wafer: 1.27 million
- Chip technology: 90 nm
- Package: FC-PBGA
- Solders: Sn 0.7%Cu & Sn 2.0% Ag

The following product construction was used for the reliability test:

Transfer Technology
- BLM: Sputtered TiW / Cu, Plated Ni / Cu
- Capture pads nominal diameter: 98 um & 110 um

Reliability data using this test vehicle is described in Figure 9. One of the Accelerated Thermal Cycle (ATC) tests was purposely run without underfill. The intention of this test was to determine the fatigue life of the Sn/0.7% Cu versus the Sn/2.0% Ag alloys. The SnCu alloy had a better fatigue life than SnAg. This is as expected, since SnAg has a higher yield strength and hardness than SnCu. This points out an advantage of C4NP, in that the solder alloy can be easily changed. The material properties of certain lead free solders can be a concern in some applications, especially with the migration to lower K, more fragile, back end dielectrics. C4NP allows rapid changeover between any low melting solder alloys.

Test	Conditions	Duration	Quant	Status	Comment
DTC	DTC: -55/125C	1 Kcyc	45	1250cyc	No fails
ATC	ATC -55/85C	3Kcyc	46	2500cyc	No fails
ATC	ATC -15/125	3Kcyc	47	1500cyc	No fails
ATC	ATC 0/100	3Kcyc	50	2000cyc	No fails
ATC	ATC 0/100C (No UF)	300cyc	28	Completed	SnAg has 50% more fails
HAST	HAST 130C/85%	96 Hrs	21	Completed	No fails
THB	THB 85C/85%/3.6V	1Khrs	21	Completed	No fails
HTS	HTS 125C	1Khrs min	21	Completed	No fails
HTS	HTS 150C	1Khrs min	21	Completed	No fails
HTS	HTS 170C	1Khrs min	21	Completed	No fails
LTS	LTS -65C	1Khrs min	21	Completed	No fails after 1500hrs
C4 EM	EM 125C,150C/0.5A,0.70A	6 months	82	530hrs	No delta Rs >5%
Tin whiskers*	T/C: -55/85C	1Kcyc	61	Completed	No electrical fails
Tin whiskers*	THB 60c/93%RH	1Khrs	34	Completed	No electrical fails
Wettability	IJ Wett	1 week		Bromont Rpt	No nonwetts, anomalies
Group B			24	Compl	Subset DTC,THB
C/A	IJ FA	10 days x 4			
Total	Total HW		485		

Figure 9: 200 um pitch reliability test results.

The test results show that there are no failure mechanisms attributed to C4NP identified in any of the testing to date. This is as expected, since the primary factor influencing reliability is expected to be the UBM construction, as opposed to the solder deposition method. The solder deposition integrity is verified by the shear strength comparison to electroplating shown in Figure 8. Figure 10 shows excellent UBM integrity and intermetallic formation, resulting from the Ni plated UBM and lead free C4NP deposited solder. In summary, C4NP is expected to demonstrate equivalent reliability performance to electroplating, and the data to date supports this hypothesis.

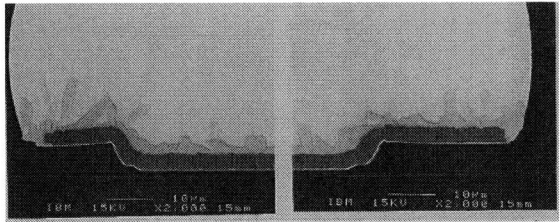

Figure 10: 150 um pitch UBM/intermetallic structure.

C4NP Production Considerations

As is the case with any new technology, high volume manufacturing is a significant aspect in the development of C4NP technology. The concept of separating the bump formation process from the bump transfer process enables a new approach to volume manufacturing for solder bumping, including lead free.

In today's volume manufacturing operations of integrated device manufacturers or packaging service providers, the UBM and bumping process steps are part of one inseparable process sequence on the wafer, thus limiting the flexibility regarding where bumping may be performed. By utilizing C4NP, volume manufacturers can prepare the UBM as part of the back end of line process in the semiconductor fab. The transfer of solder bumps to the wafer can then be performed in the best geographic location from a production and logistics standpoint.

For high volume manufacturing, high yield and low cost are critical concerns. C4NP has shown excellent yield performance in early manufacturing. A unique aspect of C4NP is that the molds can be inspected and repaired (if necessary) prior to solder transfer, thus enabling the optimization of solder bumping yield on the wafer.

In early manufacturing feasibility testing, a total of 75 mold fills for 200mm wafers over a six week pilot period have been performed. Fully automated inspection was used to characterize the defects after fill as well as after solder transfer. The bumping defects on wafer of the C4NP bumping process, not including any UBM related yield loss, was less than 10ppm.

Other advantages of C4NP include the simplicity of the process. C4NP is similar to solder printing, in that the alloy can be easily changed. However, since there is no flux containing paste involved, much tighter bump pitches are achievable than with screen printing. The material cost for C4NP is lower than for electroplating or screen printing, since pure solder is used. It is not converted to a solder paste or chemistry, which adds cost. The solder usage for C4NP is more efficient than electroplating, since solder is deposited on the UBM pads only. No solder is wasted on plating thieves. The cycle time for C4NP bumping of wafers is hours, as opposed to days for other bumping processes. This is because molds can be filled and inspected in parallel to the UBM formation operations. Finally, C4NP reduces process cost and improves bump quality since no liquid flux is used at the solder transfer step. This eliminates the possibility of void formation due to flux entrapment, and eliminates any flux cleaning operations.

Chip Scale Package Capability

The C4NP process has been demonstrated to be compatible with the greater bump pitches (400 to 500 um) associated with WLCSP. Glass molds were fabricated to produce the largest solder volume possible and still maintain the 25 um minimum specified distance between the cavities in the glass. Mold fill of these large cavities was successful. Figure 11 shows a 500 um pitch glass mold filled with solder.

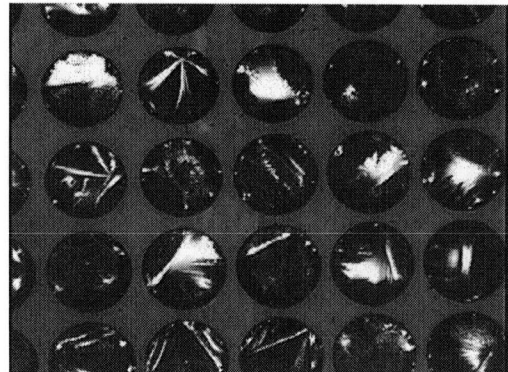

Figure 11: 500 um pitch CSP solder filled mold

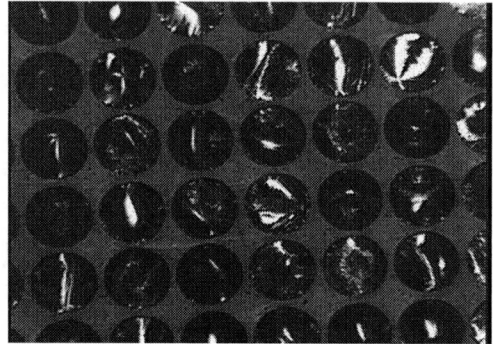

Figure 12: depicts a 400um pitch mold filled with solder

A wafer test vehicle was not available for solder transfer from these molds. However, solder bump dimensions were calculated as shown in Figure 13.

Pitch	500µm	500µm	500µm
Base Pad Diameter	250µm	300µm	400µm
Maximum Solder Volume	24X10E6 µm³	24X10E6 µm³	24X10E6 µm³
Bump Height	315µm	297µm	251µm
Bump Diameter	365µm	373µm	410µm

Figure 13: CSP calculated solder dimensions

These calculations show that traditional CSP type solder dimensions are achievable with C4NP.

CONCLUSION

The elimination of leaded solder alloys for flip chip bumping has clearly become one of the most actively pursued technology solutions in the semiconductor packaging industry. IBM's C4NP bumping technology is enabling a new method of applying lead free or leaded solders on 300mm and smaller wafers. The use of bump molds allows the separation of the bumping process from the UBM process and enables the use of any solder. With this separation, bumping cycle time is rapid, since molds can be prepared in parallel with the UBM formation instead of sequentially. Since the solder is molten, the form factor is irrelevant. Bumping pitches from very fine FCiP to courser WLCSP become possible with one cost effective process. The formation of entirely new solder or non-solder material combinations becomes possible with C4NP. Finally, since no liquid flux is used, cost is reduced and quality is improved.

These critical properties of C4NP, in combination with its efficient use of materials, its high yield and low cost capabilities make C4NP a viable alternative to existing bumping technologies.

ACKNOWLEDGEMENTS

The authors would like to thank the teams at IBM and Suss MicroTec, specifically the published and unpublished work of: Peter Gruber and Da-Yuan Shih of IBM TJ Watson Research Center in Yorktown Heights, N Y.

Guy Brouillette, David Danovitch, Jean-Luc Landreville, Valerie Oberson, and Suzanne Boutin of IBM Microelectronics in Bromont, Canada.

Barry Hochlowski, David Naugle, James Busby and Chris Tessler of IBM Microelectronics in East Fishkill, NY.

Jeffrey Friot of Suss MicroTec.

REFERENCES

1. J. Lau, Low Cost Flip Chip Technologies, McGraw-Hill Book, New York, 2000, Ch. 2, pp. 43-94.

2. P.A. Gruber, et al., "Low Cost Wafer Bumping", IBM J. Res. & Dev. Vol. 49 No. 4/5, July/September 2005.

3. B. Hochlowski, D. Naugle, P.A. Gruber, "Low Cost Wafer Bumping using C4NP", Future Fab Article, January 2005

4. Unpublished Report: IBM Systems and Technology Group, Dr. R. Levine, Presentation in Asia, September 2005.

5. K. Ruhmer et al., "C4NP: New Solder Bumping Technology – Low Cost and Lead Free", IMAPS Flip Chip Advanced Technology Workshop, Austin, TX, June 2005.

6. D. Danovitch, P. A. Gruber et al., "IMS-Injection Molded Soldering", IMAPS 2000.

7. P.A. Gruber, DY Shih et al., "Injection Molded Solder Technology for Pb-Free Wafer Bumping", ECTC 2004.

8. K. Ruhmer et al., "C4NP: Lead-Free and Low Cost Solder Bumping Technology for Flip Chip and WLCSP", Pan Pacific conference 2006.

AN INTEGRATED DEEP SILICON ETCH/ DIRECTIONAL PHYSICAL VAPOR DEPOSITION PROCESS FOR THROUGH-WAFER VIA APPLICATIONS

G. Reynolds, C. Constantine, S. Lai, K. Mackenzie, R. Westerman, D. Johnson, C. Johnson and R. Benz
Oerlikon USA, Inc.
St. Petersburg, FL, USA

J-B Chevrier
Oerlikon France
Montbonnot, France

J. Weichart, S. Kadlec, M. Elghazzali, H. Hirscher and H. Auer
OC Oerlikon Balzers Limited
FL-9496 Balzers, Liechtenstein

ABSTRACT

Through-wafer via technology has received much attention recently for stacking of multiple thin dice. Metal studs, formed by etching vias through thinned Si wafers and subsequently filling them with metal, plug into sockets on the die below, thereby forming electrical interconnects between the two dice.

Copper is the preferred metal for through-wafer interconnects due to its low resistivity and desirable electromigration characteristics. It is widely used in wafer packaging and also in on-chip interconnects. In conventional packaging applications such as under-bump metal, copper films are typically deposited by conventional sputtering. However, high aspect ratio through-wafer vias require a more directional metal flux that can be provided by Oerlikon's Advanced Directional Sputtering.

In this paper, we describe an integrated deep silicon etch/ advanced directional sputtering process developed to provide enhanced Ta/ Cu sidewall coverage in the vias. The etch process produces a tapered via in a Si wafer with a slope of 60-80° and no overhang. Etch rates greater than 7μm/min. and excellent selectivity to oxide and photoresist are observed. SEM cross-sections of the resulting etch profiles before and after metallization are presented in combination with performance data of Oerlikon thin wafer processing capabilities.

Key words: deep silicon etch, directional sputtering, through-wafer vias.

BACKGROUND

For decades, integrated circuit manufacturers have sought to increase the performance and reduce the cost of their products by shrinking the gate length, thereby increasing the density of devices on a wafer. In 1965, Gordon Moore noted that the complexity for minimum component costs had increased at a rate of roughly a factor of two per year and postulated that this rate was likely to continue in the short term [1]. Later, this came to be known as "Moore's Law", though it is often corrupted and stated as "the number of transistors on an integrated circuit doubles every 18 months," (or, in some cases, every 24 months). The drive to sustain this rate of miniaturization and complexity has become increasingly expensive in recent years. In an attempt to keep up with Moore's Law, state-of-the-art CMOS fabricators are now processing 300mm wafers and more recently, there have been calls for the industry to move to 450mm diameter Si wafers in the future [2]. In addition, the many challenges associated with patterning and filling features with critical dimensions of 220nm and smaller make "more-Moore" a daunting prospect for all but a very few companies.

This has led many manufacturers to look at other ways to improve the performance and reduce the cost of their products. The term "more-than-Moore" was coined to describe package-level integration and System-in-Package (SiP) technologies, whereby multiple dice are incorporated into a single package. These dice can have different functions that cannot be easily combined on a single wafer using the System-on-Chip (SoC) concept. Conversely, multiple dice with similar or identical functions can be stacked on top of one another to create a multi-chip package (MCP) with the same footprint on the printed circuit board substrate as a single die. In recent years, MCPs containing different forms of memory (DRAM, SRAM and Flash) have found increasing application in mobile electronic devices such as cell phones where space is at a premium.

The stacking of multiple dice on top of one another creates a need for more effective ways to connect them electrically. Conventional wire-bonding techniques are not ideally suited to connect multiple identically-sized dice arranged in a vertical stack; other methods have been proposed and investigated for this purpose. One example is 3D interconnect technology [3], where the stacked dice are connected through vias on the dice themselves. This technology has the added appeal that interconnects between the multiple dice in the package are shortened and system speed is increased: many of the gains offered by "more-Moore" advances could be realized more simply and cost-effectively with 3D interconnects. A key requirement for the effective implementation of 3D interconnects is the development of through-wafer via technology: deep vias are etched through the silicon wafer which are then lined with an insulator and filled with metal [4].

In order to etch deep vias through silicon wafers, a process often referred to as the 'Bosch Process' [5, 6] and that originated in the MEMS industry is employed. This is a multi-step anisotropic reactive ion etch (RIE) process that alternates between etch and deposition steps. During the deposition steps, octofluorocyclobutane (C_4F_8) promotes fluoropolymer formation on all surfaces. During the etch steps, sulfur hexafluoride (SF_6) is used in combination with substrate bias, to etch the silicon preferentially on horizontal surfaces where energetic ion bombardment removes the fluoropolymer and exposes a fresh silicon surface each cycle. The fluoropolymer formed during the deposition steps passivates vertical surfaces, inhibiting lateral etching and thereby allowing high aspect ratio features to be defined into silicon substrates at high rates [7]. For through-wafer vias, these features are lined with a chemical vapor deposited (CVD) dielectric, e.g., SiO_2 from a tetraethyl orthosilicate (TEOS) precursor, and filled with a metal plug.

Copper is the preferred metal for through-wafer interconnects due to its low resistivity and desirable electromigration characteristics. It is well-established as an interconnect material in back-end processing: wire-bonders have used gold and copper wires to connect dice, and most under-bump metallization schemes for flip-chip applications are based on copper. Since 1997, it has been the preferred metal for high performance multi-level interconnects. In these front-end applications, a copper seed is deposited by directional sputtering onto a tantalum or tantalum nitride liner, the feature is filled with electroplated copper and the surface is planarized by chemical-mechanical polishing (CMP) [8].

Metal deposition for traditional packaging processes such as under-bump metal has been done by physical vapor deposition (PVD) methods – evaporation and sputtering. These are clean, mature, cost-effective techniques, known to be compatible with subsequent electroplating. However, in high aspect ratio features, conventional sputtering might not provide adequate bottom and sidewall coverage: enhanced directional techniques could be necessary. Indeed, some authors have used chemical vapor deposited (CVD) titanium nitride liners and CVD copper seed layers to ensure sufficient step coverage in deep, steep-sided features [9]. The drawback to using CVD techniques is that CVD metal deposition is usually a more expensive process than PVD, and in the case of CVD copper, there are reports of delamination at the liner-seed interface during CMP. In this work, we describe an integrated sloped etch/advanced

directional sputtering approach designed to ensure robust filling of deep through-wafer vias.

DEEP SILICON ETCH (DSE)

In semiconductor processing in general and packaging processes in particular, cost is a critical factor. Even at the high etch rates of silicon characteristic of the Bosch Process, it would take a long time to etch through a wafer – a standard 200mm silicon wafer is in the range 700-750µm thick. Consequently, deep vias are etched only part way through the wafer: only to a prescribed depth that is approximately equal to or slightly more than the final thickness of the die. After the through-wafer vias are filled with metal, subsequent wafer thinning is carried out by grinding, chemical etching, atmospheric plasma thinning, CMP or a combination thereof. Despite this, throughput is still extremely important: it is essential to maximize etch rate while minimizing the formation of sidewall scallops that tend to form at the highest attainable rates [7]. Figure 1 shows the relationship between the etch rate and the sidewall scallop depth for a deep silicon etch process in an Oelikon Versaline™ etch module equipped with fast response pressure control.

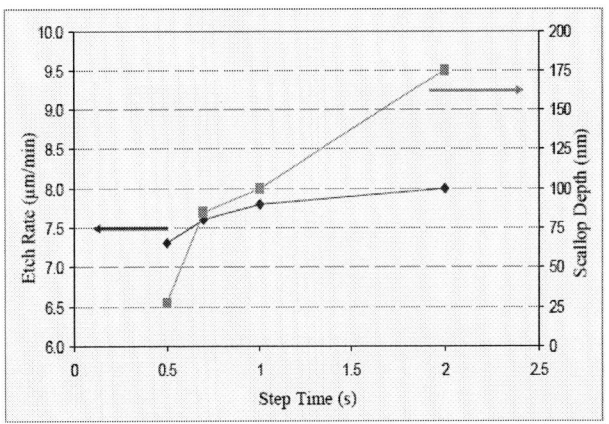

Figure 1. Etch rate vs. scallop depth as a function of step time for the Bosch Process (Oerlikon Versaline™ module).

Another important parameter that must be considered when designing through-wafer vias is the area that they occupy within the die. While vertical vias take up the least die area, they are the most challenging to fill, especially if one desires to use PVD liners and seeds. The purpose of this work was to determine how much taper is required to ensure robust filling of deep vias etched in silicon using enhanced PVD liners and seed layers.

An Oerlikon Versaline™ high density plasma etch system was used to etch a series of deep sloped vias in photoresist-patterned 150mm diameter silicon wafers. Figure 2 shows the essential features of the plasma etch system. An inductively coupled plasma (ICP) source operating at 2MHz generates the high-density plasma ($>10^{11}$ cm^{-3}). A separate RF bias supply powers the wafer chuck. Efficient cooling of the wafer during the etch process is essential to prevent reflow or reticulation of the photoresist pattern. To meet this requirement, the wafer chuck is cooled to temperatures of -10°C or less and utilizes an integrated electrostatic chuck (ESC) or mechanical clamp with helium as the transfer medium to cool the backside of the wafer.

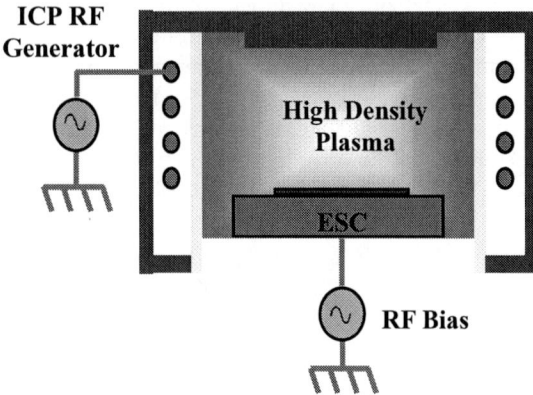

Figure 2. Schematic of DSE Chamber used for this work.

The sloped silicon process is based on a plasma chemistry of SF_6 and C_4F_8. With SF_6 alone, the silicon via is etched isotropically. The addition of C_4F_8 results in the formation of a fluoropolymer on the sidewall of the via, thereby reducing the lateral etching of the silicon. By adjusting the balance between SF_6 and C_4F_8, it is possible to tailor the slope profile of the via from vertical down to a slope angle of about 60°. Other key process parameters such as RF bias, ICP power, chamber pressure, and flow rates of C_4F_8 and SF_6 are used to optimize both the etch rate and slope profile. With this process, etch rates of greater than 7µm/min are achievable and typical selectivities to oxide and photoresist are >100:1. Figures 3(a)-(e) show examples of etched features with slopes ranging from vertical to approximately 60°. The Versaline™ system allows tight control of the slope profile – less than ±5% variation has been demonstrated on 200mm wafers.

PVD DEPOSITION

Tantalum is used as a liner in copper vias as it acts both as a diffusion barrier to, and to promote smooth nucleation of, copper thin films. Tantalum and copper were deposited in separate Oerlikon ClusterLine™ 200 PVD modules using advanced directional sputtering techniques. The Oerlikon ClusterLine™ 200 is a multi-chamber cluster tool that can be equipped with dual cassette load locks and up to six process modules. In this work, the copper seed-layers were deposited at room temperature in a specially adapted chamber with an increased wafer-to-target spacing of approximately 30cm and an enhanced unbalanced magnetron source. A schematic of this apparatus is shown in Figure 4.

This apparatus combines the features of long throw sputtering and ionized PVD. In long throw sputtering, the increased wafer-to-target spacing causes the material that is sputtered from the target at low angles to deposit on the

shields, instead of on the wafer where it would create overhang at the top corners and upper sidewalls of the features. This overhang can shadow the bottom and lower sidewalls of high aspect ratio features and result in low step coverage at those locations. If the copper seed-layer is not sufficiently thick, voiding during subsequent electrofill can occur. Only material sputtered close to normal to the target surface reaches the wafer. This directional flux penetrates deep into high aspect ratio features, improving bottom and lower sidewall coverage. Gas scattering is minimized by operating at low pressures (< 1mtorr).

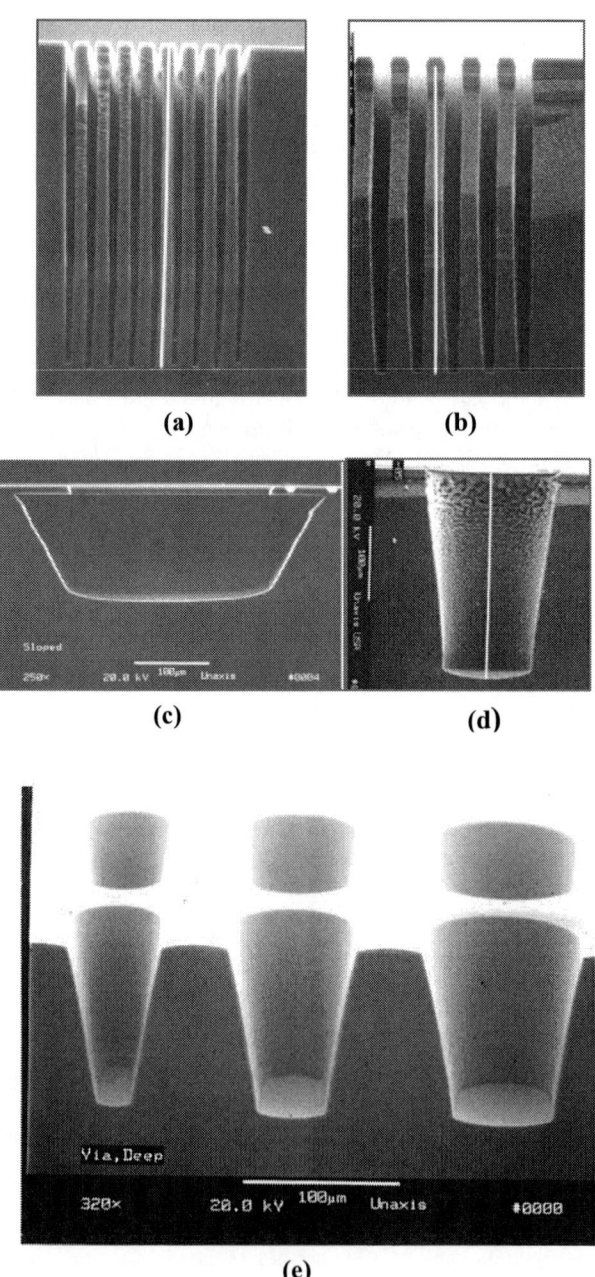

Figure 3. Uniform high and low aspect ratio features with controllable slope and high etch rates: (a) ~50:1 AR trenches, 2.5µm x 180µm; (b) ~22:1 AR trenches, 20µm x 442µm; (c) 60° profile via, ~140µm deep; (d) 83° profile via, ~280µm deep; (e) ~80° profile vias, ~130µm deep.

The enhanced unbalanced magnetron source creates a high-density plasma close to the target surface where metal and argon ions are created. Magnetic field lines extend out from the magnetron and loop through and close to the wafer (see Figure 4). At the low pressures used here, electron-gas scattering is reduced, and the electrons are able to follow the magnetic field lines through the tubus. In similar fashion, ions are guided from the target through the tubus and to the wafer by the electrons due to ambipolar diffusion. Ions arriving at the wafer are accelerated across the wafer sheath perpendicular to the wafer and in this way, highly directional metal ions penetrate deep into high aspect ratio features. The ion flux to the wafer can be further increased by application of RF bias to the wafer and the use of external DC solenoidal coils to augment the axial field lines of the magnetron. Though desirable for refractory metals such as Ti and TiN where it serves to ion peen the growing films, improving film density and lowering film resistivity [10], significant ion bombardment of PVD copper films can increase surface adatom mobility and produce anomalous grain growth. This, in turn, leads to the appearance of agglomerated copper beads or droplets, especially on the lower sidewalls of deep vias where coverage is least [11]. For this work, RF bias was not applied to the wafer and the DC coils remained off during copper deposition.

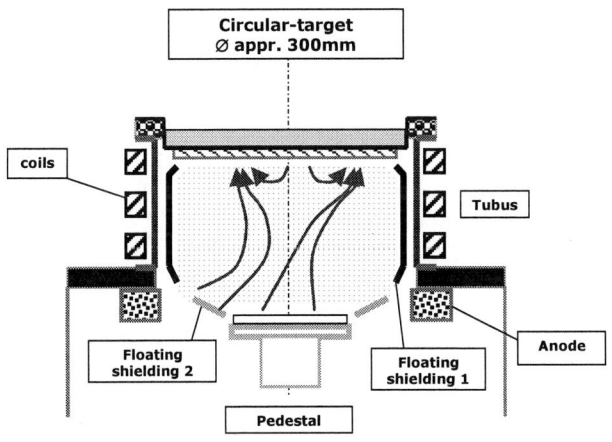

Figure 4. Schematic of Oerlikon Advanced Directional Sputtering module for ClusterLine™ 200.

Flat-field thicknesses of ~150nm tantalum and ~1.1µm copper were deposited on 150mm wafers with deep vias whose slopes ranged from ~60-80° (Figure 5). After deposition, we studied cross-sectional SEM images of the metallized vias to ascertain bottom and sidewall coverage and decide which features were best suited for electroplating. Even in the deepest, highest aspect ratio features with the steepest sidewalls (~80°), the bottom coverage looked to be close to 100% for both Cu and Ta: transmission electron microscopy (TEM) images confirmed these observations {see Figures 6(a) and (b)}. Figures 6(a) and (b) were prepared using a focussed ion beam (FIB) to create a wedge-shaped membrane from material at the bottom of the via. Figure 6(c) was taken from the lower

sidewall of a different feature. These images showed that the Ta films were continuous, dense and very smooth both in the bottom and on the sidewall. These are very desirable properties for a diffusion barrier in order to prevent copper from diffusing laterally into the silicon where it will destroy the die.

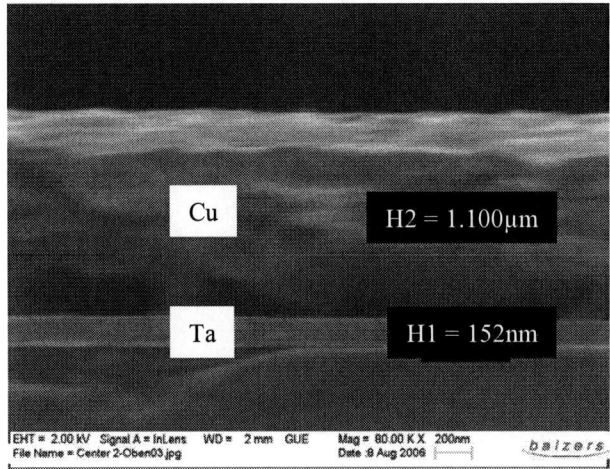

Figure 5. SEM cross section of as-deposited Ta and Cu films on the wafer surface.

ELECTROFILL

The metal coverage results looked so encouraging that we chose only the wafers with 80° sloped vias for subsequent electroplating experiments. Due to the large depth of the features (~100µm), it was necessary to run several plating experiments in order to completely fill the features (Figure 7). However, even partially filled features showed no evidence of voiding on the sidewalls that might be caused by insufficient copper step-coverage.

To test the robustness of our PVD coverage, we generated a non-optimized sloped etch profile with an extremely scalloped sidewall with near vertical and horizontal surfaces, much like a staircase structure. This should present a greater challenge for our enhanced PVD process. Again, the sidewall coverage was close to conformal, even on the vertical portions of the stair steps, and electrofill showed no voiding (see Figures 8 and 9).

The preparation of samples filled with electroplated copper for SEM cross sections poses some difficulty: cleaving at normal temperatures usually results in the copper tearing out of the feature. In this work, we employed two techniques to try to overcome this. First, we chilled the sample with liquid nitrogen and cleaved as normal. This technique required multiple attempts to produce a satisfactory sample for imaging. Figure 7 was prepared thusly. The second technique involved cleaving the wafer with a cleaving tool, polishing back to the center of the feature and removing smeared copper with an acid etch. Figure 8 was prepared thusly. Figure 9 was prepared by milling through the sample with an FIB. For the large vias imaged here, this latter technique proved to be very time-consuming and costly.

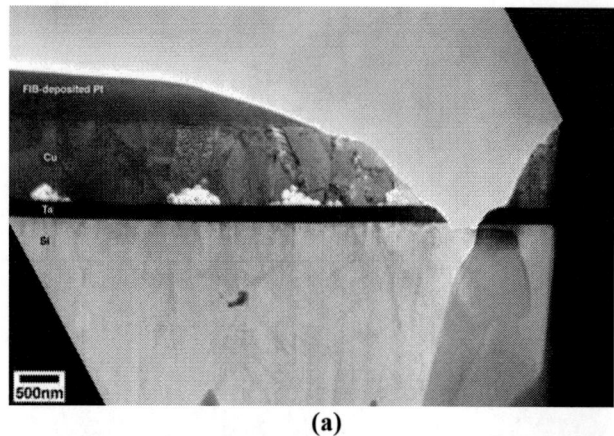

(a)

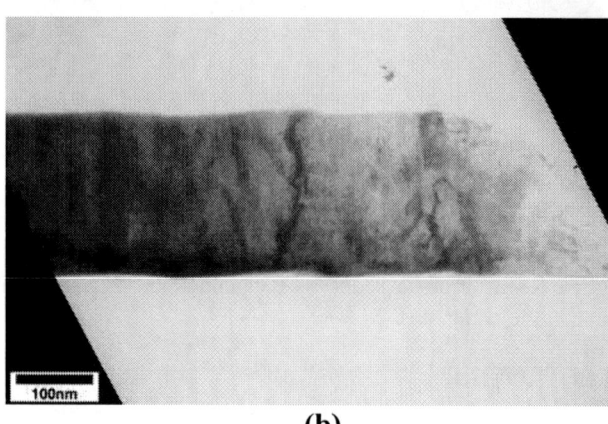

(b)

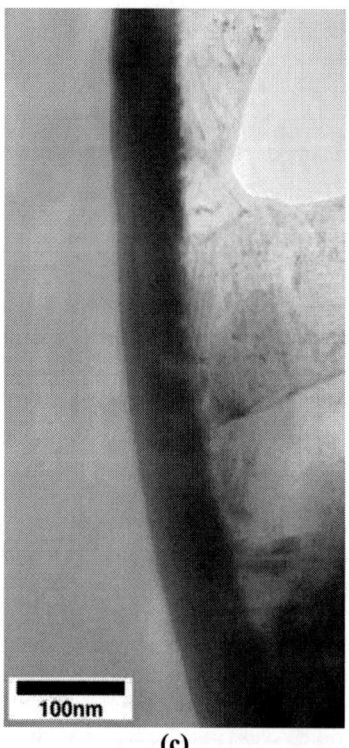

(c)

Figure 6. (a) TEM image of the bottom of a deep via in silicon; (b) Higher magnification image of Ta film in Figure 6(a); (c) TEM image of Ta on lower sidewall of feature.

Figure 7. SEM cross-section of filled electroplated deep vias in silicon.

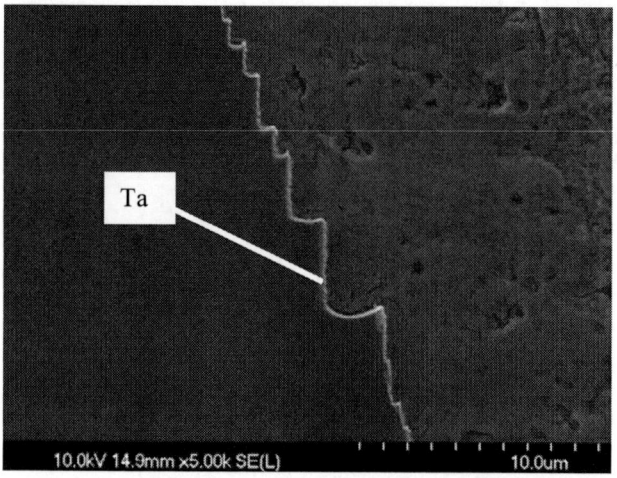

FIGURE 8. SEM cross-section of electrofilled staircase structure etched so as to challenge the sidewall coverage of our enhanced PVD technique. (Visible pits are the result of an acid etch used to remove smeared copper during sample preparation.)

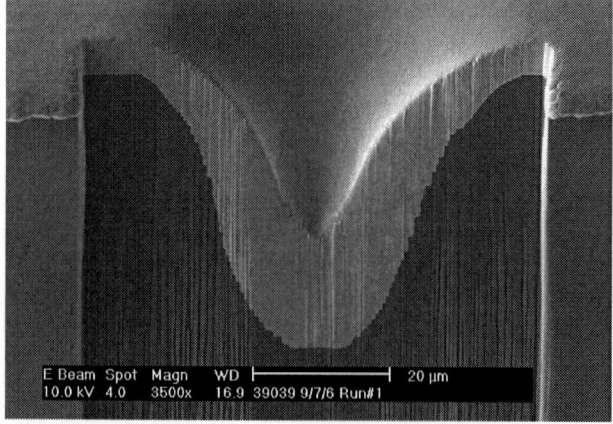

FIGURE 9. FIB cross-section of staircase structure etched so as to challenge the sidewall coverage of our enhanced PVD technique. No voiding is seen.

THIN WAFER PROCESSING

After the through-wafer vias have been etched, lined with dielectric and metal and filled by electroplating, the wafers are thinned. In some cases, especially when dissimilar die will be stacked together such as is typical in SiP applications, it is often necessary to deposit a metal redistribution layer onto the backside of thinned wafers so that the through-wafers vias will connect with contacts on the die below. The Oerlikon ClusterLine™ 200 and 300 cluster tools can be equipped with many features that are designed to handle and deposit on thin wafers:

1) A smart mapping system with a custom-designed sensor to prevent slide-out and cross-slotting in the wafer cassettes.
2) Metal cassettes that prevent the sharp edges of thin wafers from cutting into the cassette and provide improved stability that is essential for handling thin wafers. They facilitate using a reduced number of wafer slots and allow for extremely bowed wafers in the cassette.
3) Smooth, soft-touch elastomeric pins and pads to avoid scratches, particles, *etc.*, on the front side of the wafer.
4) Robot speed and acceleration is optimized for thin wafer handling.
5) Four support pins and an increased pin-lift height in the process modules and handler.
6) The lift pins and end-effectors have an edge gripping feature to minimize edge contact and center the wafers.
7) Clampless or electrostatic chucks avoid mechanical clamps that can chip the edges of fragile thin wafers.
8) *In-situ* 'on-the-fly' wafer re-alignment between process modules prevents mishandled wafers and reduces wafer breakage and loss.
9) Low energy processing and advanced stress control: thin wafers are very sensitive to stress which is controlled by limiting the temperature excursions during processing and other methods such as pulsed DC sputtering.

In production, the ClusterLine™ 200 can handle 150mm diameter thin wafers down to 70μm thick and 200mm diameter wafers down to 100μm thick without the need for a carrier. Maximum allowable wafer bow is 6-8mm, depending on the amount of wafer warpage. Throughputs of up to 500 wafers per day are currently in production.

Successful thin wafer handling and deposition depends not only on the deposition equipment and processes but also on the equipment and processes upstream. Oerlikon is partnered with several other semiconductor equipment companies who specialize in other thin wafer processing steps.

CONCLUSION

In this work, we have demonstrated an extremely robust process for fabricating through-wafer vias using proven and well-established sloped deep silicon etch processes and sputter deposition. Future work will investigate the feasibility of filling deeper, higher aspect ratio features with more vertical sidewalls to test the limits of this technology.

ACKNOWLEDGEMENTS

The authors gratefully acknowledge Mr. Peter Kohler of OC Oerlikon Balzers Limited for assistance with the analysis of these wafers and Mr. Fred Clayton of Freescale for providing un-etched, patterned wafers. TEM analysis was performed by NanoTEM, Inc.

REFERENCES

[1] G.E. Moore, "Cramming more components onto integrated circuits," Electronics **38**, pp. 114-117 (1965).

[2] J. R. Lineback, "Sematech hunts for collaborative path to 450mm," SolidState Technology, (January 2005).

[3] A. Rahman and R. Reif, "System Level Performance Evaluation of Three-Dimensional Integrated Circuits", IEEE Transactions on VLSI **8**, pp. 671-678 (2000).

[4] S. Spiesshoefer, Z. Rahman, G. Vangara, S. Polamreddy, S. Burkett and L. Schaper, "Process integration for through-silicon vias," J. Vac. Sci. and Technol. A **23**, pp. 824-829 (2005).

[5] F. Laermer and A. Schilp, "Method of anisotropic etching of substrates," U.S. Patent No. 5, 498, 312 (1996).

[6!] F. Laermer and A. Schilp, "Method of anisotropically etching silicon," U.S. Patent No. 5, 501, 893 (1996).

[7] S. Lai, "Plasma Etching," in Semiconductor Manufacturing Handbook, ed. H. Geng, McGraw-Hill, pp. 12.1-12.25 (2005)

[8] D. Edelstein, J. Heidenreich, R. Goldblatt, W. Cote, C. Uzoh, N. Lustig, P. Roper, T. McDevitt, W. Motsiff, A. Simon, J. Dukovic, R. Wachnik, H. Rathore, R. Schulz, L. Su, S. Luce and J. Slattery, "Full Copper Wiring in a Sub-0.25μm CMOS ULSI Technology," IEEE Technical Digest (IEDM), pp. 773-776 (1997).

[9] Y. K. Ko, B. S. Seo, D. S. Park, H. J. Yang, W. H. Lee, P. J. Reucroft and J. G. Lee, "Additive Vapor Effect on the Conformal Coverage of a High Aspect Ratio Trench Using MOCVD Copper Metallization," Semi. Sci. Tech. **17**, pp. 978-982 (2002).

[10] F. Cerio, J. Drewery, E. Huang and G. Reynolds, "Film properties of Ti/TiN bilayers deposited sequentially by ionized physical vapor deposition," J. Vac. Sci. and Technol. A **16**, pp. 1863-1867 (1998).

[11] L. J. Friedric, D. S. Gardner, S. K. Dew, M.J. Brett and T. Smy, "Study of the copper reflow process using the GROFILMS simulator," J. Vac. Sci. and Technol. B **15**, pp. 1780-1787 (1997).

STUDY OF Ni-P/Pd/Au AS A FINAL FINISH FOR WAFER

Kazuki Yoshikawa, Toshiaki Shibata, Masayuki Kiso, and Shigeo Hashimoto
C. Uyemura & Company, Ltd.
Osaka, Japan

Don Gudeczauskas
Uyemura International Corporation
Southington, CT, USA
dgudeczauskas@uyemura.com

ABSTRACT

Thermal diffusion of Ni-P/Pd/Au deposits for wire bonding reliability was studied. Palladium was used in an attempt to decrease nickel diffusion into the surface gold deposit. Transmission Electron Microscopy, FE-SEM, Auger analysis and wire bond analysis were used to measure the reliability of the finishes discussed. The test results show the controlled Ni-P/Pd/Au deposit to exhibit excellent reliability for wire bonding.

Key words: Wire bonding, Palladium , Ni/Pd/Au, ENEPIG, ENIG

INTRODUCTION

It is known that palladium deposits are excellent barriers to prevent nickel from diffusing into gold. In this paper, electroless palladium deposits were analyzed by TEM and thermal diffusion of each element of Ni, Pd, and Au from the electroless Ni-P/Pd/Au deposit was examined. From this study, it was confirmed that the palladium deposit could prevent diffusion of underlying nickel to gold even when the palladium deposit is comparatively thin (about 0.06μm) and provides excellent wire bonding reliability.

EXPERIMENTAL AND RESULTS

Coupons used consisted of Al-Cu (0.5%) sputtered wafers (1mm thick) for thermal diffusion test. For wire bonding reliability tests, 1μm thickness of Al-Si(1%)-Cu(0.5%) sputtered wafers which have 150μm square pads by Si-N were used. Using plating chemicals commercially available from C. Uyemura & Co., Ltd. the coupons were plated. Table 1(a) shows the Ni-P/Au (hereinafter called ENIG) plating process. Table 1(b) shows the Ni-P/Pd/Au (hereinafter called ENEPIG) plating process.

Table-1(a): ENIG plating process

Process	Chemical	Temp.	Time
Cleaner	MCL-16	50 deg C	3min.
Rinse			
Acid rinse	30%nitric acid	21 deg C	30sec.
Rinse			
1st Zincate	MCT-17	21 deg C	10sec.
Rinse			
Acid rinse	30%nitric acid	21 deg C	1min.
Rinse			
2nd Zincate	MCT-17	21 deg C	35sec.
Rinse			
Electroless Ni	NPR-18	80 deg C	25min.
Rinse			
Immersion Gold	TDS-20	75 deg C	10min.
Rinse			
Electroless Gold	TMX-16	50 deg C	14min.*
Rinse			
Dryer			

*: Thickness will follow the dipping time

Table-1(b): ENEPIG plating process

Process	Chemical	Temp.	Time
Cleaner	MCL-16	50 deg C	3min.
Rinse			
Acid rinse	30%nitric acid	21 deg C	30sec.
Rinse			
1st Zincate	MCT-17	21 deg C	10sec.
Rinse			
Acid rinse	30%nitric acid	21 deg C	1min.
Rinse			
2nd Zincate	MCT-17	21 deg C	35sec.
Rinse			
Electroless Ni	NPR-18	80 deg C	25min.
Rinse			
Electroless Palladium	TFP-30	50 deg C	5min.*
Rinse			
Immersion Gold	TDS-40	75 deg C	10min.
Rinse			
Electroless Gold	TMX-16	50 deg C	14min.*
Rinse			
Dryer			

*: Thickness will follow the dipping time

Plating thicknesses of Ni = 5μm, Pd = 0.06μm, Au = 0.02 μm were basically adopted. Pd and Au deposit thickness were adjusted as required. It was assumed that the ENEPIG deposit characteristics would greatly depend on the feature (uniformity and existence of defects) of palladium and gold deposits. Therefore, the feature of the Pd and Au deposits were confirmed by TEM, the results of which are shown in Fig. 1. By this TEM analysis result, it was confirmed that the Palladium deposit is an amorphous, smooth, and uniform layer. Since the Au deposit has the crystalline plane observed, it was confirmed that the Au deposit is crystallized.

Fig. 1: TEM analysis results of ENEPIG film

In order to confirm the deposit surface morphology, the gold surfaces with and without heat treatment at 175°C for 16 hours were observed by FE-SEM, the results of which are shown in Fig. 2.

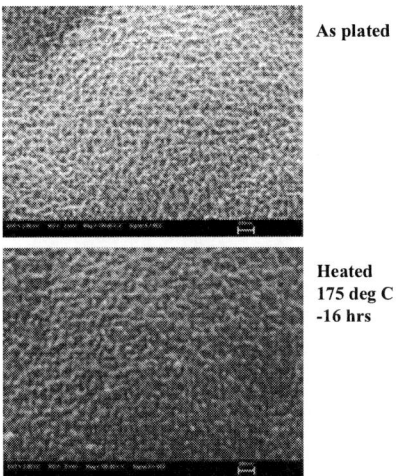

Fig. 2: FE-SEM image of gold surface

The gold surface morphologies shown in Fig. 2 coincide with the cross-sectional image of the gold layer by TEM analysis and indicate few defects, and it is confirmed that by heat treatment, the gold grain size is increased. Next, in order to confirm diffusion of each element by heat treatment, Ni=5μm / Pd=0.06μm / Au=0.02μm deposits and Ni=5μm /Au= 0.2μm deposits were heated at 175°C for 16 hours and the gold top surface was compared and analyzed by AES. As a result, it was clarified that by inserting the palladium deposit as thin as 0.06μm between the EN deposit and Au layers, diffusion of nickel to the gold deposit can be prevented and the results are shown in Fig. 3.

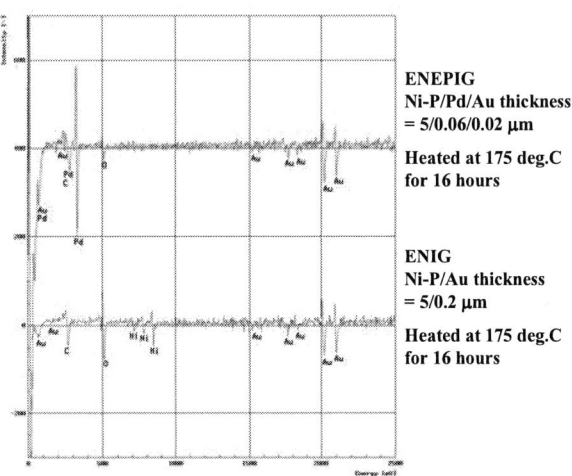

Fig. 3: Comparison of ENEPIG and ENIG in AES analysis after heated at 175°C for 16 hours

Furthermore, two types of deposits with gold thicknesses varying such as Ni=5μm / Pd=0.06μm / Au=0.02μm and Ni=5μm / Pd=0.06μm / Au=0.2μm were checked for diffusion of palladium to gold using AES-depth analysis after heating at 175°C for 16 hours, the results of which are shown in Fig. 4. As shown in Fig. 4, it was confirmed that palladium diffuses comparatively uniformly in the depth direction of the gold deposit and the Pd content in the gold film varies dependently on the gold film thickness. It is considered that the change in gold surface morphology by heat treatment shown in Fig. 2 is attributed to alloying of gold and palladium shown in Fig. 4.

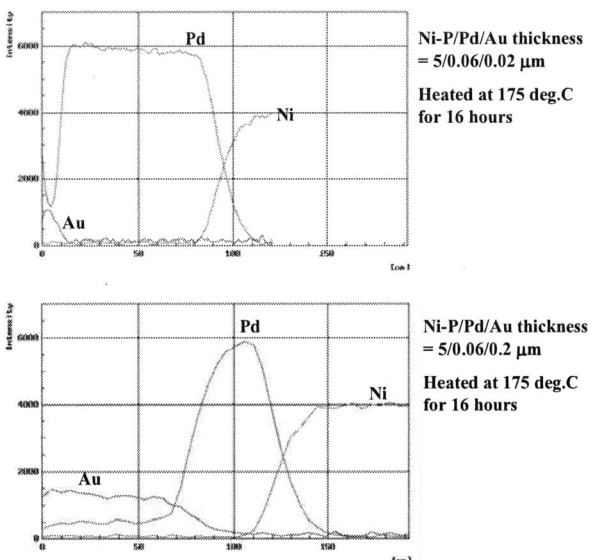

Fig. 4: Results of Au and Pd diffusion by AES analysis

Evaluations of electroless ENEPIG deposits and wire bonding characteristics were performed for wafer application.

Ni-P/Pd/Au and Ni-P/Au layers were deposited on 150μm square pads and 1st bonds were performed on them. The bonding bumps were sheared by a shear tool and shear strength and failure mode were observed.

1st wire bonding conditions are shown in Table 2, shear conditions shown in Table 3, and the failure mode shown in Fig. 5. (Mode A: Tear between bonding ball. Mode B: Tear between bonding ball and the gold surface. Mode C: No adhesion between bonding ball and the gold surface.)

Table 2: 1st bonding test condition

Model	TPT HB16
	Semi-auto
Capillary part #	B1014-51-18-12(PECO)
Frequency of U.S.	60HZ

Au Wire size	25μm
Bond power	200mW
Bond time	30mS
Bond force	500mN
Stage temp.	175°C

Table3: Bond shear test condition

Model	Dage series 4000
Bond tester	BS250
Shear speed	30μm/sec.
Tool high	5μm

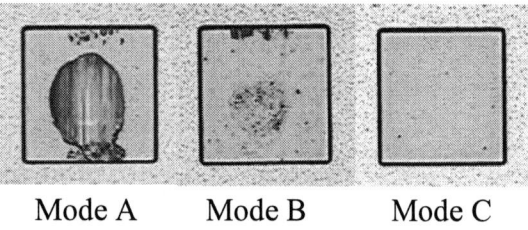

Mode A   Mode B   Mode C

Fig. 5: The failure mode of 1st bond shear test results

Firstly, coupons used were prepared with Au thicknesses of 0.04μm, 0.1μm, 0.2μm, 0.4μm, and 0.5μm for ENIG deposits. 1st bonds were evaluated by the use of coupons before and after heating at 175 deg.C for 16 hours.

The 1st bonding evaluation results shown in Fig. 6. As shown in Fig. 6, with the ENIG film, excellent wire bonding reliability was obtained with the Au thickness greater than 0.4μm, however it is confirmed wire bonding reliability decreases with the Au thickness less than 0.2μm. These characteristics coincide with estimated results from the Fig.3 data from Au surface analysis (Ni detection).

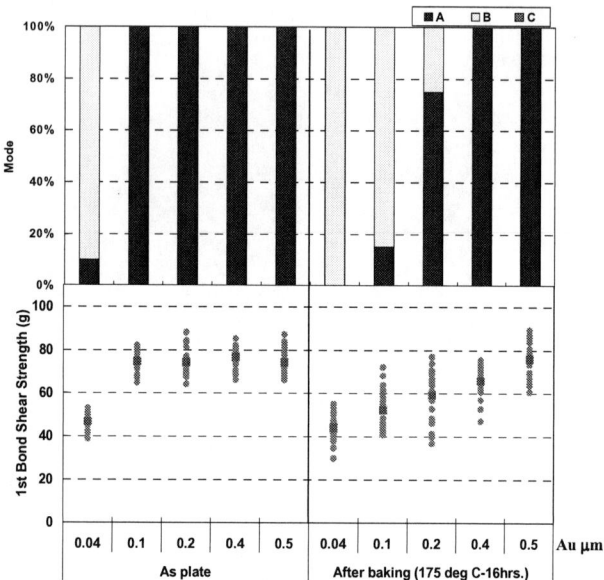

Fig. 6: 1st bond shear test results on ENIG film

Coupons used were prepared by combining Pd thicknesses of 0.06μm, 0.1μm, 0.2μm, and 0.3μm with gold deposit thicknesses of 0.02μm, 0.1μm, 0.2μm, and 0.4μm for the ENEPIG film. 1st bonds were evaluated by the use of coupons before and after heating at 175 deg.C for 16 hours. The 1st bond evaluation results are shown in Fig. 7.

As shown in Fig. 7, even if 0.02μm of Au thickness and 0.06μm of Pd thickness are implemented, excellent wire bonding reliability was obtained with the ENEPIG deposit. The characteristics coincide with estimated results from the Fig.3 data from Au surface analysis (no Ni detection).

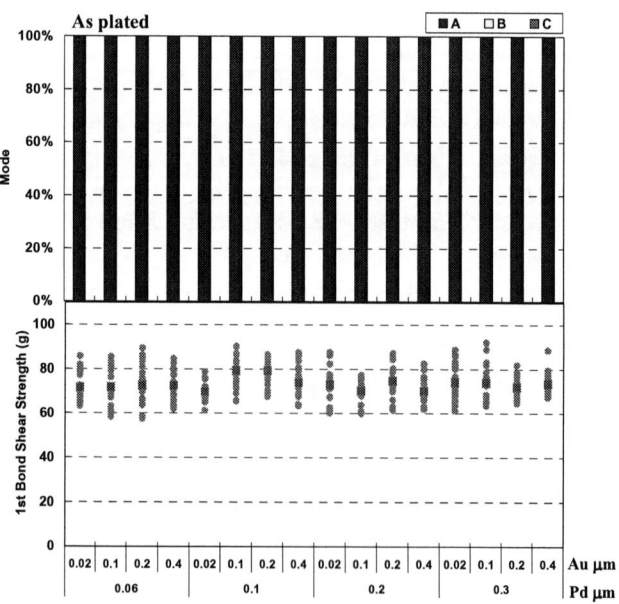

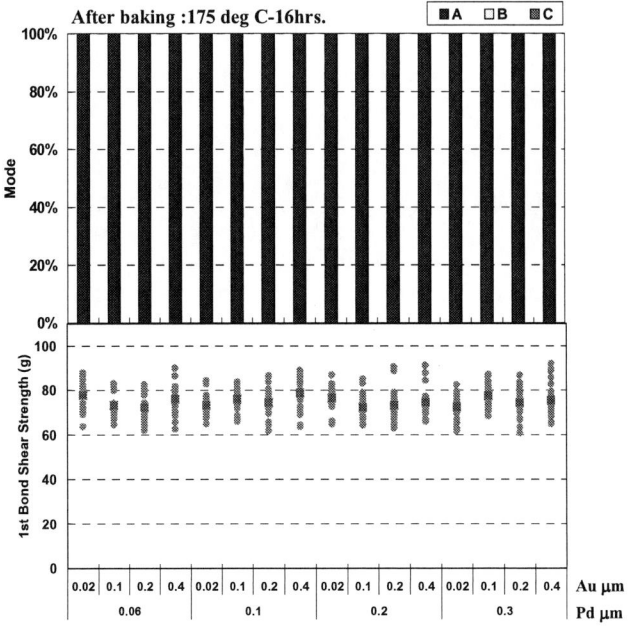

Fig. 7: 1st bond shear test results on ENEPIG film

It was confirmed the wire bonding characteristics depended on Au thickness and heat treatment with Ni-P/Au film. However, we did not confirm similar results with the Ni-P/Pd/Au film. Therefore, in decreasing the bond power to 130mW, 1st bonding characteristics were observed with each thickness of Pd and Au film shown in Fig.8.

As shown in Fig.8, it was confirmed thicker Au showed better wire bonding characteristics without dependence on the Pd film thickness under low bond power.

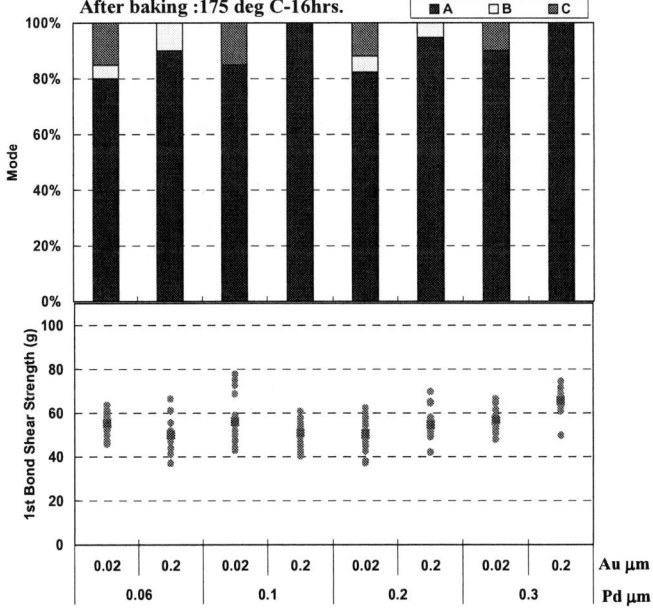

Fig. 8: 1st bond shear test results on ENEPIG film (Bond power 130mW)

SUMMARY

In this report, heat diffusion of each element in the ENEPIG film was investigated and it was found that diffusion of Ni to gold can be prevented by insertion of an approximate 0.06µm thick electroless palladium layer between Au and EN deposits.

Based on the results, even if the gold thickness of Ni-P/Pd/Au deposit was 0.02µm, excellent wire bonding characteristics were obtained. It was confirmed thicker Au showed better wire bonding characteristics without dependence on the Pd film thickness under low bond power. Excellent wire bonding characteristics were obtained by depositing thicker Au with Ni-P/Au deposits which could better withstand thermal aging.

ADVANCED PLASMA PROCESSING TECHNIQUES FOR IMPROVING DESCUM AND OTHER WLP PROCESS PERFORMANCE

Scott D. Szymanski
March Plasma Systems
Concord, CA, USA
sszymanski@marchplasma.com

ABSTRACT

Wafer level packaging (WLP) technology has been advancing by leaps and bounds over the past few years. As with all emerging technologies, the manufacturing challenges for building WLP devices can be great; moreover, the packaging and assembly processes are not always convergent, meaning that a solution that works for one type of WLP process may not be suitable for other processes. The challenge is to find solutions that are flexible enough to improve a wide range of WLP processes in order to be cost-effective.

The WLP industry has found that a variety of commercially-available plasma treatments exist to address many of these challenges. Plasma processing can remove photo resist (PR) residue after development (a process commonly referred to as "descum" in the semiconductor industry), remove organic, metal or oxide contamination, generally clean the surface of the wafer before the next step in the process, roughen and activate the chemical bonds at the surface of the wafer in order to improve bondability, increase the wettability of the surface of the wafer, and increase the uniformity of liquid flow (such as liquid photo resist material) over the surface of the wafer.

Moreover, data show that plasma technology is a convergent technology, where both established and emerging WLP processes can be improved in existing plasma systems (by changing, for example, plasma chemistries, flow rates, process pressures, power settings, and chamber configuration, etc.), and without the high costs usually associated with traditional front-end equipment capital equipment.

Plasma processes can directly, and in some cases, dramatically improve the performance and reliability of the finished WLP package. This discussion will cover how plasma treatments can improve the performance of several common WLP processes, such as descum. In this way, plasma treatment is rapidly changing from an enhancing technology to an enabling one.

Key words: Wafer level packaging, plasma treatment, cleaning, descum.

WAFER LEVEL PACKAGING PROCESS FLOW

For this example of a typical WLP process flow, we will not consider any steps required for redistribution; that is, in this case we will assume the solder bumps are created directly above the wafer's original bond pad locations. We will also assume that the wafer's final passivation layer (for example, silicon nitride or benzocyclobutene, also referred to as "BCB") has already had vias created in it to expose all the bond pads. In this case, a WLP manufacturing process flow involves the following steps:

(1) Under bump metal (UBM) deposition over the entire passivated surface of the finished wafer.
(2) Resist coating the entire wafer.
(3) Exposure of the resist so that the areas above the bond pads can be removed, while leaving resist intact over the rest of the wafer.
(4) Development of the resist to remove the resist as described above.
(5) Descum.
(6) Plating or depositing solder bump material into the open areas of the resist.
(7) Stripping away the remaining resist.
(8) Etching the UBM to remove all the UBM that is not covered and protected by the solder bump material.
(9) Reflowing the bump material into spherical solder balls that can be mounted to a suitable WLP substrate after dicing.
(10) Final inspection of the completed bumped wafers, where the wafers are sorted into good wafers, bad wafers and/or wafers that require re-work to be acceptable.

The good wafers can then proceed to subsequent dicing steps, and the individual die can be flipped and mounted to a substrate in order to complete the packaging process.

This WLP manufacturing process flow is shown in Figure 1 below.

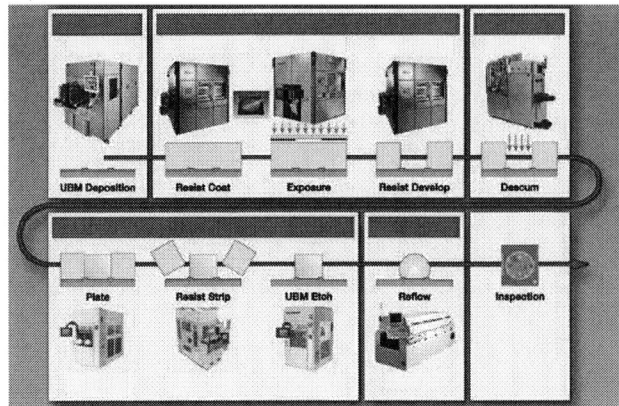

Figure 1. Manufacturing process flow for a typical WLP application, which shows the 10 major processing steps. Note that the Descum process is Step 5 (top right corner).

The described WLP manufacturing process flow is straightforward, and often steps can be performed by competing technologies or processes (For example, resist stripping can be done by wet/liquid processes, by dry/plasma-based processes, or by a combination of the two). WLP process steps can also be performed by a variety of commercially available semiconductor processing equipment from both the Front- and Back-end equipment industries. Using "off-the-shelf" technologies and equipment from is critical in order to minimize the cost of manufacturing WLP devices. Both technology and cost drivers are taken into consideration when choosing the correct manufacturing processes and equipment for a particular WLP application.

PLASMA TREATMENTS TO IMPROVE THE SOLDIER BUMP

When performing wafer level packaging, it is critical to have a bump material deposition process (Step 6) such that the bump material makes complete contact with the previously deposited UBM and also has maximized adhesion to the UBM layer. Unfortunately, even at the geometries typically involved in with today's WLP processing, the resist develop process (Step 4) can leave a resist residue along the sides and bottom of the photo-defined vias that prevents complete contact between the bump material and the UBM. This residue is known as "scum" and it is highly desirable (if not absolutely required) to remove this scum before the next process step. Scum is removed in a step referred to as "descum" (Step 5), and dry plasma treatments are an attractive solution for the descum step, as long as they do not remove too much of the desired resist but simply clean out the undesirable scum material. Secondly, it is desirable to increase the bondability of the UBM surface to the bump material as much as possible. Plasma treatments are well known for their ability to increase the adhesion ability of a surface (without causing damage to said surface) by a process called "surface activation" [1]. The ability of certain types of plasma to make a microelectronic device's hydrophobic

(non-wettable) surface into a hydrophilic (highly-wettable) surface is also well documented [2, 3]. In fact, post-treatment improvements can be easily demonstrated almost anywhere by using a simple water dropper and an inexpensive, commercially-available Contact Angle Meter (CAM).

What is especially noteworthy and of particular interest to us in regard to WLP processes is that an increase in the hydrophilic characteristic of a surface thanks to plasma treatment correlates to a significant increase in both pull and shear strengths [4], which is highly desirable in the manufacturing of WLP devices, because it is better to have the solder bump bond as strongly as possible to the underlying UBM layer. The benefit is greater reliability and fewer failures of the solder bumps, and therefore, of the finished package as well.

PLASMA-BASED DESCUM PROCESS

For a plasma process to be suitable for WLP descum applications, the process must remove the scum thoroughly but not remove too much of the desired resist material because the resist material is used to make sure a sufficient quantity of solder bump material is placed directly over the bond pad and UBM layer. This can be difficult because the scum and the desired resist material can have similar selectivity in a given plasma etch environment, although the desired resist is usually more difficult to etch than the undesirable scum material.

In any case, the resist etch rate and uniformity of the resist etch across the surface of the wafer are both key aspects of performance for any type of descum equipment. The etch rate of the bulk resist material is used because it is nearly impossible to measure the etch rate of scum material in a meaningful and repeatable way. Thus, the etch rate of the bulk resist material serves as a proxy for the etch rate of the scum for descum processes.

For uniformity measurements, it is usual to exclude the very edge of the wafer from measurements. It is common to exclude a region from 3 to 10 mm wide from the edge of the wafer.

It is important to note that the descum process is similar to, but significantly different from ashing/stripping processes. Both descum and ashing/stripping processes remove resist material. However, in ashing/stripping processes, bulk resist material is completely removed ("stripped" away) by using a high-power dry plasma system or a highly reactive chemical bath (or by a combination of these two methods). Resist ashing/stripping is therefore best defined as a "bulk" process. On the other hand, descum is better defined as a "fine" process, because both good etch rate control and good etch uniformity over the entire surface of the wafer is required for a successful descum process. This is typically achieved by using lower power and less flow of the reactive chemicals compared to an ashing/stripping process.

In addition to etch rate and uniformity, it is also important to ensure that the process is fast enough to keep up with the customer's required manufacturing levels. The descum step must have a high enough throughput and not cause a bottleneck in the production line. Increasing the throughput of a descum process can be accomplished several ways. For example, the customer may choose a descum process that is inherently fast and meets the customer's manufacturing requirements. Another way to increase throughput is to load multiple wafers into a single processing chamber and processing the group of wafers together, which is sometimes referred to as a "batch" process. Still another method is to add additional process modules to meet required throughput levels. This approach is possible with batch-type systems, but is especially common for systems that process one wafer at a time in each process module. Such systems are commonly referred to as "single-wafer" systems. Adding extra process modules to single-wafer processing systems can be very desirable, because it allows a single material handler system to feed multiple processing stations, which decreases handler system idle time while increasing throughput at a much lower cost compared to buying additional systems.

REQUIRED PERFORMANCE CRITERIA FOR DESCUM PROCESSES

When defining acceptable performance for a WLP descum process, it is most important to take into consideration the three readily measurable criteria described above:
(1)   Etch rate of the bulk resist material.
(2)   Uniformity of the etch across the entire surface of the wafer excluding the edge exclusion zone.
(3)   Throughput of the process on a wafers-per-hour (WPH) basis.

The following target values are somewhat arbitrary, and actual required values will vary from customer to customer, but these values are consistent with market research and customer interactions carried out by March Plasma Systems:

Etch Rate: An etch rate of bulk resist of $\geq 3,000$ angstroms per minute is desirable, because it allows sufficient time for plasma processing plus the overhead time required for each wafer (in a single-wafer system) in order to reach the required throughput level.

Uniformity: A uniformity of $\leq 10\%$ within each wafer with a 5 mm edge exclusion zone allows the customer to maximize the number of die that can be taken from each wafer while ensuring that die from every point on the wafer will receive sufficient descum treatment.

Throughput: On a wafers-per-hour basis, throughput for a single-wafer system with one process module should be

>60 WPH or nominally one wafer per minute. A system with two process modules should therefore reach 120 WPH unless the system is somehow handler-limited. Note that the throughput value includes all the overhead time required for processing the wafers.

To ensure that the etch rate and throughput mentioned above are both reasonable and achievable, we assume that a successful descum process removes about 1,500 angstroms of the bulk resist material. This amount of resist removal completely removes the scum material, but does not significantly affect the performance of the remaining resist, because the resist thickness is nominally 1 to 4 microns (10,000 to 40,000 angstroms) thick before the descum step. The time to etch 1,500 angstroms is about 30 seconds at the aforementioned etch rate, and the material handling consumes approximately 20 seconds per wafer (i.e. 10 seconds to move a wafer into the chamber plus 10 seconds to remove the wafer). This leaves 10 seconds for the system to pump down before processing and also to vent after processing. The total cycle time is 60 seconds per wafer, which equates to a steady-state throughput of 60 WPH.

EXPERIMENTAL SET-UP AND METHOD

A commercially-available plasma processing system with a chamber volume of approximately 4 liters was used. The RF generator and vacuum pumping system were self-contained within the chassis of the system, as were the system's mass flow controllers (MFCs) and automatic matching network. The system was configured with a 600 watt 13.56 MHz radio frequency (RF) generator; a 1,000 watt RF generator was also available (but was not used for this set of experiments). The plasma system was configured with two MFCs, each rated at a maximum flow rate of 500 standard cubic centimeters per minute (sccm). Semiconductor-grade Oxygen (O_2) was used as the process gas for these experiments. The wafer stage (also known as the wafer "chuck") was water-cooled using a closed-loop chiller unit and the temperature setpoint for all wafer runs was 25 degrees Celsius. The wafer chuck was configured for 8 inch (200 mm) standard-thickness (approx. 700 microns thick) reclaimed silicon wafers. The wafers in this experiment were uniformly coated with approximately 3.5 microns of commercially-available resist material and then soft-baked to ensure the resist was hard enough for material handling purposes. Wafers were hand-loaded into the chamber for these experiments; however, an integrated automated wafer handling system is available for this plasma system.

Four process conditions in terms of RF power setting and steady-state process pressure were evaluated as part of this experiment. The four conditions are shown in the figure below.

Condition #1: High Power, High Pressure	Condition #3: Low Power, High Pressure
Condition #2: High Power, Low Pressure	Condition #4: Low Power, Low Pressure

Figure 2. The four process conditions in terms of RF power setting and steady-state process pressure that were evaluated as part of this descum experiment.

The within-wafer results for the process condition with the best results are shown in the Experimental Results section of this paper.

To confirm wafer-to-wafer repeatability, which is arguably as important as within-wafer uniformity, the process condition with the best results was repeated with eight additional wafers (for a total of nine wafers). Those results are also shown in the Experimental Results section of this paper.

All wafers were processed for 180 seconds (three minutes) of RF-on time, in order to mitigate process ramp-up and loading effects. The etch rate per minute was then calculated.

Pre- and post-etch data was collected with a commercially-available stylus-type profilometer system. Before processing, patterns were created at several areas on the wafer by removing areas of resist down to the bare silicon wafer. A pre-etch measurement was taken of the height of the resist step. After processing, a post-etch measurement of the height of the resist step was taken. The resulting delta height value was then calculated.

Height measurements were taken at 9 points over the surface of each wafer. Uniformity for each wafer was calculated by taking the difference between the maximum value minus the minimum value, then dividing by two times the mean value, and then multiplying the resulting value by 100% in order to get a percent uniformity value. Uniformity between the nine wafers was calculated by the same formula, using the mean value of each wafer as that wafer's data point.

EXPERIMENTAL RESULTS

The experimental condition which yielded the highest etch rate and best uniformity was Condition #1: High Power, High Pressure. This result is not unexpected at all, because high power processes typically create more reactive plasma species compared to low power processes, and high pressure plasma processes contain a greater quantity of reactive chemicals compared with low pressure processes. The within-wafer results for Condition #1 are shown in the figure below.

Maximum Etch Rate	3,626 Å/min.
Minimum Etch Rate	3,348 Å/min.
Mean Etch Rate	3,506 Å/min.
Within-wafer Uniformity (Max-Min/2*Mean)*100%	3.98%

Figure 3. Within-wafer results for experimental Condition #1: High Power, High Pressure, which yielded the highest etch rate and the best uniformity.

To confirm wafer-to-wafer repeatability of this descum etch process, the same process conditions were repeated for an additional eight wafers, for a total of nine wafers. As with the first experiment, data was collected from 9 points on each wafer, and the mean etch rate value for each wafer was calculated. The mean etch rate results for Condition #1 for nine wafers are shown in the figure below.

Wafer Number	Mean Etch Rate
Wafer #1 (Original)	3,506 Å/min.
Wafer #2	3,462 Å/min.
Wafer #3	3,529 Å/min.
Wafer #4	3,423 Å/min.
Wafer #5	3,615 Å/min.
Wafer #6	3,598 Å/min.
Wafer #7	3,477 Å/min.
Wafer #8	3,604 Å/min.
Wafer #9	3,611 Å/min.

Figure 4. Mean etch rate results for a total of nine wafers for experimental Condition #1: High Power, High Pressure. Wafer #1 was one of the original wafers run in the first part of this experiment.

Wafer-to-wafer uniformity was calculated using the same formula as within-wafer uniformity. The results for Condition #1 for the nine wafers are shown in the figure below.

Maximum Etch Rate	3,615 Å/min.
Minimum Etch Rate	3,423 Å/min.
Mean Etch Rate	3,536 Å/min.
Wafer-to-wafer Uniformity (Max-Min/2*Mean)*100%	2.71%

Figure 5. Wafer-to-wafer uniformity results for a total of nine wafers for experimental Condition #1: High Power, High Pressure.

Although we did not define a success criteria for wafer-to-wafer uniformity above, a wafer-to-wafer uniformity value of $\leq 3\%$ should be acceptable for most WLP applications. This value is based on market research and customer interactions by March Plasma Systems.

CONCLUSION

Using a commercially-available RF plasma treatment system, we were able to achieve successful results for the WLP descum process, in terms of etch rate, uniformity within-wafer, uniformity wafer-to-wafer, and throughput. A summary of these results is shown in the figure below.

Criteria	Requirement	Result	OK/NG
Mean Etch Rate, Single Wafer	$\geq 3,000$ Å/min.	3,506 Å/min. (1st wafer)	**OK**
Mean Etch Rate, Multiple Wafers	$\geq 3,000$ Å/min.	3,536 Å/min. (9 wafers)	**OK**
Uniformity, Within-wafer	$\leq 10\%$, 5 mm edge exclusion	3.98% (9 points)	**OK**
Uniformity, Wafer-to-wafer	$\leq 3\%$	2.71% (9 wafers)	**OK**
Throughput, Steady-state	≥ 60 WPH, per module	65 WPH @3500 Å/min. etch rate	**OK**

Figure 6. Successful results for the WLP descum process. This figure shows a summary of descum criteria, requirements, and experimental results.

OTHER PLASMA PROCESSES TO IMPROVE WLP DEVICE MANUFACTURING

The focus of this discussion has been primarily on the descum process in WLP device manufacturing. However, dry plasma treatments can improve a number of other WLP manufacturing steps. In the interest of brevity, data will not be presented in this particular paper, but a short list of these processes is listed here for the reader to consider. Specific applications or plasma technology questions may be directed to March Plasma Systems for response.

Other common dry plasma treatments for WLP device manufacturing:

(1) Incoming Wafer Cleaning: In order to remove organic, halogen (i.e. fluorine), metal oxides and other contamination from the surface of the wafer as it transitions from front-end fabrication to back-end device packaging.

(2) Cleaning Metal Pads and Lines used for Wafer Level Packaging or Redistribution: Especially pads/lines made from aluminum, gold, thin gold, copper or more exotic "sandwiched" metal stacks that are very fragile.

(3) Surface Conditioning of the Wafer: In order to increase wettability of the resist liquid, which allows the resist to flow faster over the entire surface of the wafer (improving throughput) and coat more uniformly across the wafer.

(4) Surface Treatment of the Wafer: In order to increase the adhesion of dielectric layers, such as benzo-cyclobutene (BCB) or silicon nitride (SiN_x).

(5) Treatment of the Wafer Surface after Wet Resist Stripping: In order to insure that all of the resist material has been completely removed. This process is also referred to as "post-wet resist strip dry plasma cleaning."

For example, plasma process 5 above could be done in lieu of a high-power ashing-type resist strip process, in cases where the cost of the high-power ashing equipment is prohibitive or where the high-power ashing process might damage the wafer/devices due to excessive plasma energy or high processing temperatures.

REFERENCES

[1] Getty, J., Ph.D., "How Plasma-Enhanced Surface Modification Improves the Production of Microelectronics and Optoelectronics," Chip Scale Review, January/February issue, 2002, pp. 72-75.

[2] Zhao, J., Ph.D., and J. Getty, Ph.D., "Plasma for Underfill Process in Flipchip Packaging," Proceedings of IMAPS-Taiwan Tech. Symposium, Taipei, Taiwan, June 24-25, 2005, pp.3.

[3] Zhao, J., Ph.D., J. Getty, Ph.D., and D. Chir, "Plasma Processing for Enhanced Underfill," Chip Scale Review, July issue, 2004, pp. 1, 3-4.

[4] Zhao, J., Ph.D., and J. Getty, Ph.D., "Plasma Treatment of Pre-Underfill Flip Chip Devices," Proceedings of IMAPS Austin Tech. Symposium, Austin, Texas, 2005, pp.4-5.

USING THE 2D MACRO CD METROLOGY PACKAGE TO MEASURE CD LINES

Rajiv Roy, Matt Wilson, and Chris Hawes
Rudolph Technologies
Bloomington, MN, USA
rajiv.roy@rudolphtech.com

OVERVIEW

Automated Visual Inspection (AVI) systems are used extensively in Advanced packaging and Memory Outgoing QA areas. Recently customers have begun using these tools in-line to monitor process. An interesting use of AVI has been the use of CD measurement. CD measurement in the area of advanced packaging was done by a microscope. With AVI tools becoming faster and cost-effective CD measurements are also being done by these tools.

ADVANCED PACKAGING - PPL

An important trend in the area of wafer bumping is Post-Passivation Layer (PPL) processing. Examples of PPL include ReDistribution Lines (RDLs) for I/O rerouting, and the integration of passives at the chip/package interface. PPL processing involves adding multiple thin-film dielectric and sputtered or plated metal layers on top of the chip to enhance device functionality, as shown in Figure 1. Greater performance is achieved as integration takes place at the package level and the form factor shrinks. Performance of these passives are directly proportional to the geometry,

hence the need to measure line widths to assure device performance.

In addition, as more device manufacturers use RDLs to allow for greater flexibility, inspection processes need to capture the measurements of these critical dimension (CD) lines. Instead of relying on manual inspection, or purchasing additional equipment, manufacturers can collect information about CD lines using existing advanced macro inspection systems.

The 2D Macro CD Metrology Package on the NSX® Series and 3Di™ Series systems provides a tool for measuring straight or curved CD lines in real time during advanced macro inspection. Capturing line measurements without impacting throughput, the tool also yields repeatable results so that manufacturers can rapidly detect any issues in RDL processes, minimizing impacts to yields.

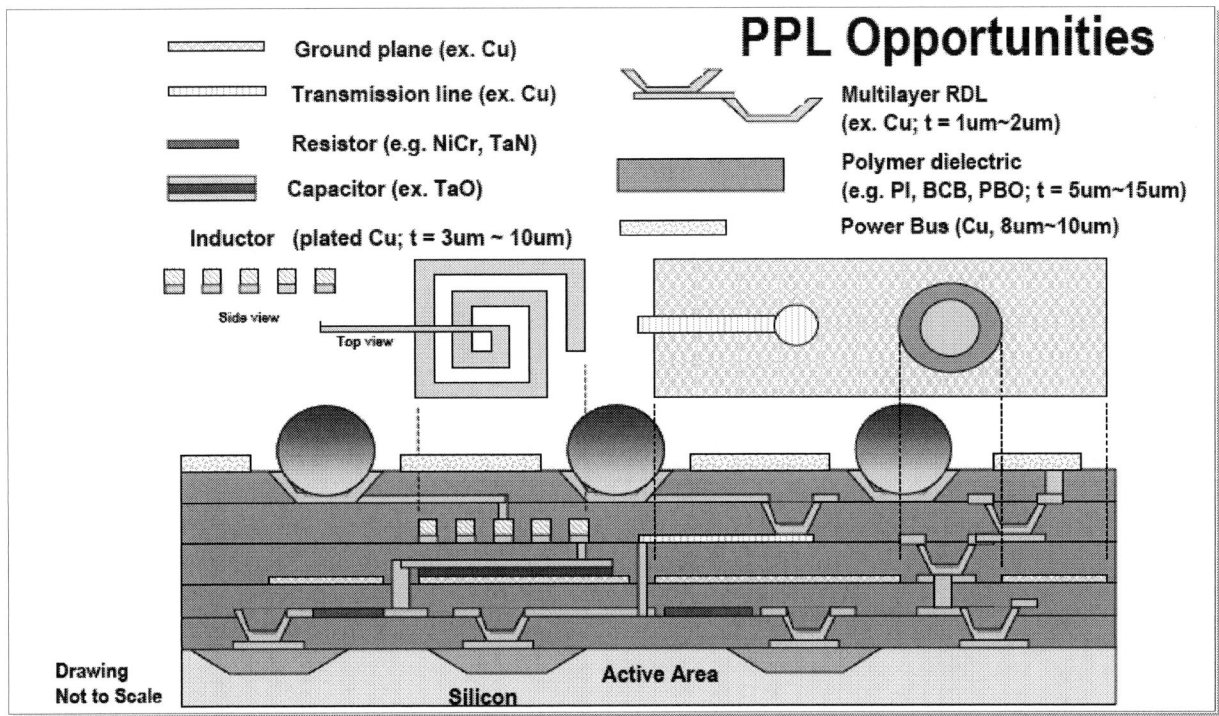

Figure 1: Post-Passivation Layer

MEMORY – RDL

Memory is increasingly being shipped into stacked-die applications. Today a bulk of these die are wire-bonded. Clearly if memory has been typically assembled in a Lead-On-Chip configuration then, the bond-pads have to be re-distributed to the edge from the center of the die (Figure 2). This gives flexibility to the Memory operations to choose to

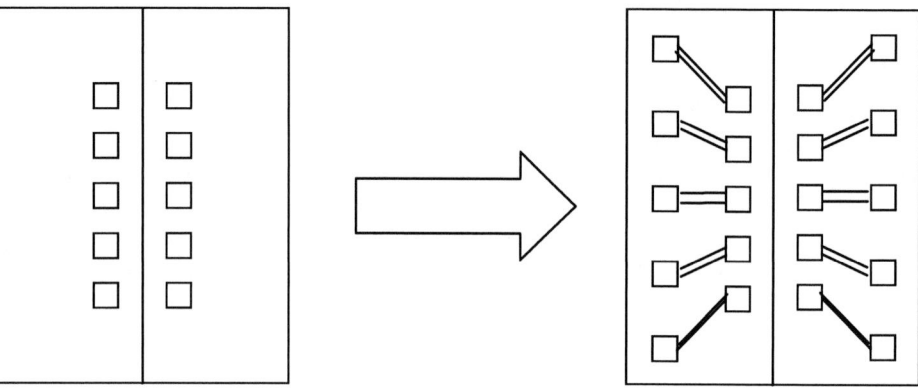

Figure 2: Lead-On-Chip Memory redistributed for Stacked Die applications

allocate devices to either traditional packaging or to stacked packaging. Increasingly these die are operated at >500Mhz where performance can be affected by the RDL dimensions. RDL is being used extensively to measure the CD width to monitor process performance.

HOW THE 2D MACRO CD METROLOGY PACKAGE WORKS

The 2D Macro CD metrology package uses a binary thresholding approach that is localized to within a specified region of interest (ROI), as shown in Figure 4. Each ROI contains a feature of interest that will be isolated from the background, binarized, and then measured. The feature is typically a bump, a via, or a metal line—and in all cases, has a gray value that is statistically different than the rest of the image within the ROI. Once the feature has been binarized, the Metrology Package can count pixels in nearly any direction across the feature and report the results in microns.

In real time, the system measures features — whether they are bumps, CD lines, or vias—using ferrets, at specified angular increments. For example, to measure the average diameter of a bump, engineers setup the angular increment of the ferret to be a minimum of every 45 degrees, as shown by the red line in Figure 3.

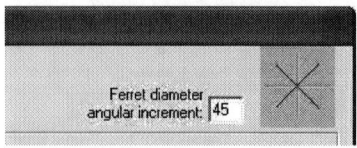

Figure 3: 45 Degree Ferret

Each ferret (or measurement) can be enabled or disabled, and the software will report the average of whichever measurements have been enabled for that ROI.

Figure 4 illustrates five ROIs on a wafer image. The vertical yellow line that is in the middle of each ROI marks the ferret setting.

Figure 4: Post-Passivation Layer

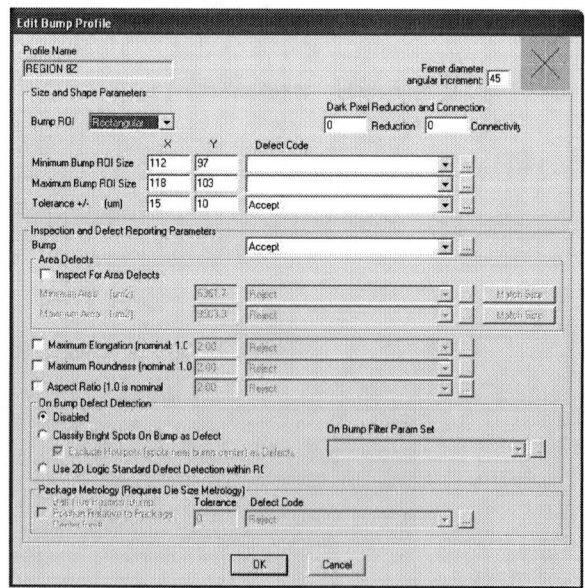

Figure 5: Ferret Settings on CD Lines

THE REPORTING CAPABILITY

The system reports on all features that are measured in several ways. A Bump Summary report, included in the standard Crystal Report template, includes a summary of all of the measurements in the form of averages, minimums, maximums, standard deviations, and coefficient of variance on both a per die basis, as well as a per wafer basis.

Detailed results on each measurement are available in a Raw Bump Data report that can be automatically generated at the end of inspection.

Regardless of the type of report that is generated, they all refer to features as bumps. Similarly, the system assumes that any feature measured by the 2D Macro CD Metrology package is round. Consequently, when measuring square or rectangular features it is important to manage the use of ferret diameter definitions appropriately so the information in the report will make sense for the feature you are measuring.

SETTING UP THE SYSTEM TO MEASURE CD LINES

Before the setup takes place, the resolution and accuracy of the measurements must be determined. This decision will drive the magnification setting, which in turn will determine whether the measurements can be made during a standard 2D inspection, or whether a separate inspection pass will be required to capture the measurement of each feature.

In many process control situations, it will be necessary to measure line widths at the highest magnification (20X) setting which will result in a measurement accuracy of +/- 0.5µ. At this resolution it is frequently more cost effective to measure up to five sample areas on a wafer, instead of inspecting the whole wafer.

When performing an inspection at 20X, the first step is to define the inspection region by overriding the inspection area for the inspection pass. This is important so the system does not attempt to define inspection sites over a large area.

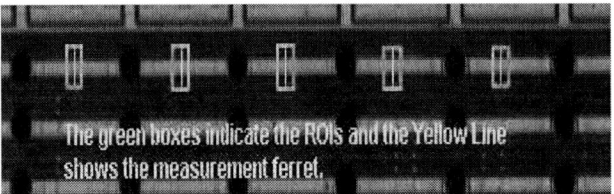

Figure 6: Ferret Settings on CD Lines

Ideally, the size of the inspection area will be less than one field of view (FOV) at 20X (0.6mm x 0.8mm).

When setting up the ROIs, the engineer must first create a measurement profile to define the nominal range of the measurements to be taken, as shown in Figure 6. This also includes enabling or disabling the appropriate ferrets to ensure that the width or height of features is measured correctly. The ROIs cannot typically be located automatically, so it will be necessary to manually create the ROIs and place them in the correct location on the wafer.

Once the ROIs are placed correctly, a binary threshold must be calculated. This threshold will determine which pixels are included in the measurement and which pixels are ignored. This is a critical setting that will ultimately determine the accuracy of the measurement.

APPLICATION OF CD LINE MEASUREMENTS IN FABS

A leading semiconductor manufacturer uses its advanced macro inspection system to inspect the width of RDLs, as well as polyimide openings. Using a multi-pass inspection process, the system first inspects whole reticle masks across the entire wafer surface at 1X. In the second pass, the 2D Bump function captures the width of CD lines that are approximately 20µ wide, in five areas on the wafer at 20X.

Figure 7 shows an image of one ROI, or inspection site. Note that in this inspection site, two ROIs have been defined so that the system captures two measurements at a time. The width of these lines and openings provide a process control mechanism within the RDL process to ensure the consistency of the metal lines.

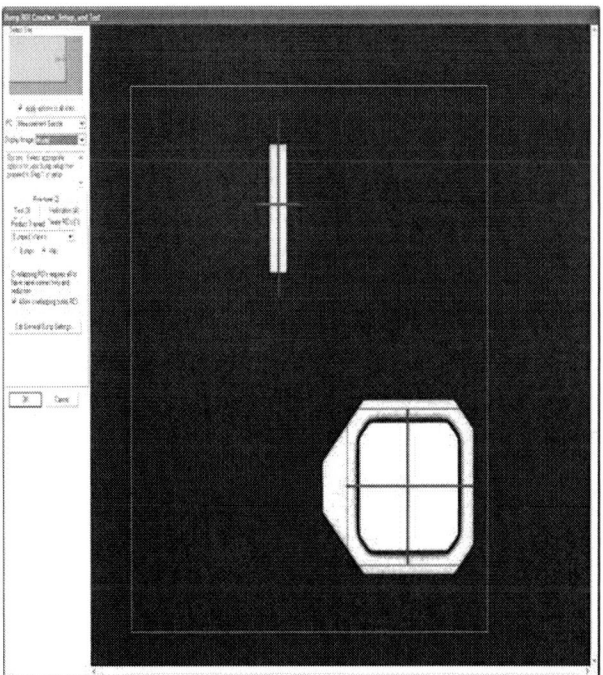

Figure 7: Ferret Settings on CD Lines

Measurements for the above example were repeatable to within less than 0.2µ.

PERFORMANCE DATA

Data was also taken on 5 dies to measure accuracy and repeatability.

The interesting aspect about the data is that it is offset in one direction and hence with an offset correction, data could be dialed in to minimize error.

Die	Manual 20X measurement (u)	Automated result	Delta
Die 4, 4	18.5	21.02	+1.57
Die 4, 3	19	21.85	+2.85
Die 1, 2	19.5	22.27	+2.77
Die 6, 2	19	21.15	+2.15
Die 4, 1	19.5	22.58	+3.08

Die	Manual 20X measurement (u)	Automated After offset correction
Die 4, 4	18.5	18.54
Die 4, 3	19	19.37
Die 1, 2	19.5	19.79
Die 6, 2	19	18.67
Die 4, 1	19.5	20.1

There were repeatability tests conducted on the same 5 dies.

Average repeatability was 0.19 3 sigma.

Die	Repeatability (3 σ)
Die 4, 4	0.18
Die 4, 3	0.31
Die 1, 2	0.22
Die 6, 2	0.19
Die 4, 1	0.07

CONCLUSION

Rather than investing in additional inspection devices, or requiring an out-of-process inspection step, the 2D Macro CD Metrology function on the NSX and 3Di Series systems can measure CD lines in real time, delivering consistent results without impacting throughout. Engineers can experiment with the 2D Macro CD Metrology function to determine whether it can capture useful line information in process.

SURFACE CLEANING FLIP CHIP WAFERS FOR TEST AND ASSEMBLY IMPROVEMENTS

Terence Collier
CVInc.
Richardson, TX, USA

ABSTRACT

Flip chip wafers are dirty compared to the wafers that are processed in the front end fab. The wafer and bumps have layers of carbon contamination, residues and non-desirable oxides that are unfounded in the front end. These contaminants not only can impact assembly yield but reduce the overall test yield as well. Proper cleaning helps recover yield loss, reduce cycle time and rework.

Keywords: Flip chip, cleaning, test, yield enhancement, CRES

BACKGROUND

One technique for wafer bumping involves electroplating solder into a defined via opening. The photoresist is removed, flux added and the solder reflowed to form a sphere. In another technique the solder is screen printed directly onto the wafer followed by reflow to form the bump. Both processed involve large amounts of flux. The electroplating process can also leave behind photoresist residue. When polyimide or BCB is added an additional source of carbon contamination can remain on the bumps.

Cleaning these carbon residues is critical for stable test yield and assembly. It has been demonstrated that proper cleaning of these residues improves probe card life and first pass yield at test. Improved cleaning methodologies provide an added gift by reducing overall process cycle time. In some cases the customer has gone from 50 touchdowns (insertions) between cleans to over 400.

Most commercial flux and wafer cleaners are marginally effective at removing contaminants. Some commercial cleaners can make the wafer look worse than the starting wafers with no cleaning. It has also been demonstrated that oxide reduction can eliminate the need for flux during assembly reflow. As such, having a cleaner that can remove bump oxides as well as flux allows a lead free reflow profile without rapid (quench) cooling. The rapid cooling is used to reduce the amount of oxidation which forms on the solder sphere and can be problematic with some older equipment.

A History of Dirt and Grime

The flip chip process begins at the wafer fab where things are typically clean. A wafer is provided with vices corresponding to some standard I/O layout. Pads are typically pure aluminum or some hybrid aluminum alloy containing silicon or copper. During the later metallization and passivation steps, the pads can be subject to processes that lead to contamination, corrosion and residual amounts of photoresist residues.

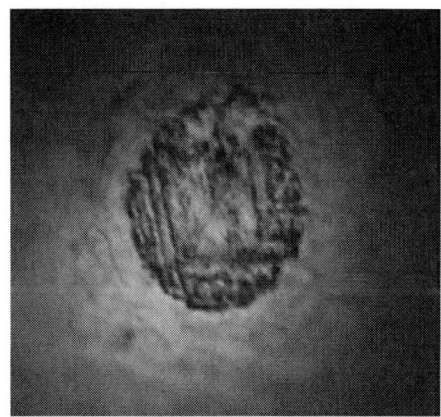

Figure 1. Pre Probed Bump Showing Residue

"Dirty pads" on die lead to difficulty during UBM processing as well as long term device reliability. Performance can be improved by removing the various native and non-native oxides, fluorides, etch residues, photoresist and other debris that might surface. While the focus of this paper is mainly what happens after UBM and bumping, it should be known that the author did investigate pre UBM cleaning as well and found many reliable aqueous cleaners without having to us dry clean recipes. These are cleaners that will remove the debris on pads to aid in UBM deposition. One unexpected benefit discovered is that the cleaners used in this project worked effectively at cleaning bumps as well as both aluminum and copper bond pads.

Vertical Test – How to Probe Flip Chips

Flip chip test is unlike probing on a traditional bond pad. Traveling in the z-direction, the probe card has vertical instead of horizontal scrub motion as with bond pads on wafers. The tester directs the probe card to position relative to the bump just making electromechanical contact. Additional overtravel is then added to penetrate the barrier layer to make stable electrical contact with the device undergoing test.

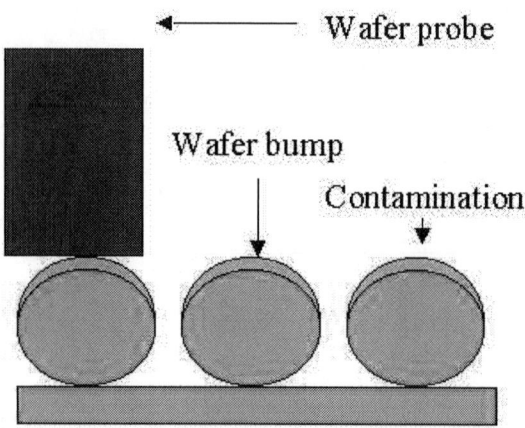

Figure 2. Vertical Type Probing of a Flip Chip

Additional overtravel is required when excessive contamination layers are present. This additional overtravel can lead to coplanarity issues and damage the underlying structures including UBM. Bump damage, coplanarity specifically, has a direct impact on assembly yield and is demonstrated when the flip chip is reflowed to the substrate during final assembly.

Problems also occur when a foreign substance is deposited on the bumps after continuous probing. This material can be incorporated into the matrix and lead to poor intermetallic and voids at the interface. Each could lead to an electrical test failure, a time zero package failure, or a long term reliability risk.

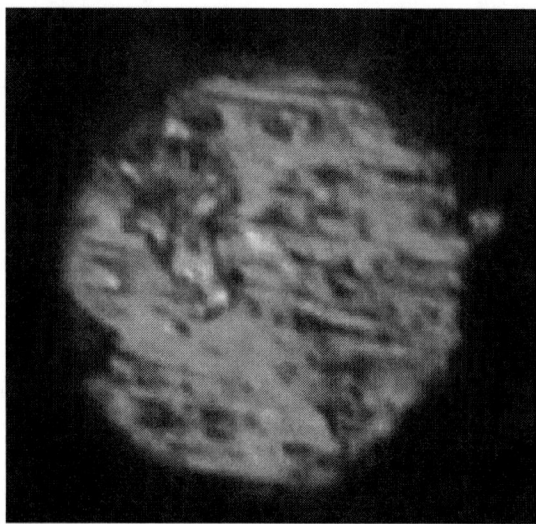

Figure 3. Probe Tip Debris

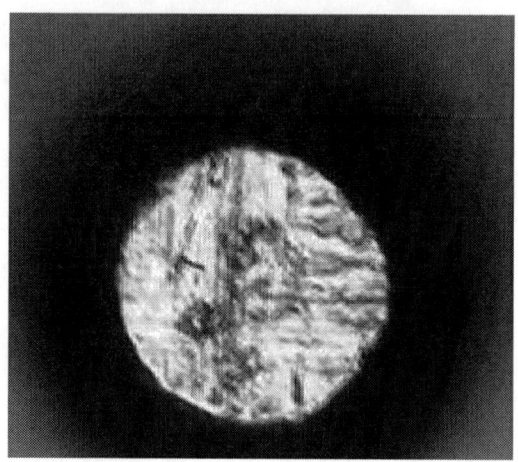

Figure 4. Post Probe Dirty Bump

Contact Resistance

Metals have low bulk and surface resistance typically less than 50 milliohms. The contact area, typically the surface of the pad, might have higher than normal surface resistance due to contaminants and corrosion. This contact resistance (CRES) can increase the normally low resistance between the outside world by up to a few megaohms in some cases. Since the electrical contact to the outside world is made at this interface, contact resistance is a critical parameter. Increased CRES impedes electrical contact between the device and tester. Reducing CRES improves yield and extends the life of hardware and PM costs.

Increased contact resistance will affect test signals and measurement, especially in applications using terminated transmission lines.[1] For example, device outputs driving 50-ohm transmission lines to a tester with 50-ohm receiver termination loads will expect a connection as:

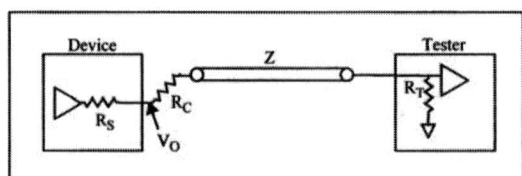

Figure 5. Contact Resistance Circuit

If the receiver termination (R_T) and the transmission line impedance (Z) are 50 ohms, the contact resistance (R_C) is 0 ohms, then the transmission line properly terminates with a load reflection coefficient of $(R_T - Z) / (R_T + Z) = 0$. The device would produce $V_O * R_T / (R_T + R_C) = V_O$ at the tester receiver input.

The probe contact resistance is near the device output and effectively increases the output source resistance. If the device driver control maintains V_O at the output pad, probe contact resistance values of 5 to 50 ohms would cause measured voltage levels to vary 8% to 50% from the true device output. Output current levels (I_{OL}/I_{OH}) would also be lower than expected.

92

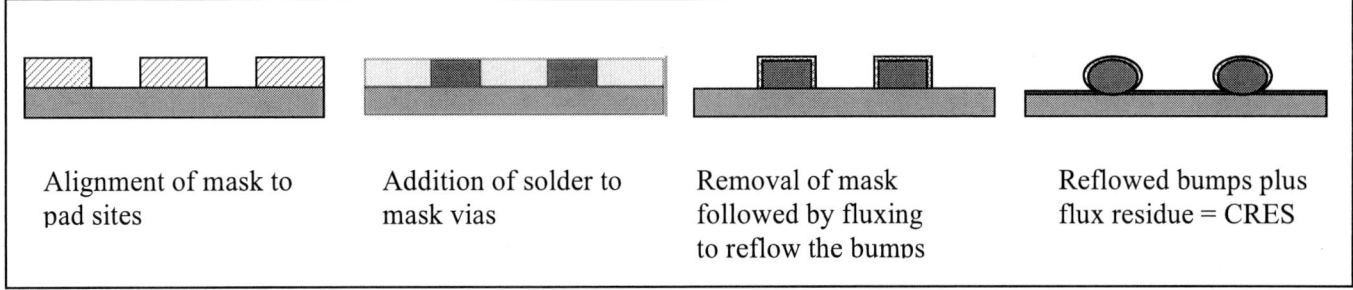

| Alignment of mask to pad sites | Addition of solder to mask vias | Removal of mask followed by fluxing to reflow the bumps | Reflowed bumps plus flux residue = CRES |

Figure 6. Overview of a Typical Flip Chip Bump Fabrication Process

CLEANING

The discussion on CRES demonstrates the criticality of clean wafers and bumps. Fabrication of CSP packages typically proceeds as wafer fab, electrical verification, UBM, CSP fabrication, CSP assembly and final package test. CSP fabrication stages can lead to dirty bumps. While there are a number of bumping techniques, screen printing and electro deposited processing are discussed in this application.

The process to reflow the bump requires a flux material to facilitate the melt and reflow. While the flux is a necessary material, residues remain even after cleaning processes to remove the material. A number of recipes and flux removal solutions are available but the results can all be the same. Residue from the flux, cleaner, interactions between the flux and cleaner, or the cleaner and final rinse solution can all leave debris on the bump and wafer.

CVI studied six materials to look at the impact of cleaning on improving test and assembly yield. Four materials were from the BPS series of products offered by ACT. BPS 100, BPS 125, BPS 130, and BPS 135. A competitor product and IPA was also evaluated. The fluxes studied were a number of clean, no clean, tac and paste with flux. For this paper, the results presented are from using off the shelf 63/37 SnPb paste with flux.

The paste was applied to a 1 inch square aluminized wafer that was made solderable with electroless nickel gold (ENIG). Paste was applied and placed in a reflow oven per the reflow profile specified.

After the reflow the sample was inspected and pictures taken. The image below shows one such slide after reflow. A large amount of flux residue and many microspheres are apparent in the center portion.

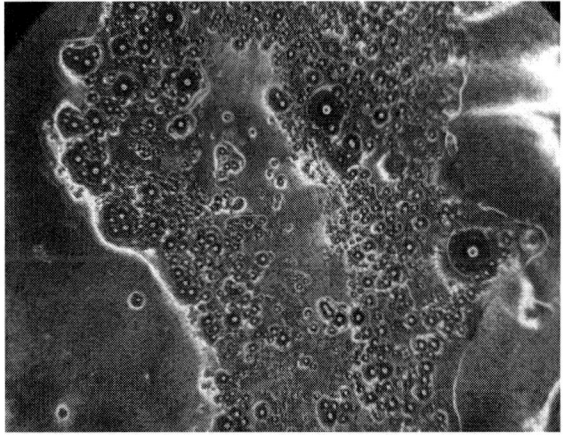

Figure 7. Solder as reflow with no cleaning.

The samples were then dipped in a 150ml beaker containing either IPA, BPS or the control cleaner. The beakers were placed in an ultrasonic bath for 30 seconds. It was shown that IPA is only marginally effective at removing large amounts of flux. Similarly the control of the shelf cleaner was only marginally effective leaving a large amount of white residue as seen in figure 8.

One BPS product, BPS 125, is presented in figure 9. This material was highly effective at removing the flux (as was the other BPS materials) and left no residue. Some oxidation was noted but that was apparent on all the samples and seems to be a by-product of the flux during reflow. BPS135 was also used to evaluate removal of polyimides. When left in a BPS solution for over 20 minutes, the polyimide on one set of flip chip die was effectively removed without leaving residue.

BPS103, 100 and 130 were also found to effectively remove photoresist residues. The author was surprised at the effectiveness of the IPA compared to some of the cleaners sold as cure alls. In the experiment IPA performed much better than the control material that is high volume mover.

93

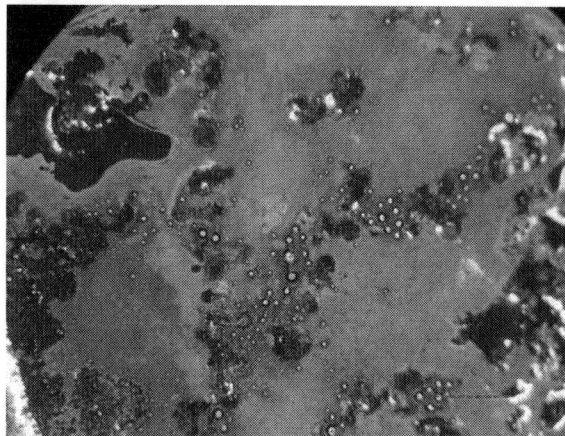

Figure 8 Commonly sold flux remover

The off the shelf cleaner used in figure 8 also left a heavy white residue/salt on the sample upon rinsing with DI water. Many micro solder spheres remain and the salts formed were difficult to remove with IPA afterwards. The author did note that some of the BPS materials were able to remove the residue and clean the layer with the 30 ultrasonic dip process.

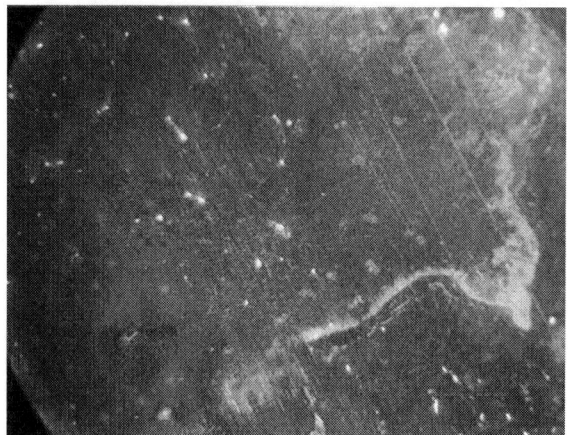

Figure 9. BPS 125 after 30 seconds.

Better images and process capability were obtained by looking at flip chip die. This is the real application and an apples to apple comparison helps show how effective the BPS series was at removing both flux and PR residue.

Figure 10. Die with large amounts of flux

Figure 11. Post cleaning

Figures 10 and 11 show the delta of an effective cleaning solution. These are singulated die. Electrical test on such die would not contaminate the probe needle as quickly leading to spike in CRES and false test failures.

Figure 12. Pre flux cleaning

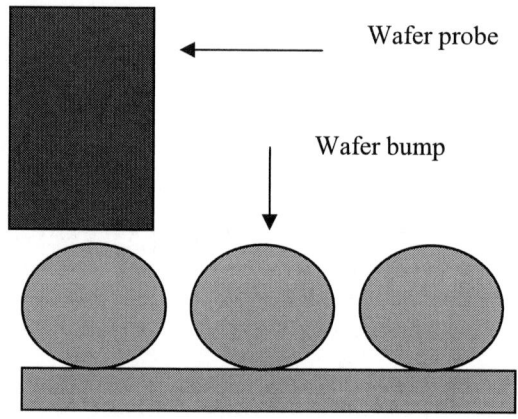

Figure 13. Probing on clean bumps

Figure 13 is analogous to the situation. The bumps here are clean and free of residue. These clean bumps will improve test yield and throughput. Similarly bumps free from residues and contaminants will yield better during assembly as well.

Figure 14. Close up of cleaned bumps

CRES II

CRES was discussed early. It's helpful to discuss again after reviewing the bumps. As previously stated, contact resistance, or CRES, is the measured impedance between the probe tip (or socket) and the electrical output of the DUT. The typical impedance range of the metals used for probes, sockets, pads and bumps have values in the milliohm range.

With continuous probing or insertions, contaminants build up causing the impedance to rise beyond time zero values. This rise in impedance results in poor contact, electrical fails (opens, speed related fails, Vth, Io, etc), and reduced life of probe hardware.

The test hardware must be cleaned each time the CRES reaches a critical value. Without cleaning, increased test time, process bottlenecks and assembly issues become drains on resources and profit. Unfortunately the cleaning reduces the life of the probe hardware. Implementing an effective cleaning solution helps eliminate the root causes of contact resistance.

Figures 15, 16 and 17 demonstrate the pre and post effects of cleaning. Figure 15 shows the wafer as is with no cleaning. The contact resistance quickly goes above the 20ohm cutoff leading to false failures. The data in Figure 16 demonstrates cleaning with BPS can improve CRES directly impacting yield, cycle time and indirectly back end packaging (coplanarity). Figure 17 are the results on a wafer lot. Stable CRES and process control have been achieved. This improvement would also reduce PM costs, engineering hours and hardware cost (probe cards and sockets).

CONCLUSION

It has been demonstrated that dirty bumps and wafers can lead to increased contact resistance (CRES) during test. The contaminants that are present during test can likewise carry over to assembly and cause yield issues there as well. One particular concern is the coplanarity issues cause as a result of excessive overtravel. By using the BPS series of materials the author has shown that flux, photomask, polyimide and residues from other cleaners can be effectively removed to promote improvements at test and assembly.

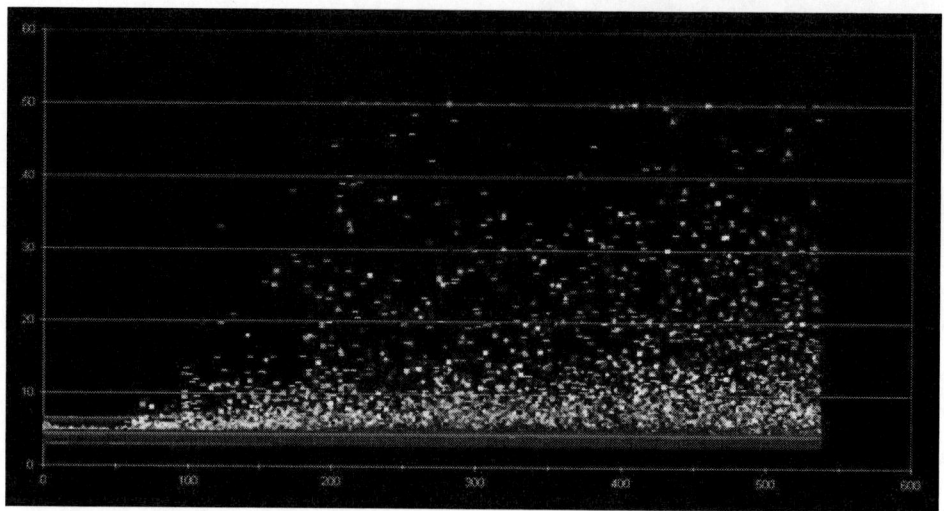

Poor CRES results at time zero. Impedance goes out of control near 100 touchdowns.

Figure 15.   Poor CRES as a result of inadequate cleaning

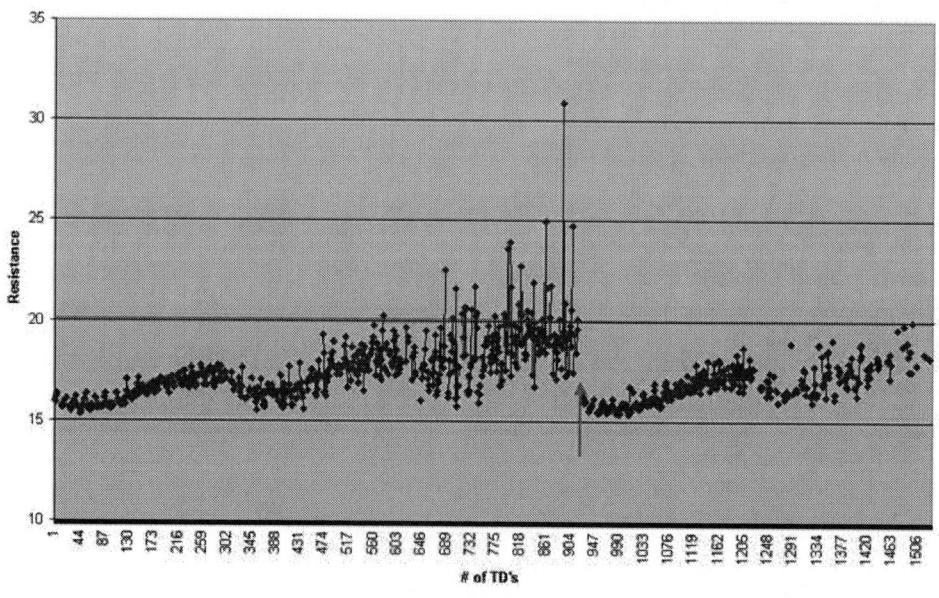

The above plot demonstrates the desired performance! Hundreds of touchdowns between cleans!

Figure 16.   Stage 1 cleaning improvement demonstrates a drastic improvement. The 20ohm limit was initially hit at 120 insertions.

First 13 wafers

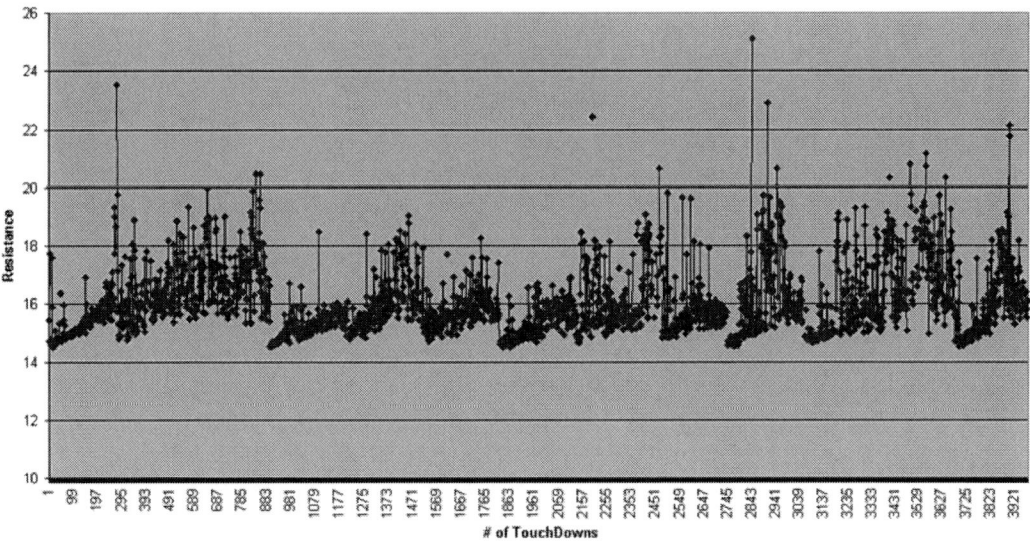

Figure 17. Production floor runs on very dirty wafers show that control can be maintained. These wafers initially yield less than 10%. Proper cleaning brought the yield over 80%.

NON LITHOGRAPHIC MICROCELL PLATING
FOR INTEGRATED PASSIVES AND RDL

P. Moller and M. Fredenberg
Replisaurus Technologies AB
Kista, Sweden
patrik@replisaurus.com

P. Leisner
Acreo AB, Jönköping
Science Park, Jönköping

M. Ostling
Royal Institute of Technology, IMIT
Kista, Sweden

ABSTRACT

ECPR (ElectroChemical Pattern Replication) is a fabrication process for microscale metal patterns, characterized by high accuracy, high throughput and cost efficient production. Using ECPR, the cost of metallization for microelectronics can be reduced significantly compared to using lithography based processes. The technology utilizes a master electrode for electrochemical pattern replication, which enables direct metallization with high deposition rates, short cycle times, and fewer equipment modules.

ECPR provides metallization on most substrates such as silicon wafers, ceramic substrates and flexible or rigid organic substrates. The technique currently enables pattern transfer of copper structures down to 500/280 nm line/space with uniform material distribution and minimum line width variations.

Examples of applications suitable for ECPR metallization are wafer redistribution, integrated passives, ultra-fine pitch pillar bumping, advanced substrates/interposers and power-ICs.

Key words: Metal printing, microcell plating, copper metallization, electrochemical pattern replication.

INTRODUCTION

Continuous demand for further miniaturization, increased functionality and cost reduction in combination with reliability requirements drives the development of novel packaging concepts and efficient manufacturing processes in the field of advanced packaging. Integrated passives, such as on-chip inductors for RF and wireless applications, is one example where high accuracy pattern definition, and thickness uniformity can enable new advanced designs and functionality. The recently introduced Electrochemical Pattern Replication technique

(ECPR) [1] is a non lithographic microcell plating process that enables direct replication of metal patterns. ECPR offers improved accuracy and a simplified process scheme compared to traditional metallization based on lithography and electroplating. The elimination of polymer patterning also gives a more environmentally friendly production and enables significant cost and cycle time reductions. [2]

Accurate CD control, submicron resolution [3] and uniform thickness, with no pattern dependency, arising from the uniform current distribution in the microcell plating, are two intrinsic advantages associated with ECPR. This paper reports recent result from ECPR plating of integrated passive devices, such as copper inductors.

THE ECPR PROCESS

In the ECPR process, a template (master electrode) consisting of an electrically conducting electrode layer and one or more patterned layers of electrically insulating materials, is pressed against a substrate with an electrolyte applied between the two surfaces (Fig. 1a).

When put in contact, excessive electrolyte is forced away from the master electrode and substrate interface and local electrochemical micro cells, filled with electrolyte, are formed according to the cavities in the pattern of the master electrode (Fig. 1b). When an external potential is applied over the master electrode and substrate surfaces, electrochemical material transfer takes place inside each local electrochemical cell. Metal is dissolved into ions at the anode and transported through the electrolyte in each micro cell and deposited on the cathode (Fig. 1c). Metal patterns can be either electrochemically etched or plated, depending on the polarity of the applied potential. [6,7]

When using ECPR plating (Fig. 1), prefill of anode material is done prior to each plating cycle, or less frequently if thicker layers can be allowed.

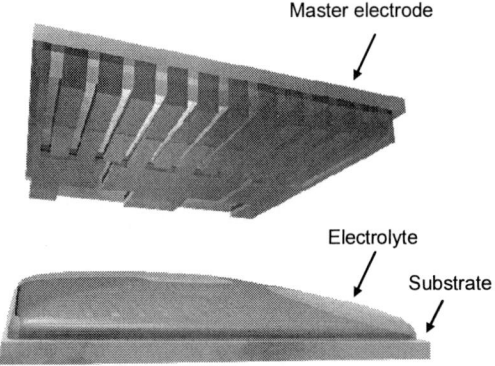

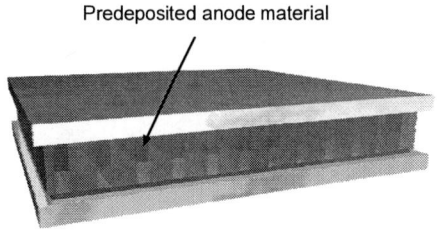

Figure 1a. Master with anode material and substrate with seed layer and electrolyte on.

Figure 1b. Master and substrate in contact.

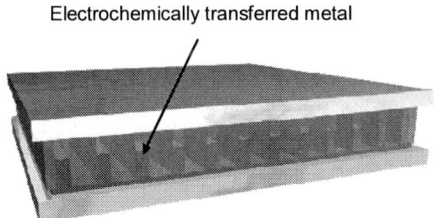

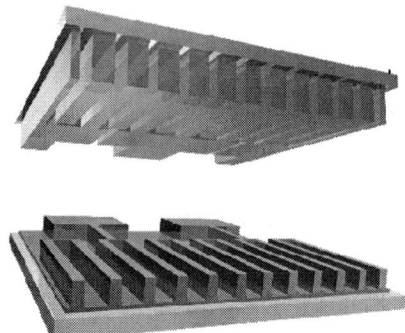

Figure 1c. Electrochemical metal transfer.

Figure 1d. Separation, followed by seed etch.

The purpose of the master electrode is both to accommodate an electrode surface in each micro cavity and to accurately define each active electrode area on the substrate by a conformal master-to-substrate contact. A conformal contact results in an accurate contact pattern on the substrate, corresponding to the master pattern. The technique offers good control of the sidewall angle of the metal patterns, close to 90° perpendicular to the substrate surface, if desired.

ECPR PLATING RESULTS
With a uniform contact surface, each insulating structure prevents a well defined substrate area from electrochemical reaction, resulting in high resolution replication in the regions that are not blocked by the insulating layer. (Fig. 2,3)

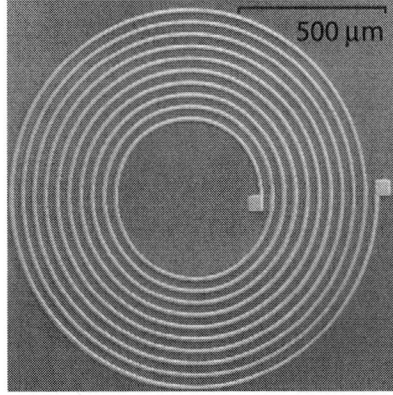

Figure 2. Copper coil fabricated on silicon substrate by ECPR with 10 μm line width and 3 μm copper thickness.

Successful demonstration of 500 nm lines with 280 nm spaces (Fig 3) using a demo equipment in Kista/Sweden indicates that there are no fundamental limitations reducing feature sizes further and that the ECPR technology can target also nano scale applications.

99

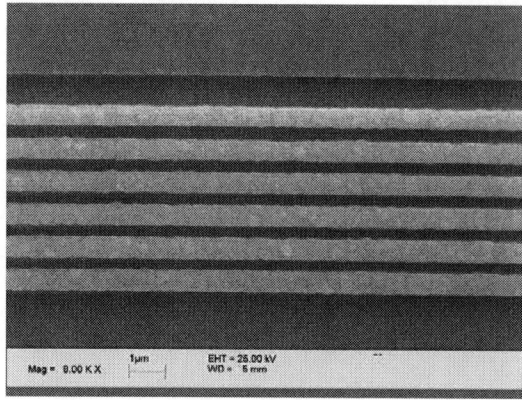

Figure 3. 500 nm copper lines with 280 nm spacing

Accurate replication of 500nm thick metal patterns was demonstrated by the process and it was shown that CD (Critical Dimension) variations in the replicated pattern were controlled entirely by the variations in the master electrode and that the variations introduced by the electrochemical pattern transfer itself had insignificant impact.

REDUCED COMPLEXITY AND COST

ECPR offers a cost/time efficiency advantage that derives from the ability to reuse the master electrode to define the pattern instead of having to pattern a resist template with lithography on each substrate. Hence, the following steps of the traditional photolithography based metallization process:

- Metal seed layer sputtering/PVD
- Photo resist spin/spray coating
- Exposure through photo mask
- Resist development
- Descum
- Electroplating
- Resist strip
- Metal seed layer etch

can be replaced with:

- Metal seed layer sputtering/vaporization
- Prefill of metal in the master electrode
- ECPR transfer of the metal from the master
- Metal seed layer etch

Predeposition of metal in the master and ECPR patterning of the substrate are done in the same equipment, and the total number of equipment modules in a manufacturing line can be reduced from eight to four. This means a significantly lower investment in a manufacturing line for a certain throughput. It also enables a reduction of clean room area, installation and maintenance costs, operating costs as well as chemicals and resist costs. It is also likely that the total manufacturing downtime will be reduced when having a process line with less number of equipment modules since the risk for machine breakdown will be lower. ECPR has shown the ability to electroplate copper with speeds up to 5 µm/min [1,2]. The cycle time for producing copper patterns with heights of 4-15 µm, will then be 1-3 minutes plus handling time, which is significantly less than the cycle time that can be achieved with the standard processes. Therefore, it is possible to get a higher throughput and a significantly lower production cost per wafer with ECPR compared to traditional through-mask plating processes.

UNIFORM MATERIAL DISTRIBUTION

One of the main technical advantage of ECPR metallization is the ability to get a uniform plating height, independent of the pattern density. This is hard to achieve with standard lithography based electroplating. In a normal electroplating cell the ratio of anode-to- cathode area can vary over the substrate depending on the density of the structures. Areas with low active-area density (i.e. regions that are more densely populated with insulating resist) will receive a higher current density than areas with high active-area density due to the nature of electrolytic current distribution that causes bending of the current lines in response to the pattern of the resist blocking the electrode surface of the substrate. This results in different deposition speeds in areas with different active-area density and hence leading to an uneven material distribution and height variations [4,5]. The problem is normally addressed by using advanced agitation equipment, adding organic additives to the electrolyte and taking the active-area density into account when designing the pattern. Additional dummy patterns may be added to the design and repeated process optimization is typically needed when changing between two different designs. Despite these efforts, non uniform material distribution is still common in normal electroplating processes, with various results depending on the pattern design.

The ECPR setup has the advantage of always having a 1:1 anode-to-cathode area ratio since the electroplating cells area formed by the cavities between the master and the substrate (Fig. 1). In this way the potential field lines do not bend across any isolating material and it is possible to get the same current density and uniform material distribution over the substrate area despite varying active-area density.

APPLICATIONS

ECPR metallization can be used for numerous applications in the advanced packaging space. Demanding applications requiring highly uniform thickness of metal over the wafer surface and minimum CD variations can especially benefit from ECPR, but the cycle time and cost per wafer advantages do apply to most applications where copper layers with a few microns thickness are desired on wafer sized substrates. Typical applications of interest are wafer redistribution, integrated passives, ultra-fine pitch pillar bumping and power IC metallization. Also advanced substrates, such as interposers, MCM/MCP-substrates and HDI Flip Chip substrates have been metallised using the process. The current capabilities of producing structures with line/space of 500/250 nm are well ahead of the requirements for these applications as well as for the next few years according to the Technology Roadmap for Assembly and Packaging [6].

Integrated passives, such as on chip inductors (Figure 4) and capacitors, are target applications where the combination of exceptional thickness and CD control offered by ECPR can enable more advanced designs and increased functionality.

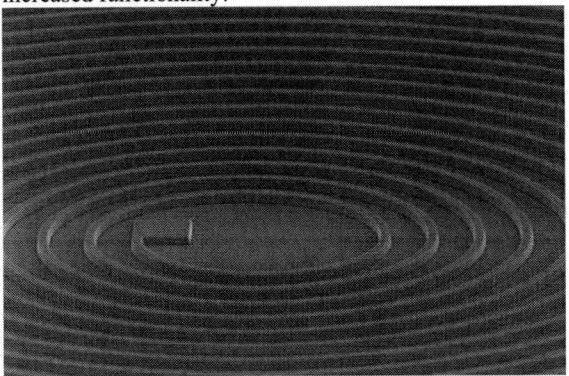

Figure 4. Printed on-chip copper inductor with 5μm line width and 5μm thickness.

CONCLUSIONS

ECPR utilizes a reusable patterned master electrode to enable direct metallization without any polymer processing and hence short cycle times can be achieved. The production cost for advanced metallization applications can be significantly reduced due to high throughput in combination with less need for equipment, clean room and operating resources. The ECPR process also offers a major advantage, concerning problems with uneven material distribution, over traditional through-mask plating. The present capabilities of ECPR points out interesting application areas within advanced packaging metallization. ECPR is unique in being able to do direct metal replication without polymer processing.

REFERENCES

1. Moller P., Fredenberg M., Wiwen-Nilsson P., in AESF SURFIN / Interfinish 2004, proceedings (2004).
2. Fredenberg M., Möller P., Recent Progress in the Development of ECPR (ElectroChemical Pattern Replication) - Metal Printing for Microelectronics, ECS 208 Fall Meeting, Los Angeles, October 19, 2005.
3. Möller P., Fredenberg M., Dainese M., Aronsson C., Metal Printing of Copper Interconnects Down to 500 nm using ECPR – ElectroChemical Pattern Replication, Micro- and Nano Engineering 2005, Microelectronic Engineering 83 (2006) 1410–1413.
4. Mehdizadeh S., Dukovic J. O., Andricacos P.C., Romankiw L. T., Journal of The Electrochemical Society, Vol 139, No. 1, January 1992.
5. Mehdizadeh S., Dukovic J., Andricacos P.C., Romankiw L. T., Journal of The Electrochemical Society, Vol. 140, No. 12, December 1993.
6. International Technology Roadmap for Semiconductors, 2005 Edition, Assembly and Packaging.

ASSEMBLING OPTICAL DEVICES UTILIZING WAFER LEVEL TECHNOLOGY AND CHIP ON BOARD PROCESS TO ENABLE HIGHER YIELDS AND REDUCED COSTS

Yehudit Dagan, Giles Humpston, and Michael J. Nystrom
Tessera Inc.
San Jose, CA, USA
ydagan@tessera.com

ABSTRACT

Device manufacturers using a chip on board (COB) process to attach image sensors or other optoelectronic devices to a circuit board have long sought means by which to increase yields and therefore reduce overall costs. While COB has certain advantages over traditional packaging, such as short signal path and small low profile, the process results in lower yields than competing technologies due to contamination of the optical sensor during the assembly process. Preventing contamination becomes an even greater issue for higher resolution (mega-pixel) image sensors, where the effect of contamination from dust particles is compounded by the small size of the pixels. Furthermore, as future camera modules become more complex, requiring additional assembly, tuning and calibration, unprotected image sensors will be exposed to dust particles over a longer time span.

The author will present the concept of integrating wafer-level packaging of optical devices with COB assembly processes. The resulting technology, SHELLCASE® CF, permits attaining high assembly yields regardless of the imager resolution, which in turn reduces the cost of the final product.

INTRODUCTION

The market of camera phones (cellular phones equipped with cameras) emerged in 2000 and has grown rapidly since then. In 2005, close to 390 million camera phones were shipped, and that number is forecast to double to approximately 800 million by 2009. This growth is due to a combination of market expansion together with the increased penetration of camera phones, which is expected to reach close to 80 percent by 2009. On most new cell phones, cameras have become a standard feature, much like color screens. The volume, form factors, reliability requirements and cost pressures in the camera phone application are also driving the packaging and assembly needs of camera modules.

A camera module in a cellular phone consists of, at a minimum, a lens assembly and an image sensor attached to a laminate. The number of lens elements varies based on the requirements of the optical design. Additionally, this module may combine other functions through the integration of an image processor and memory devices.

Assembly processes for cell phone camera modules has proved to be extremely challenging compared with standard integrated circuit assembly processes. With more than 90% of defects being optical rejects caused by particles contamination, these assembly processes must entail strict particle control regimes starting from facility control through material selection and labor training. This paper will discuss these challenges of assembling image sensors for camera phones and offer an innovative method to address these challenges.

BACKGROUND

The electronic devices at the heart of digital imaging are solid state image sensors, which build upon today's integrated circuit technology. A solid state image sensor is effectively an array of miniature solar cells, converting a broad spectrum of light into an electronic signal. Each light sensing division in the sensor is called a pixel. For a given picture, the number of pixels determines the quality of the digital picture. If the picture is divided into a small number of pixels, each pixel must be larger and the quality of the picture is lower. The higher the number of pixels, the smaller is each pixel and consequently, the better the picture quality. Each pixel needs to be assigned a location and color. The location of the pixel is represented by its X-Y coordinates in the grid. Since all colors can be broken down into Red, Green and Blue components, the color of a pixel is obtained by placing a filter on top of the pixel that only allows certain wavelengths to pass. The most common pattern of coloring the pixel array is called the Bayer Filter Pattern (See Figure 1).

Figure 1: Bayer Filter Pattern on a pixel array (Courtesy www.aluminumstudios.com).

The two fundamental types of solid state image sensors prevalent in the market today are charged coupled devices (CCD) and complementary metal oxide semiconductor (CMOS) devices. In CCD devices the entire area of each pixel is sensitive to light, requiring signal processing outside of the imager. The ratio between the active area of the pixel and the pixel area is referred to as the Form Factor of the sensor. CCD devices are typically of close to 100% form factor. CCD devices provide the advantages of high sensitivity and low noise. However, since these sensors are produced by a non-standard silicon process, they tend to be of higher cost. In addition, other electronic components for signal processing cannot be integrated into the non-standard silicon, leading to multi-chip solutions. CMOS image sensors are produced using standard silicon processes, allowing integration of signal processing electronics within the device. CMOS sensors are typically about 80% form factor. This standard process also results in lower cost and system power for CMOS image sensors; a distinct advantage for mobile imaging applications. The drawbacks of these sensors are higher noise and lower sensitivity. A primary reason for the lower sensitivity is the reduced size of the light sensitive region, since a portion of each pixel is occupied by the integrated signal processing electronics.

To compensate for the reduced light sensing region within each pixel, additional optical elements, namely micro lenses, are placed over the image sensor. Micro lenses, positioned above each pixel, focus incident light onto the light sensing region, increasing the light gathering ability of the sensor (Figure 2). Customarily micro-lenses are made of polymeric material which is very susceptible to contamination. As a result, if particles are created during the assembly process of the image sensor or during a subsequent test, they stick to the micro-lenses which in turn are very hard to clean.

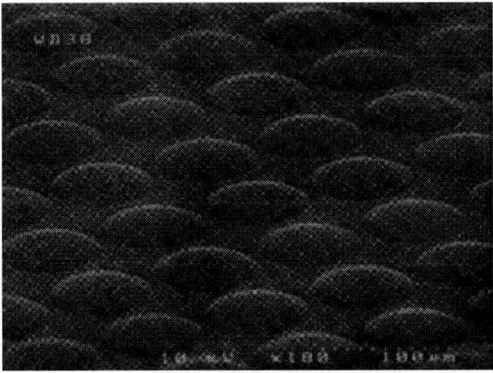

Figure 2: Micro lenses increase the light gathering ability of the light sensing areas of individual pixels (Courtesy www.phy.hw.ac.uk).

Integrating wafer level technology with COB
The main technology used today for assembling image sensors in camera phones is Chip-on-Board (COB). In this technology, the image sensor is mounted directly on a printed circuit board (PCB) or flexible circuit, and traditional wire bond techniques are used for electrical contacts (See Figure 3). The principal deficiency of this technology, which will become increasingly important as pixel size decreases, is particle contamination. The exposed micro lenses and hence pixels are at risk of obscuration by particles until the module assembly is completed.

As pixel sizes decrease to permit higher resolutions and smaller die, this problem worsens. The smaller the pixel, the smaller the size of a critical particle that creates a reject. Use of dirty processes such as dicing, die mounting and wire bonding become more difficult with smaller pixels leading to progressively higher yield loss due to particle contamination. In order to protect the sensor from contamination, a stringent clean room environment needs to be maintained until the camera module assembly is complete.

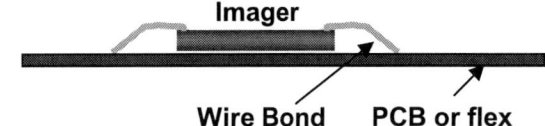

Figure 3: Schematic of typical chip-on-board mounting of image sensor.

In order to overcome this challenge, additional investment would need to be made in the COB assembly lines to make them cleaner, which is complicated as many are already operating at Class 10 or better. Fortunately there is another way: utilize the fundamental capability of wafer level technology of die encapsulation at the initial stage of processing.

SHELLCASE® CF, a new wafer level technology recently introduced to the market, is the optimal combination of wafer level processing with wire bond technology. In this technology, the image sensor is encapsulated by a glass cover at the initial stage (See Figure 4a), keeping it protected from contamination throughout the assembly processes. The glass cover is elevated from the active area to allow the use of micro-lenses, thus improving the sensitivity of the imager. The bond pads are then exposed to allow standard wire bonding at later stage (See Figure 4b). The wafer is then thinned and devices are singulated into individually protected image sensors which can then be mounted on a board using standard wire bond technology (See Figure 4c).

Figure 4a: Imager is protected by a glass cover from the initial stage.

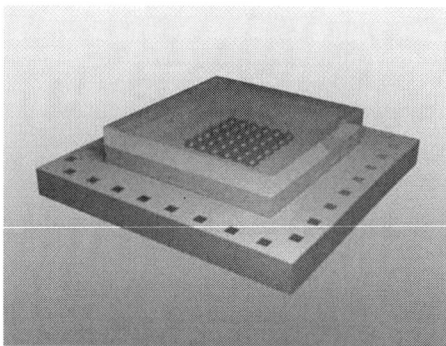

Figure 4b: Bond pads are exposed for wire bonding at later stage.

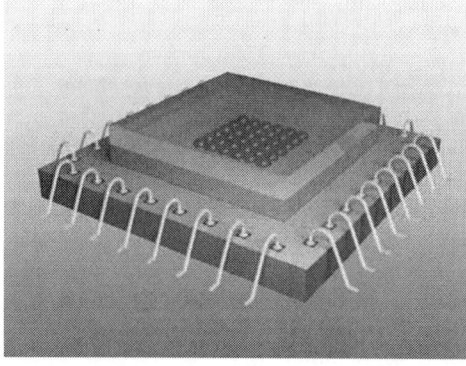

Figure 4c: Protected image sensor can now be mounted on a board using standard wire bond technology.

Using Shellcase CF, the micro lenses and pixels are encapsulated at the beginning of the process, minimizing their exposure to particles. The bond pads can be exposed to allow wire bond assembly to the printed circuit. Particles produced during dicing, die attach and wire bonding processes can be removed more easily from a polished glass wafer than a semiconductor wafer with its complex topography. In addition, any particles that do land and remain over the image sensor area are less noticeable in the electronic image because they are removed from the focal plane. A group of pixels of decreased brightness is visually much less obvious and less distracting than a single black

pixel in an image. Table 1 lists the reliability specification of the Shellcase CF technology.

Test	Test Conditions	Results
Moisture Sensitivity Level 2	125C, 24hrs 60C / 90% RH, 168hrs 3 Pb-free reflow	Pass
High Temperature Storage	150C, 1000hrs	Pass
Thermal Cycling	-40C / 125C, 1000 cycles	Pass

Table 1: Reliability spec for the SHELLCASE® CF.

Wafer Level Optical Testing
Customarily, before the image sensor wafer leaves the FAB, electrical tests are conducted for preliminary analysis of the imager performance. Since these tests do not detect all optically related defects, module makers are facing the need to assemble defected image sensors into the modules only to find out at the end of the module assembly process that the sensors are defected. This methodology incurs additional cost on the module makers as well as prolongs time to market. One of the ways to overcome this hurdle would be to perform wafer level optical testing prior to module assembly.

Shellcase CF, being a wafer level technology, allows module makers to perform full optical testing of the imager prior to module assembly. Once the image sensor is encapsulated by the glass cover and the bond pads are exposed, the sensors are ready for optical testing.

The optical test can be performed either immediately after the bond pads are exposed or after the devices are singulated into individually protected imagers, since the devices are not connected electrically throughout the process. In addition, since the devices are still in wafer form, optical testing can be done simultaneously to multiple devices thus reducing the test cost per die.

CONCLUSION
Image sensors for camera phones improve constantly. This is mainly reflected by increasing pixel counts and decreasing pixel dimensions. Therefore, particulate contamination, which is the major cause of yield loss when packaging image sensors, will get worse as the pixel dimensions decrease and pixel counts increase.

Providing wafer level glass protection to the image sensors is a cost-effective solution as it overcomes the issue of particulate contamination by applying a glass lid on top of the image sensor at the initial stage of packaging. Exposing the bond pads for wire bonding at later stage also allows module makers to perform full optical testing at the wafer level prior to module assembly thus reducing cost and cutting down on time to market.

Integrating wafer level technology with COB assembly processes facilitates high assembly yields, regardless of the image sensor's dimensions, resolution and pixel size. At the same time, this approach leverages the flexibility and cost base of COB die attach and interconnect processes.

REFERENCES

[1] Mark Christensen, "Market Trends, Applications and Analysis for Multi-die Packaging." 2006 Electronic Product Miniaturization Symposium, Tokyo, 17 January 2006

[2] Asif Chowdhury, Amkor Technology Inc. "Camera Module Assembly and Test Challenges", Semiconductor International, 2/1/2006

HYBRID WAFER-LEVEL PACKAGING FOR RF-MEMS APPLICATIONS

J. Iannacci[1,2], M. Bartek[2], J. Tian[2], S. Sosin[2], A. Akhnoukh[2], R. Gaddi[1] and A. Gnudi[1]

1) ARCES-DEIS Università di Bologna, Bologna, Italy
2) HiTeC-DIMES, Delft University of Technology, Delft, the Netherlands
j.iannacci@ewi.tudelft.nl, jiannacci@deis.unibo.it

ABSTRACT

In this work, we present our progress in development of a wafer-level packaging (WLP) solution suitable for RF-MEMS devices and enabling hybrid co-integration of additional IC dies. The proposed approach is based on a capping high-resistivity silicon substrate that is wafer-level bonded to an RF-MEMS device wafer. The capping substrate contains electroplated through-substrate copper vias for signal redistribution to an area-array solder bumps and recesses for accommodation of RF-MEMS devices. Optionally, through-substrate cavities or recesses are realized for hybrid integration of additional IC dies. Several aspects of this packaging solution are studied. First of all, the packaging process flow details are briefly described. Then the options of achieving a hermetic/semi-hermetic sealing of the packaged cavities are investigated together with description of the corresponding process flows. Furthermore, co-integration of RF-MEMS devices and CMOS control circuitry based on hybrid-packaging approach is discussed. For this purpose, two solutions where the capping substrate facilitates the placement of CMOS dies are introduced. Finally, package electromagnetic characteristics are analyzed and the technological degrees of freedom (e.g. via diameter, capping substrate thickness, etc.) are used to optimize the package radio-frequency (RF) behavior. The experimental results of the first fabricated sample of capped test structures (50Ω transmission lines and shorts) are presented and compared to the simulations.

Key words: Wafer-Level Packaging, RF-MEMS, RF-performance.

INTRODUCTION

The wafer-level packaging (WLP) approach to packaging of radio-frequency micro-electro-mechanical systems (RF-MEMS) has proven to be a viable option mainly due to its capability to provide for more benefits than just the basic packaging functions and the protection of fragile parts from the harsh environment (i.e. shocks, contaminations, etc.) [1]. Indeed, the WLP based e.g. on a capping protective substrate can be exploited as an interface in between functional parts fabricated using different processes and which need to be then co-integrated into one electronic package i.e. hybrid-packaging or system-in-package (SiP) approach [2]. Moreover, when dealing with radio-frequency (RF) range, the requirements of a certain application in terms of parasitic effects introduced by the package-related components may be rather strict [3]. Hence, it is inevitable that the design of the package-related parts (e.g. capping substrate with signal redistribution) must be performed with the same care required by the actual RF-MEMS devices that are to be protected. In this work, an extensive electromagnetic optimization of all the technological degrees of freedom (dofs) for a given packaging process is performed in order to reduce as much as possible the package parasitics.

WAFER-LEVEL PACKAGING PROCESS

The fabrication is realized with the facilities available at the DIMES Technology Centre (Delft University of Technology, the Netherlands). The process is based on the etching of through-vertical vias in a high resistivity silicon (HRS) substrate by means of the Bosch DRIE process (Deep Reactive Ion Etching). Recesses are etched on the back wafer side in order to accommodate the MEMS devices (see Figure 1).

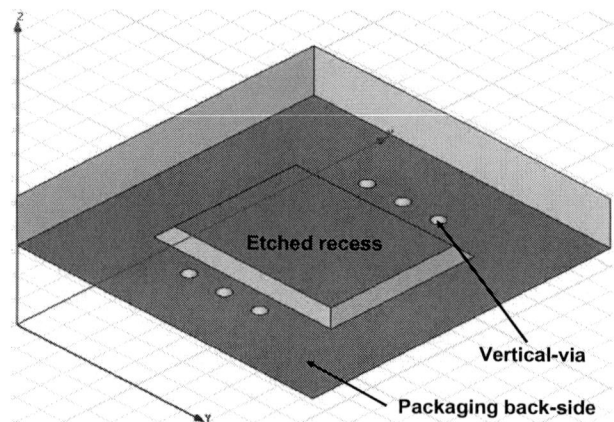

Figure 1. Schematic view of a capping silicon substrate with etched vertical vias and a recess.

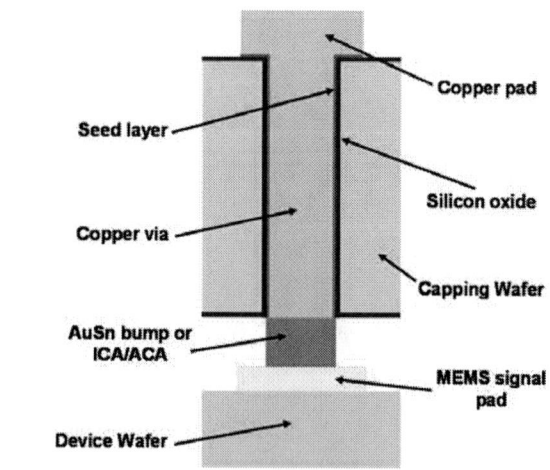

Figure 2. Schematic via cross-section.

After growing of an insulation oxide layer, a metal seed layer is sputtered on the capping substrate front side and on the vertical via sidewalls. In the subsequent electroplating steps, copper pads are plated and the vias are bottom-up filled with copper. In Figure 2, the cross-section of the resulting copper via is shown. The wafer-to-wafer bonding of the capping substrate to the RF-MEMS device wafer is performed either by ICA/ACA (Isotropic/Anisotropic Conductive Adhesive) [4,5] or by solder reflow. In the latter case, AuSn bumps have to be first electroplated in correspondence with the vias. (see Figure 2) [6]. The main difference is that the adhesive bond does not allow to maintain hermeticity while with solder reflow it is feasible. On the other hand, the ICA/ACA bonding requires a lower thermal budget (curing temperature ~120-150°C) compared to the AuSn solder reflow (~300°C).

PACKAGE HERMETICITY OPTIONS
Reliable operation of RF-MEMS devices requires their proper sealing from the environment. If hermeticity of copper-filled interconnect vias can be guaranteed, straightforward solution to achieve a (semi-)hermetic package is application of sealing rings around each RF-MEMS device or a one common sealing ring at the periphery of the entire package. The sealing rings can be formed using the same processing steps and the same material as used for electrical signal interconnect i.e. either conductive adhesive or solder alloy. In case of solder alloy, full package hermeticity can be achieved. Other option is to use a sealing ring material that is different from the one used for electrical interconnect and/or use additional fabrication steps. Even if this will increase process complexity (more masks are required, etc.), potential advantages have to be considered. The most important advantages are: a/ possibility to select most suitable materials required by their function (i.e. electrical contact vs. sealing and bonding properties); b/ possibility to optimize for the surface topology of the RF-MEMS device wafer.

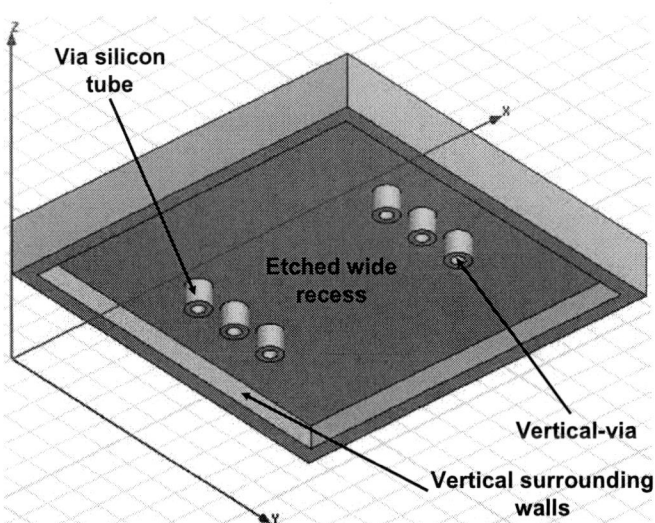

Figure 3. Schematic view of the capping substrate in the modified process. A recess is formed on the entire silicon surface except the sealing ring area and around each via.

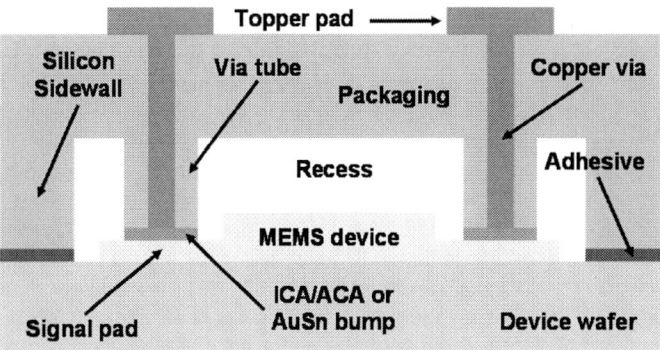

Figure 4. Cross-section of a capped MEMS devices obtained by the modified process flow discussed above.

Figures 3 and 4 demonstrate such alternative approach. In this approach, the recess on the capping wafer is formed on its entire surface except the sealing rings and a small area around each via. Also the capping substrate topology matches the interconnect height of the RF-MEMS device wafer i.e. the sealing ring plane is lower than the plane of the vias (see Figure 4). Solder alloys or polymers such as SU-8 or BCB can be used as the sealing ring material and have been demonstrated previously [7, 8].

HYBRID PACKAGING
As mentioned before, packaging technology provides means for co-integration of functional parts fabricated in different (incompatible) technologies into one functional system (hybrid-packaging). Indeed, the capping substrate can be suitably exploited to facilitate this step by providing for instance the necessary space to accommodate additional IC dies (e.g. CMOS control circuitry) to be interfaced with the passive RF-MEMS devices. Few feasible solutions have been already presented by the authors [9]. In this paper, more focus is spent on a solution for the RF-MEMS/CMOS hybridization which allows to maintain the hermetic sealing of the MEMS parts. The solutions which are described, are applied to the modified fabrication process of the capping part (shown in the previous section). However, these can also be applied to the traditional packaging solution without wide recess and via tubes.

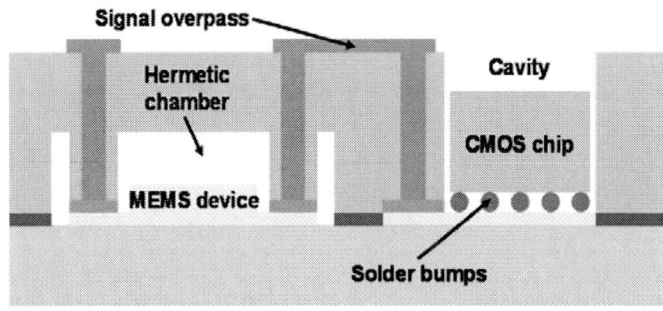

Figure 5. Hybrid-packaging solution (RF-MEMS/CMOS). The electrical interconnection between the two parts is provided by overpasses. This allows to maintain hermeticity in the cavities accommodating the MEMS parts.

After preparation of the capping substrate for the MEMS part (i.e. etching of vias and recess) the space for the CMOS dies has to be formed in the capping substrate. In the solution proposed in Figure 5, a through-substrate cavity has to be formed. An effective solution, in order to achieve this, has already been demonstrated by the authors [10]. Here, a non-continuous trench is etched (Bosch DRIE process) around the cavity perimeter while narrow supporting bridges remain to keep the bulk silicon inside the cavity. In a subsequent ultrasonic cleaning step, the anchor points (bridges) are broken and the cavity is released. In Figure 6 a SEM picture of the non-continuous etched groove around the cavity is shown, while in Figure 7 the released cavity after the ultrasonic treatment is reported.

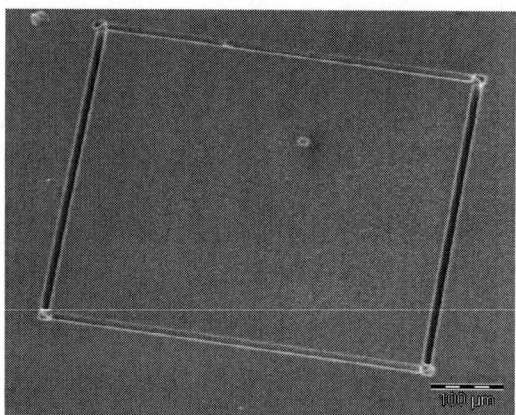

Figure 6. Narrow trench defines the to-be-formed through-substrate cavity. The supporting bridges are visible in the four corners.

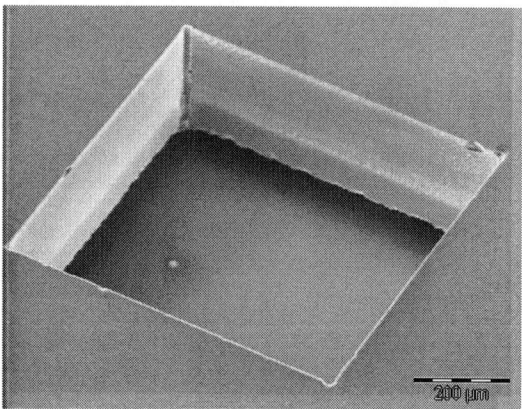

Figure 7. Same cavity as in Figure 6 after the release of the bulk silicon by means of ultrasonic cleaning.

The electrical interconnection in between the MEMS and the CMOS part is provided by overpasses (see Fig. 5). The signals of the MEMS part are brought on top of the capping substrate by means of vias and connected to other vias outside the MEMS area by uppper interconnects (overpasses). The CMOS chip is then introduced in the cavity with the solder bumps downward and is flip-chip bonded to the interconnect on the RF-MEMS device wafer.

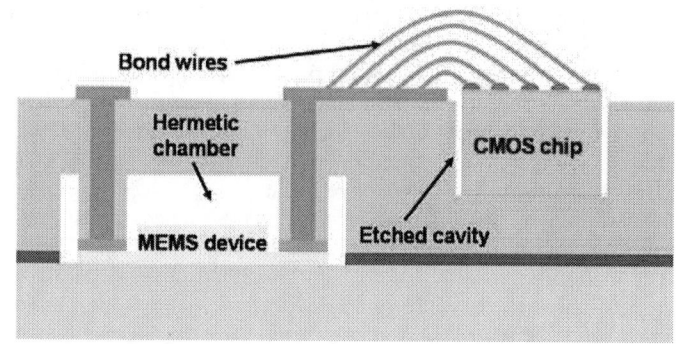

Figure 8. Hybrid-packaging solution with a shallow cavity for the CMOS chip. In this case, the interconnect is realized on the upper part of the capping substrate. The interconnection is obtained by wire bonding.

Alternative solution based on conventional wire bonding is shown in Figure 8. In this case, a recess with depth equal to the CMOS die thickness is required. The interconnect array on the device wafer is not necessary anymore since the CMOS die is placed in the cavity with the signal pads upward. The interconnection to the MEMS part is performed by means of wire bonding technique.

ELECTROMAGNETIC OPTIMIZATION
The issue of the losses introduced by packaging when dealing with radio-frequency (RF) applications has been already highlighted. In this scenario the correct understanding of the trends to be followed in designing the capping part represents a key-factor in reducing as much as possible its influence on the RF-behavior of MEMS devices.

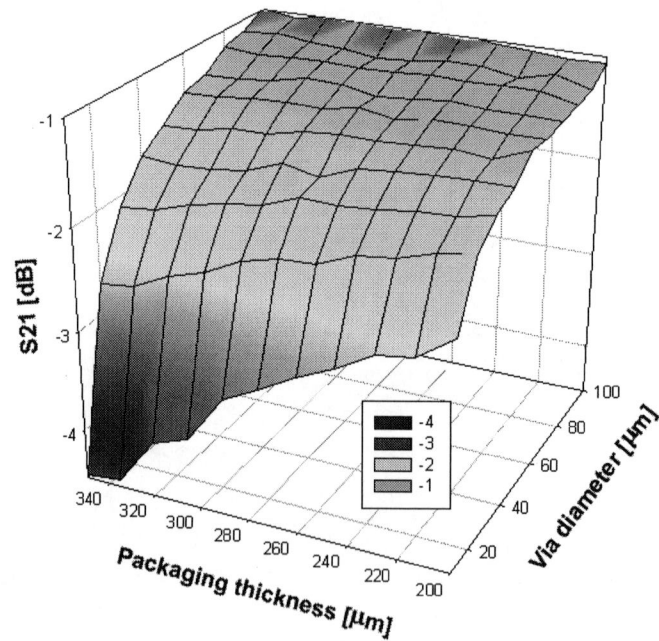

Figure 9. 3D-plot of a two parameters optimization in Ansoft HFSS 3D simulator. On the x, y and z-axis the via diameter, capping substrate thickness and transmission parameter (S21) are reported, respectively.

The authors have already pointed out the degrees of freedom (dofs) allowed by the technological process (e.g. etched recess depth, via diameter etc.) and their trend on the RF-performances of test structures (50Ω transmission lines) by means of the Ansoft HFSS electromagnetic simulator [6]. In the 3D-plot of Figure 9 the influence of two dofs on the transmission parameter (S21) of a capped transmission line is shown. The frequency chosen for the analysis is 15GHz. On the xy-plane the via diameter and the capping substrate height are reported while the S21 (expressed in dB) is plotted along the z-axis. The via diameter has been ranged in between 10 μm and 100 μm while the cap height was swept between 200 μm and 350 μm. All these values are compatible with the employed technology and have been achieved in several trials for the optimization of the process. The S21 parameter seems to be particularly sensitive to the capping substrate thickness in the lower range of via diameter (up to about 50 μm). For instance, for via diameter of 10 μm the S21 parameter ranges from -2.29 dB to -4.47 dB (variation of -2.18 dB) for a capping substrate thickness from 200 μm up to 350 μm respectively @ 15 GHz. On the other hand, in the same packaging thickness range the S21 variation is just -0.12 dB for a via diameter of 60 μm. This suggests that wide vias are preferable in the design. Indeed, it is possible to get good RF-performances even with a more thick capping part (about 300-350 μm instead of 200 μm) whose mechanical strength is definitely higher than the one of a very thin substrate. The technological advantages of the packaging solution with a wide recess and via tubes have been already discussed above. It is now interesting to investigate if this solution also introduces benefits from the point of view of the RF-performances.

For this purpose several transmission lines have been simulated in HFSS with the two different packaging solutions. The results for one of these test structures are shown below. The transmission line is 1000 μm long. The signal and ground lines width is 80 μm and 500 μm respectively while the gap is 30 μm. This structure has been simulated firstly applying a traditional packaging solution like the one shown in Figure 1 and then with a capping part similar to Figure 3. In both cases the cap substrate height is 350 μm, the via diameter equals to 40 μm and the recess depth is 150 μm. The comparisons of the S-parameters plots for the two simulations are shown in Figures 10, 11 and 12. Looking at Figure 10, the reflection parameter of the line with the packaging solution with wide recess is lower than the one with a normal recess in the frequency span from 3 GHz up to about 25 GHz. For instance, the S11 parameter for the solution with normal recess is -20.8 dB @ 8 GHz while it changes in -32.0 dB @ 8 GHz for the packaging with wide recess and via tubes. Concerning the transmission parameter the two curves are superposed up to about 6 GHz. Beyond this frequency a higher value of the S21 parameter is observable for the solution with wide recess. Indeed, the reflection parameter is -1.73 dB @ 18 GHz for the traditional solution while it is -1.10 dB @ 18 GHz for the wide recess.

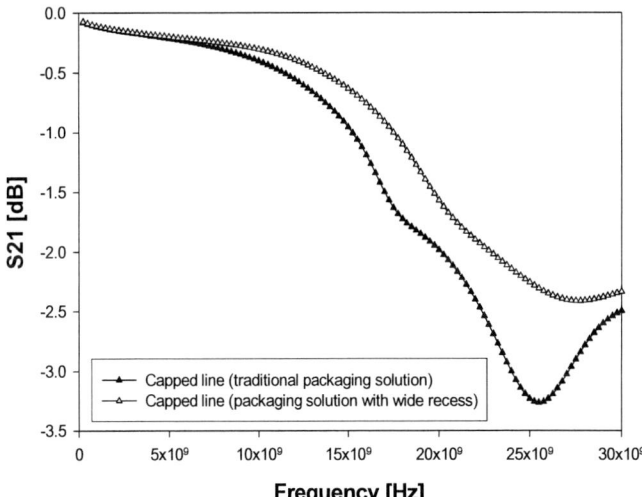

Figure 11. Comparison of the transmission parameter (S21) for the transmission line packaged with the two different technological solutions (normal and wide recess).

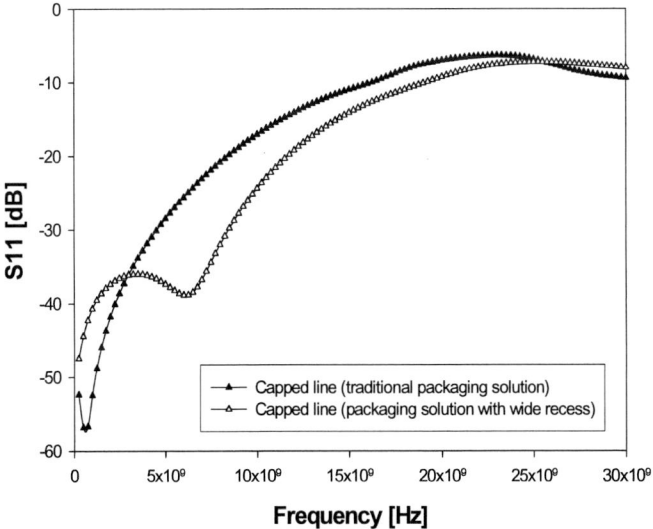

Figure 10. Comparison of the reflection parameter (S11) for the transmission line packaged with the two different technological solutions (normal and wide recess).

Finally, the Smith chart of the S11 parameters (Fig. 12) shows that the curve referred to the packaging solution of Figure 3 is closer to the centre (50Ω) with respect of the solution which employs a normal recess. The same enhancements of the S-parameters behaviour was observed for the other simulated structures.

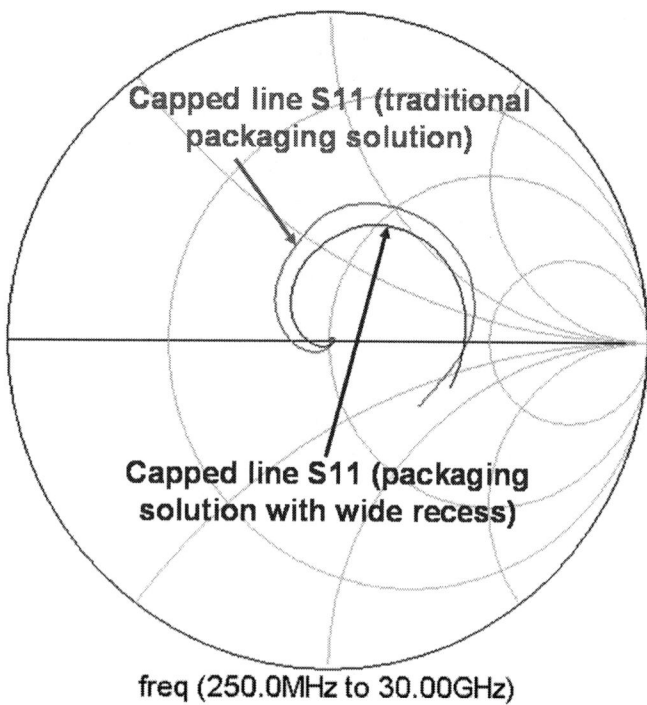

freq (250.0MHz to 30.00GHz)

Figure 12. Comparison of the S11 Smith chart for the transmission line packaged with the two different technological solutions (normal and wide recess).

EXPERIMENTAL RESULTS

The first samples obtained using the proposed packaging process have been fabricated. Instead of an RF-MEMS device wafer, test structures like various 50Ω transmission lines and shorts have been used. The utilized ICA is the CE 3103 WLV provided by Emerson & Cuming (www.emersoncuming.com). The experimental data for one of the transmission lines with and without capping substrate are shown below. The line is 1350 μm long. The signal and ground lines width is 116 μm and 300 μm respectively while the gap is 65 μm. Finally, the via diameter is 50 μm. The comparison between the measured data for the uncapped and packaged line are shown in the Figures 13, 14 and 15. In the S11 parameter plot (Figure 13) the maximum increase of it introduced by the packaging is around 17 GHz. Indeed, at this frequency the reflection parameter is -26.8 dB for the uncapped line while it becomes -16.4 dB after the cap is applied. Whereas, looking at the transmission parameter (Figure 14) it can be seen that its offset increases with the frequency. For instance, the difference between the S21 parameter for the uncapped and capped line is 0.11 dB @ 10 GHz but it changes in 0.24 dB @ 20 GHz and finally in 0.40 dB @ 30 GHz. Nevertheless, the changes in the S-parameters behavior due to the packaging are not too large.

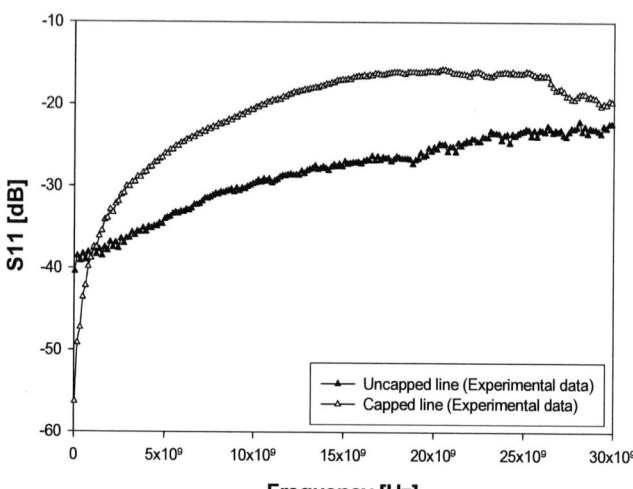

Figure 13. Comparison of the measured reflection parameter for an uncapped and packaged transmission line.

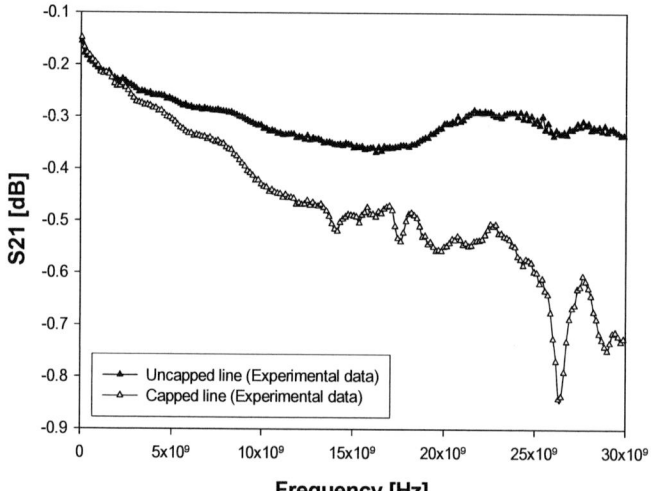

Figure 14. Comparison of the measured transmission parameter for an uncapped and packaged transmission line.

Furthermore, the results just shown should be considered encouraging since the design of this first packaging sample was not optimized for the reduction of the losses. Hence, a smaller deviation of the S-parameters related to the packaged structures with respect of the uncapped ones is expected in the next fabrication run. Indeed, in the design of the capping substrate the results of the electromagnetic optimization mentioned earlier will be taken into account.

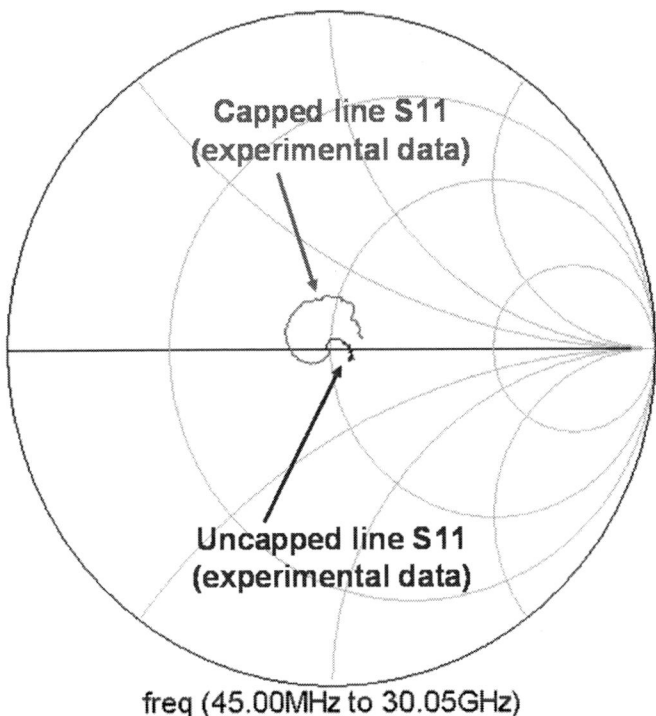

Figure 15. Smith chart of the measured reflection parameter for an uncapped and packaged transmission line.

In Figures 16 and 17 the experimental data referred to a short are reported. The plot of the reflection parameter (Fig. 16) shows that the lowering of the S11 curve due to the packaging is never too large over all the frequency range. For instance, the reflection parameter for the uncapped short is -0.43 dB @ 10 GHz while it becomes -0.68 dB @ 10 GHz for the packaged short. Despite beyond 18 GHz the two curves show several peaks (especially the uncapped short) these are still rather close to each other.

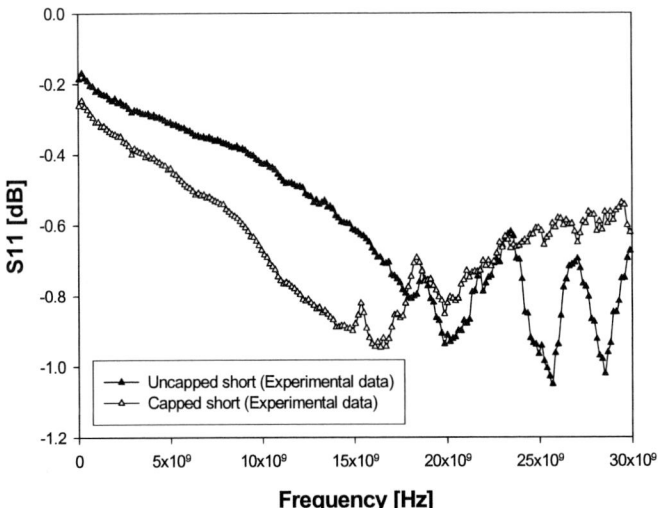

Figure 16. Experimental comparison of the S11 parameter for an uncapped and packaged short.

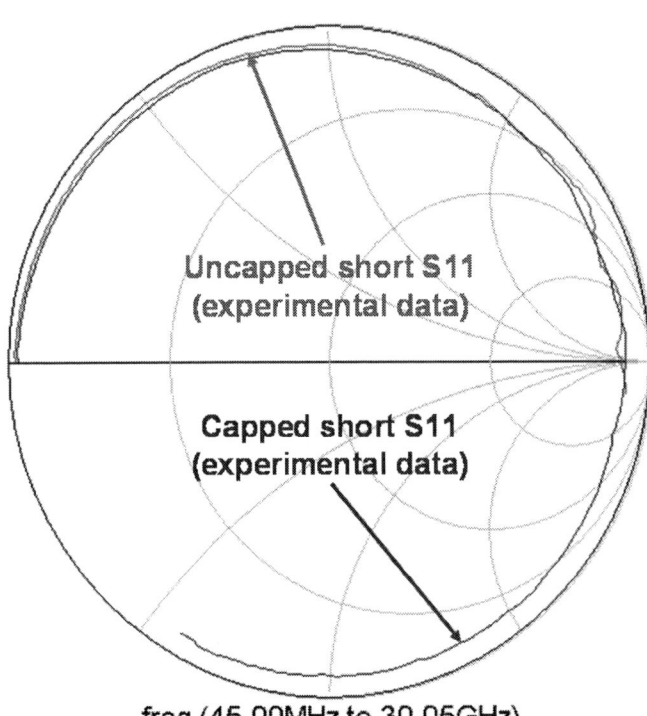

Figure 17. Smith chart of the measured S11 parameter for an uncapped and packaged short.

The Smith chart of the S11 parameter in Figure 17 reports a good superposition of the rotation for the two structures as well. The extra rotation of the capped short plot in the lower part of the Smith chart is due to the longer path of topper pads and vertical vias the RF-signal has to go through.

MEASUREMENTS VERSUS SIMULATIONS

The accuracy of the Ansoft HFSS electromagnetic simulator in predicting the S-parameters behavior of packaged structures was proven by comparing the results of simulations with the experimental data for various test structures. In Figure 18 and 19 the measured vs. simulated data for the capped transmission line mentioned in the previous section are plotted. The simulated S11 parameter shows a good agreement with the measured one (Figure 18). Indeed, the measured reflection parameter value is -20.4 dB @ 10 GHz and -16.0 dB @ 25 GHz while the simulated one is -15.9 dB @ 10 GHz and -7.42 dB @ 25 GHz respectively. Concerning the transmission parameter (Figure 19) the simulated curve is rather close to the measured one up to about 13 GHz. In this range the largest displacement in between the simulated and the measured S21 parameter is 0.13 dB @ 6 GHz. Whereas, beyond 13 GHz the simulated transmission parameter is lower than the experimental one. The largest displacement is around 24 GHz where the experimental S21 value is -0.57 dB while the simulated one is -1.32 dB.

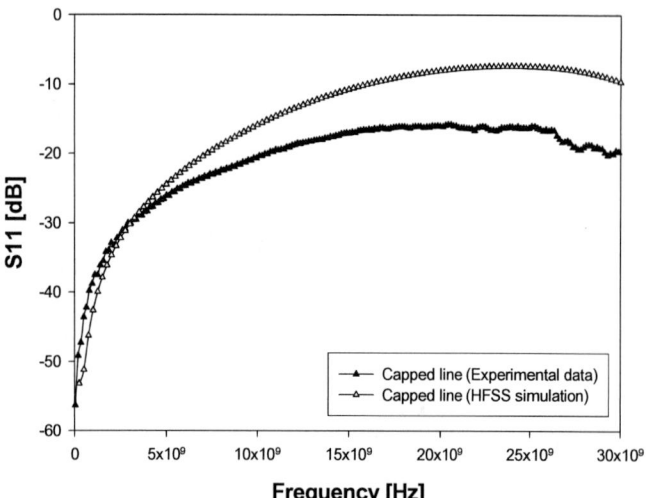

Figure 18. Comparison of the S11 parameter for a measured and simulated packaged transmission line.

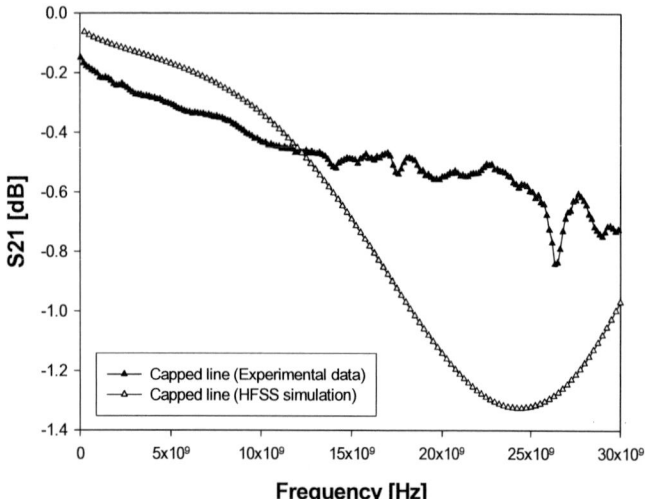

Figure 19. Comparison of the S21 parameter for a measured and simulated packaged transmission line.

Although for the S21 parameter the disagreement in between the measurement and the simulation is quite large at high frequency the HFSS prediction must be considered good. Indeed, in the first sample of packaged test structures a few process parameters were not well controlled and therefore were only estimated for the simulation purposes. In the next experimental run based on optimized values and improved process control, better fit of the simulations with the experimental data is expected.

CONCLUSIONS
In this work, our progress in development of a wafer-level packaging (WLP) solution suitable for RF-MEMS devices and enabling hybrid co-integration of additional IC dies is presented. The proposed approach is based on a capping high-resistivity silicon substrate that is wafer-level bonded to the RF-MEMS device wafer. The capping substrate contains electroplated through-substrate copper vias for signal redistribution and recesses for accommodation of RF-MEMS devices. Optionally,

through-substrate cavities or recesses are realized for hybrid integration of additional IC dies. Several aspects of this packaging solution are discussed.

Firstly, the processing flow for the fabrication of a capping substrate has been outlined. The proposed packaging solution is performed at wafer-level and the wafer-to-wafer bonding is performed by means of ICA/ACA (Isotropic/Anisotropic Conductive Adhesive) or AuSn solder reflow. A modified package structure that is more suitable for achieving (semi-)hermetic packaging is presented. Furthermore, two different approaches suitable for hybrid co-integration of CMOS dies are presented. In both cases the etching of a deep recess in the cap is necessary for hosting the active chip. In the first case, through-substrate cavity and flip-chip bonding is employed. The connection to the MEMS part is provided by through-vias and conductive pads on top of the packaging (overpasses). In the second case, the chip is accommodated in the cavity with the contacts upward and the electrical interconnections are provided by wire bonding. In both these solutions the hermeticity of the packaged MEMS part is not compromised.

Furthermore, the electromagnetic optimization of the package structure with the Ansoft HFSS electromagnetic simulator is performed. All the technological degrees of freedom (dofs) made available by the packaging process (e.g. via diameter and recess depth) have been parameterized and analyzed using Ansoft HFSS in order to reduce as much as possible the losses (capacitive and inductive couplings) introduced by the package-related components. As an example, the influence of the via diameter and package thickness on the transmission parameter (S21) are shown. Experimental data of the first capped test structures (50Ω transmission lines and shorts) are shown. The losses introduced by packaging are not too large especially taking into account that the design of the cap was performed before the electromagnetic optimization in HFSS.

Finally, the comparison of the HFSS simulations with the experimental data of a capped transmission line has been presented. Despite the fact that few process parameters were not well known in this initial experiment, the Ansoft HFSS predictions matched satisfactory the qualitative behavior of the measured S-parameters.

REFERENCES
[1] A.C. Imhoff, "Packaging technologies for RFICs: current status and future trends," 1999 IEEE Radio Frequency Integrated Circuits (RFIC) Symposium, 1999, pp. 7-10.

[2] R. Ramesham and R. Ghaffarian, "Interconnection and Packaging Issues of Microelectromechanical Systems (MEMS) and COTS MEMS," Internal Report of the NASA Electronic Parts and Packaging Program (NEPP), http://nepp.nasa.gov/.

[3] K.I. Kim, J.Mu Kim, J.Man Kim, G.C. Hwang, C.W. Baek and Y.K. Kim, "Packaging for RF-MEMS devices using LTCC substrate and BCB adhesive layer," Journal of

Micromechanics and Microengineering, Vol. 16, pp. 150-156, 2005.

[4] R. Aschenbrenner, A. Ostmann, G. Motulla and E. Zakel, "Flip Chip Attachment Using Anisotropic Conductive Adhesives and Electroless Nickel Bumps," IEEE Transaction on Components, Packaging and Manufacturing Technology, Part C, Vol. 20, No. 2, April 1997.

[5] R. Sihlbom, M. Dernevik, Z. Lai, J. P. Starski, J. Liu, "Conductive adhesives for high-frequency applications," IEEE Transactions on Components, Packaging, and Manufacturing Technology, Part A, Volume 21, Issue 3, pp. 469-477, September 1998.

[6] J. Iannacci, J. Tian, S. M. Sinaga, R. Gaddi, A. Gnudi and M. Bartek, "Parasitic Effects Reduction for Wafer-Level Packaging Of RF-MEMS, " Proceedings of DTIP 2006, 26-28 April 2006 - Stresa, Lake Maggiore, Italy, pp. 25-30.

[7] J. Tian, M. Bartek, "Low Temperature Wafer-Level Packaging of RF-MEMS using SU-8 Printing," Proc. STW Annual Workshop on Semiconductor Advances for Future Electronics and Sensors (SAFE 2005), 17-18 November 2005, Veldhoven, the Netherlands, pp. 56-59.

[8] A. Polyakov, M. Bartek, J.N. Burghartz, "Area-Selective Adhesive Bonding Using Photosensitive BCB for WLCSP Applications," Transactions of the ASME, Journal of Electronic Packaging, Vol. 127, No. 1, March 2005, pp. 7-11.

[9] J. Iannacci, J. Tian, S. M. Sinaga, R. Gaddi, A. Gnudi and M. Bartek, "Electrical Optimization of a Wafer-Level Package for RF-MEMS Applications," Proceedings of EMPS 2006, Terme Catez, May 22-24, 2006, pp. 79-84.

[10] S. Sosin, J. Tian and M. Bartek, "Hybrid Wafer Level Packaging Based on a Capping Substrate with Cavities," Proceedings of Eurosensors XX, Goteborg, Sweden, September 2006.

UTCP : 60 μm THICK BENDABLE CHIP PACKAGE

W. Christiaens, B. Vandevelde, E. Bosman, and J. Vanfleteren
IMEC/TFCG Microsystems
Zwijnaarde, Belgium
Wim.Christiaens@elis.ugent.be

ABSTRACT

This contribution describes modelling and technology for an Ultra-Thin Chip Package (UTCP), based on the embedding of ultrathin dies in flexible substrates. Chips with thickness in the range of 20 - 30 μm are packaged in between 2 polyimide layers. The result is a very thin chip package, with a total thickness of only 50 - 60 μm. Chip, PI layers and metal are so thin that the whole package is bendable.

The base substrate is a uniform polyimide layer, applied (spin coated and cured) on a rigid carrier. Then the ultrathin chip is placed and fixed, using BCB as die attach material. Next, the second spin-on polyimide layer is applied and cured. Vias to the contacts of the chip are laser drilled and the contact metal layer is sputtered and photolithographically patterned. This metal layer is providing a fan out to the contacts of the chips. Finally the whole package is released from the rigid substrate.

The resulting package can be SMD assembled on PCB or flex, or can be embedded in a stack of PCB layers, replacing for example the naked die - with the advantage that alignment constraints for the embedded package are not as severe as for the embedded die. Also the "Known Good Die" (KGD) issue is solved as the packaged chip can be tested before embedding. UTCP's have also the potential for stacking in order to obtain 3D type chip packages.

Key words: chip packaging, 3D, flexible

INTRODUCTION

In electronic circuit technology there is a clear trend towards mechanically flexible circuits, mainly for portable applications where high compactness and reduced weight are important. Furthermore chips need to be assembled in bare die format to reduce the size of the resulting circuit. Flip-chip is a quite well known technology in this respect. In order to make use of the third dimension embedding of components, especially chips is the next step. One approach is to embed the bare die in the inner layers of the rigid or flexible board and to connect the pads of the chip with the wiring of the print. Possible problems with this option are the testing of the dies before embedding (Known Good Die problem), the necessary precise placement of the bare die, and the need for very fine pitch PCB or FPC (flexible printed circuit), compatible with the pad pitch of the embedded chip. A second approach is to provide an interposer, permitting the testing of the chip before embedding and providing a contact fan out with more relaxed pitches, thus eliminating the need for precise placement and ultra high density printed circuit boards. Of course in order to be able to embed chip + interposer in the inner PCB or FPC layers, the unit itself has to be extremely thin, using ultrathin interposer layers and chips. It is precisely one of the goals of the EC funded project "SHIFT" ("Smart High Integration Flex Technologies") to develop the technology for a chip package, capable of being embedded in the flex board [1].

ULTRA THIN CHIP PACKAGE CONCEPT

The Interuniversity MicroElectronics Center (IMEC), together with the University of Ghent, has developed a new concept for packaging ultra-thin chips: the ultra-thin chip package (UTCP). The UTCP is based on embedding of ultra-thin chips (with thickness below 30 μm) in polyimide. Wafer thinning down to 20-30 μm is still not common but a number of manufacturers have already presented equipment for this technology, and the first services offer such extreme wafer thinning and dicing.

An overview of the process flow for the ultra-thin chip package (UTCP) is shown in figure 1. The base substrates are a 20 μm polyimide layer spincoated on a rigid glass carrier. For the fixation and the placement of the chips a bicyclobutane (BCB) is used as adhesive. The chip is covered with a next 20 μm thick polyimide layer. For the contacting to the chip, contact openings to the bumps of the chips are laser drilled and a 1 μm TiW/Cu layer is sputtered and photolithographically patterned. This metal layer provides a fan out to the contacts of the chips. Finally the whole package can be released from the rigid carrier.

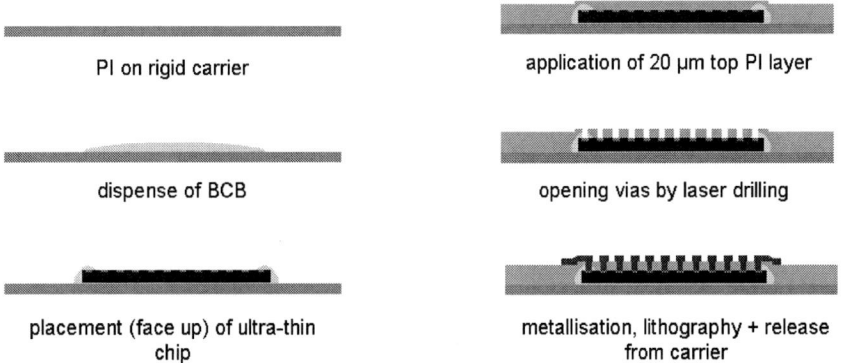

PI on rigid carrier	application of 20 µm top PI layer
dispense of BCB	opening vias by laser drilling
placement (face up) of ultra-thin chip	metallisation, lithography + release from carrier

Figure 1. Overview of the process flow of the UTCP.

PROCESS FLOW
Preparation of the substrates
The base substrate is a 20 µm polyimide layer spincoated on a rigid glass carrier. The polyimide used for this is PI 2611, HD Microsystems.

After processing the package has to be released from the rigid carrier. An easy release of the package from the rigid substrate is obtained in a special way: before spinning the first polyimide layer, the 4 edges of the square glass substrate are coated with an adhesion promoter. The consequence of this is that the first layer of polyimide adheres well to the edges of the substrates, and has marginal adhesion strength to the centre of the substrate. However the adhesion to the edges is sufficient to allow for the whole process cycle (1) through (6) as described in figure 1. After processing the package can be cut out in the area of marginal adhesion and thus peels off easily from the rigid substrate.

Chips
The used chips are silicon chips thinned down to 20-30 µm. Chip size is 5x5 mm^2, providing a daisy chain with 176 pads. The contact pitch is 100 µm. The pad metallization is an electroless Ni/Au flash, with a thickness of 3 µm.

Fixation of the chip
An adhesive material is needed for the fixation of the ultra-thin chips on the base polyimide layer. The adhesive has to be resistant to the high curing temperature of the top polyimide, PI2611 (350 °C). So bicyclobutane (BCB) and polyimide could be used as bonding material. Several polymers were already compared for full wafer bonding [2], in that study BCB bonding offers the highest bond strengths. The polymers we tested as adhesive bonding material were: polyimides PI 2610 and PI 2611 (both from HD Microsystems) and the BCB Cyclotene 3022-46 (DOW).

It is very important to prevent void formation at the bond interface. Voids can be caused by small air bubbles, trapped between the adhesive layer and the surface of the chip. These bubbles can be prevented by placing the chips in a vacuum environment during bonding. Another possible solution could be found in dispensing a well controlled amount of the adhesive. While placing the chip the dispensed adhesive flows from the middle of the surface to the edges of the chip without air at the interface.

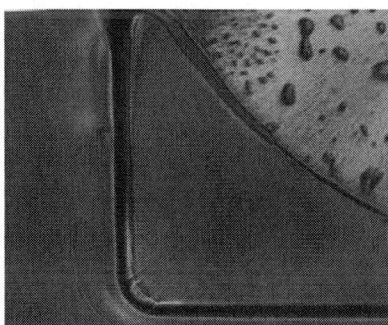

Figure 2. Air bubble under the chip, after cure of BCB.

Voids can also be created during the curing of the tested polymer: evaporating solvents create bubbles. Using BCB material, this can be solved with a pre-curing of the adhesive BCB layer before bonding: solvents can be evaporated during this pre-curing.

Void-free bonds were not realized in the tests with the polyimide PI2610 or PI2611 as adhesive materials. Voids evolve during the curing after the bond, even after a pre-cure [2]. This can be explained by the difference in curing mechanism of BCB and the tested polyimides. BCB does not produce any byproducts that are released during the curing process (if solvents are evaporated during a pre-curing). During the imidization process of the tested polyimides water is sublimated as by-product, also solvents evaporate during the curing process [3].

Also the application of a minimum bonding pressure is required to achieve void-free bonds [2]. So BCB is suitable to be used as adhesive to glue the chip on the polyimide layer: the solvents of the BCB evaporate during a pre-curing and when the chips are placed properly (in vacuum or with a dispensed BCB) void free bonds can be obtained. In next experiments we will further optimize chip placement on dispensed (pre-cured) BCB, and try to avoid voids by controlling the dispensed quantity of BCB. (This way we would not need a vacuum environment.)

Covering of the chip
After the cure of the BCB at 350 °C, the chip is fixed on the polyimide layer. Then a covering polyimide (PI2611) layer is spincoated on this fixed die, with a thickness of 20 μm. In order to obtain a good adhesion between the top polyimide layer and the (cured) base polyimide, this cured polyimide has to be pre-treated. Before the spincoating of the top polyimide layer the cured layer is first plasma-treated for 2 minutes in a CHF3/O2 plasma followed by a 2 minutes an oxygen plasma treatment.

The cross section below is made after the cure of the top polyimide: it shows the good edge-coverage of the spincoated top polyimide layer.

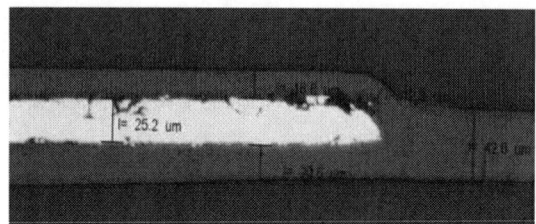

Figure 3. Cross section of embedded chip (metallization and vias not visible).

Contact opening
Contact openings to the bumps of the chips are laser drilled through the top polyimide layer. Laser drilling has emerged as the most widely accepted method of creating microvias in high-density electronic interconnect and chip packaging devices [4]. UV lasers are known for high-precision material removal and their ability to drill the smallest vias. The UV lasers have also the ability to effectively ablate metal layers, this also means that measures must be taken to avoid too much laser damage to the metallization of the bumps of the chip.

Laser drilling tests were performed with two different UV lasers: a 248 nm KrF excimer laser and a frequency tripled Nd-YAG-laser, working at 355 nm.

Three different set-ups were compared:
- excimer laser
- YAG-laser with Gaussian beam
- YAG-laser with shaped beam

The best results are achieved using the tripled YAG-laser with a shaped beam. Beam-shaping optics transform the natural Gaussian irradiance profile to a near-uniform "tophat" profile. This imaged beam removes the polyimide material uniformly across the via, without creating undesirable underlying metal damage at the centre of the imaged spot (which is difficult to control with a Gaussian beam). It is also shown that substantially less irradiance dose is required for drilling when using the reshaped beam profile. A lower irradiance dose reduces considerably the thermal load on the material and improves dramatically the overall hole quality (and reduces the debris). Due to the uniform profile of the beam also the tapering can be better controlled [5].

Via diameters with a top diameter down to 35 μm are already realized using the tripled YAG laser with shaped beam. Figure 4a and figure 4b are showing top and bottom diameter of a metallized via.

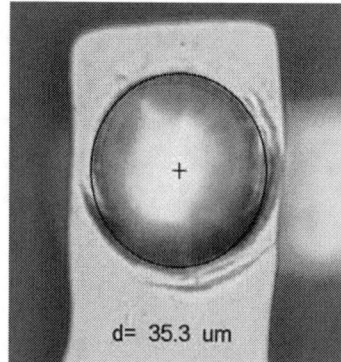

Figure 4a. Metallized via with top diameter of 35 μm.

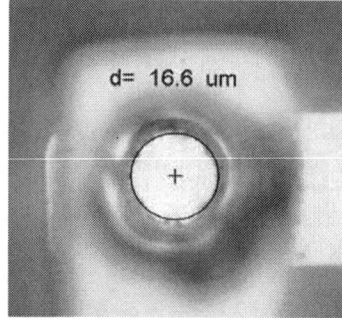

Figure 4b. Bottom diameter of same via.

Top metallization
A top metal layer is sputtered, metallizing the contacts to the chip and providing a fan-out (see figure 5).

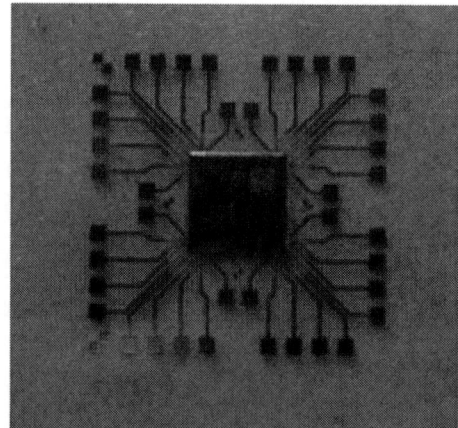

Figure 5. Fan out metallization.

The metallization is realised by sputtering a 1 μm TiW/Cu layer. In order to have optimum adhesion strength of the sputtered TiW/Cu layer on the top polyimide, reactive ion etching was tested on spincoated polyimide layers as pre-metallization surface treatment.
3 different plasma treatments were compared:
- CHF3/O2 (4/1) gas mixture
- O2 plasma
- CHF3/O2 plasma + O2 plasma

Polyimide was spincoated on glass-substrates, plasma treated and then metallised with a 1 µm sputtered TiW/Cu layer.

All the samples passed the Scotch-tape test. In order to perform a peel strength test the 1 µm copper had to be plated up. The sputtered copper was plated up to a 25 µm copper thickness and then photolithographically patterned. The measured peel strength on the samples treated with a CHF3/O2 plasma + O2 plasma was higher than 1.6 N/mm and the peel strength of the metallisation on the oxygen treated polyimide was even higher than 2 N/mm.

SOME RESULTS

After processing the whole package (polyimide layers + embedded chip + fan-out metallization) can be released from the rigid substrate. Chip, PI layers and metal are so thin, that the whole package is bendable (see figure 6 and 7).

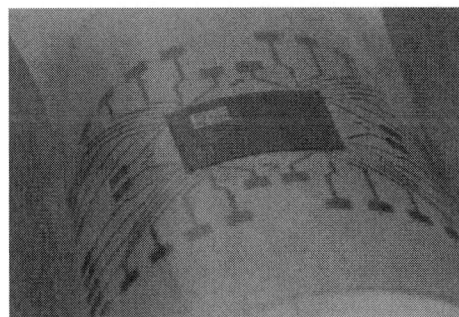

Figure 6. Bended chip.

Typical thicknesses for the different layers are:
- polyimide : 20 µm
- BCB : <5µm
- chip : 20 – 30 µm
- metallisation : <= 1µm

This results in a very thin flexible chip package (down to 50 – 60 µm thickness).

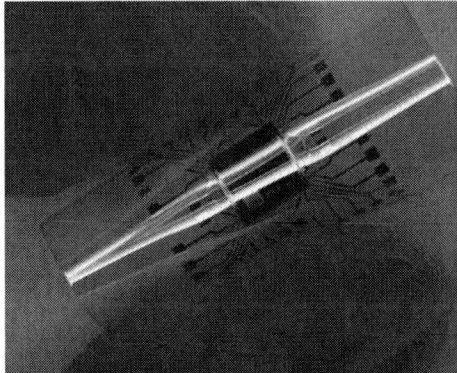

Figure 7. Showing bendability of packaged chip.

THERMO-MECHANICAL ANALYSIS

FEM has been used to calculate the stresses induced in the UTCP caused by first processing succeeded by a mechanical bending.

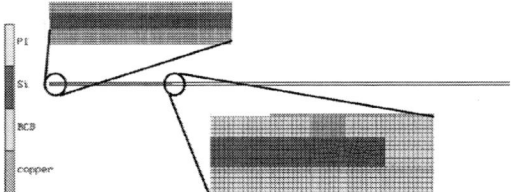

Figure 8. 2D-FEM the UTCP.

The first part of the FEM simulation is the stress calculation after processing. Both the polymer and BCB are cured at 350°C, so the whole structure (excluding the copper metallisation) gets a cooling down from 350°C to 20°C. The silicon chip (169 GPa, 2.6 ppm/°C) and the polyimide (8.5 GPa, 3 ppm/°C) have almost the same coefficient of thermal expansion (CTE) so there will be not so much induced stress. However, the thin BCB (2.9 GPa, 42 ppm/°C) layer has a much higher CTE which will cause high tensile stresses in the BCB layer itself.

Figure 9 shows the real deformation after processing. Thanks to the almost similar CTE of Si and the PI2611, the flexible package stays almost flat. In the BCB layer, a uniform "tensile" stress equal to 61 MPa is found. Near the chip edge, also tensile stress in the silicon chip is found (+74 MPa).

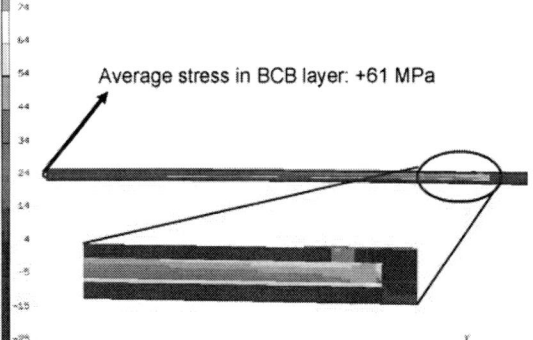

Figure 9. Deformation and in-plane stress (σ_x) in UTCP after processing.

The second part of the FEM simulation is the mechanical bending of this UTCP. As shown in figure 10, it is possible to manually bend this package with a curvature of about 5 mm without damaging the silicon chip nor the BCB layer. For the silicon chip, the maximum allowable tensile stress is around 300 MPa. For BCB, the ultimate tensile stress is 87 MPa.

For the downward bending, a curvature of 5 mm is feasible, as it compensates the tensile stress in BCB by compressive stress. The stress in the silicon chip is in the neighbourhood of the maximum allowable stress. For the upward bending, it is much more critical, as additional tensile stress in the BCB layer is induced due to the upward bending.

117

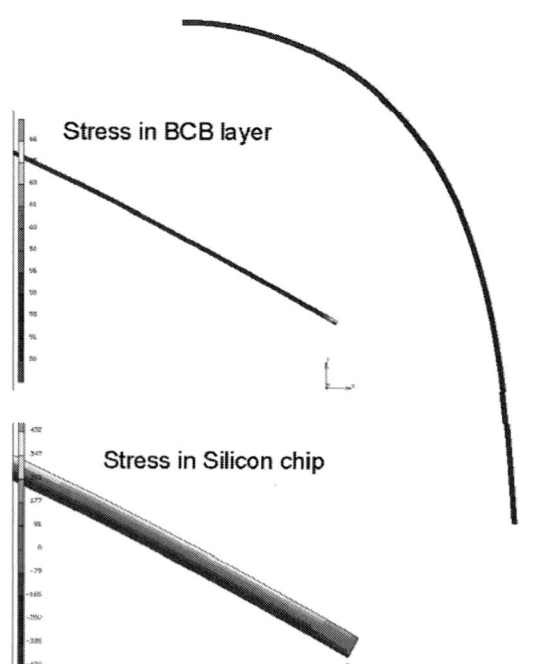

Figure 10. FEM of "downward" bending of the UTCP (curvature ~ 5 mm).

APPLICATIONS OF THE UTCP

This ultra thin package can be embedded in a stack of PCB or FPC layers or can be used as flexible chip in for example smart textile. The process flow also provides an alternative for the driver chips in certain types of flexible displays. In such displays, a base polyimide layer is applied on a rigid substrate, followed by processing of an active matrix of thin-film transistors (TFTs) [6].

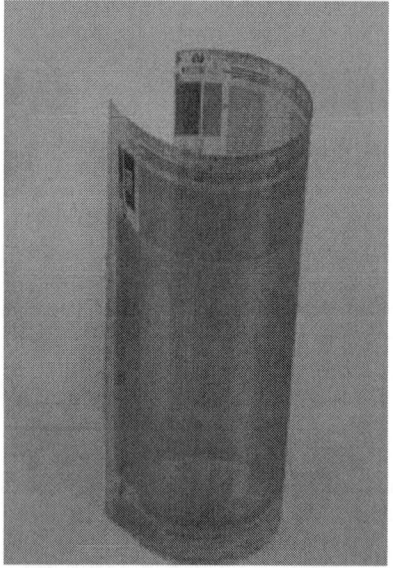

Figure 11. 50 x 50 mm a-Si TFT array on 3 μm thick freestanding polyimide layer made by the EPLaR process [6].

Each TFT cell addresses one pixel in the display. The rows and columns of the display are driven by external silicon chips, which are packaged in tape-carrier-package or chip-on-film format and attached to the display substrate using adhesives or solders. Alternatively, the ultra-thin package technology described here can be used for embedding the driver chips in the polyimide base substrate, before processing of the active matrix. This results in an enhanced flexibility of the display module and a strongly reduced number of interconnections from the display substrate to the external driver electronics.

FLAT UTCP

An alternative process flow makes it possible to produce real flat chip packages. This process is based on a photodefinable polyimide: cavities can be defined in this extra photodefinable spin-on polyimide layer and the chips are placed and fixed in these cavities. The polyimide HD-7012 (HD Microsystems) is used for this purpose. It has typical cured film thickness between 20 and 70 μm. The thickness we need is about 30 μm, so that the chip + adhesive fill the cavity.

A suitable consequence of process steps for this flat UTCP is shown in figure 12. The base substrate is a uniform polyimide layer (PI2611), applied on a rigid carrier. Next, a 30 μm photodefinable polyimide (PD PI) layer is spin coated. After the prebake, the PD PI is exposed. The unexposed regions are etched away by the developer (PA400D, rinse with PA400R), defining the cavities in the PD PI layer. After cure of the PD PI, the ultra-thin chip is placed and fixed in the cavity. The chip and the adhesive are filling the cavity. Next, the top polyimide is applied, via openings are laser drilled and the TiW/Cu is sputtered. After patterning of the metal layer the flat package can be released from the carrier.

This alternative process places the thin dies in cavities, so that the packages are flat. The advantage of these flat UTCP's is that they have also the potential for stacking in order to obtain 3D type chip packages.

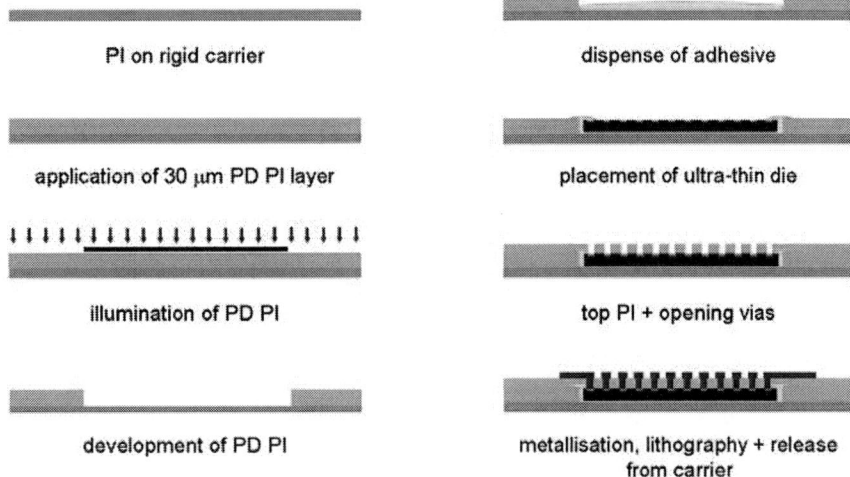

Figure 12. Process flow for flat UTCP.

CONCLUSIONS

A new concept of packaging ultra-thin chips has been developed: the UTCP. With a total thickness of the packaged die of only 50-60 μm, the UTCP is so thin that the whole package is flexible.

As vias with diameters down to 35 μm are realised using a tripled YAG laser, chips with contact pitches down to 80 μm could be packaged with the UTCP.

The UTCP can provide an interposer, permitting the testing of the chip before embedding and providing a contact fan out with more relaxed pitches.

ACKNOWLEDGEMENTS

The work was carried out under the EC IST funded project IP-SHIFT (contract number 507745). The authors also want to thank their SHIFT partners: Technische Universität Berlin (TUB) for the test wafers, Fraunhofer-Gesellschaft zur Förderung der angewandten Forschung e.V. (FhG) for the thinning of the chips and Hightec MC AG for the technical discussions.

REFERENCES

[1] www.shift-project.org
[2] F. Niklaus, P. Enoksson, E. Kälvesten and G. Stemme, "Low-temperature full wafer adhesive bonding", Journal of Micromechanics and Microengineering, Vol. 11, pp. 100-107, 2001.
[3] Information from the material data sheets.
[4] C. Dunsky , "High-speed microvia formation with UV solid-state Lasers", proceedings of the IEEE, Vol. 90, No. 10, pp. 1670-1680, October, 2002.
[5] D.M. Karnakis, J. Fieret, P.T. Rumsby and M.C. Gower, "Microhole drilling using reshaped pulsed Gaussian laser beams", proceedings of the SPIE, Vol. 4443, pp. 150-158, 2001.
[6] I. French, D. McCulloch, I. Boerefijn, N. Kooyman, "Thin electrophoretic displays fabricated by a novel process", SID Dig. Tech. Papers, pp. 1634-1637, 2005.

COPPER PANEL FABRICATION AND STACKING CONCEPT FOR VLP FB DIMMS

Peter Salmon
Peter C. Salmon, LLC
Mountain View, CA, USA
peter@petersalmon.com

ABSTRACT

A new concept for fabricating very low profile fully-buffered dual in line memory modules (VLP FB-DIMMs) is proposed. Interconnection layers are built up on copper panels using high resolution photolithography. The copper substrate provides a built-in heat spreader. 8-high die stacks are employed. A new Copper-Pillar-Well (CPW) flip chip connector is used; it provides convenient rework of all 144 memory chips in each module, without melting any solder. The same connectors enable the Advanced Memory Buffer (AMB) chip to be attached within the width limitation of 18.3 mm for the VLP module. Improved electrical, thermal and mechanical properties are potentially achievable, enabling 16 GByte modules having advanced specifications.

Keywords: FB-DIMM, CPW, copper substrate, 80Au20Sn, stacked package, rework, memory module, HDI, 3D semiconductor.

BACKGROUND AND MOTIVATION

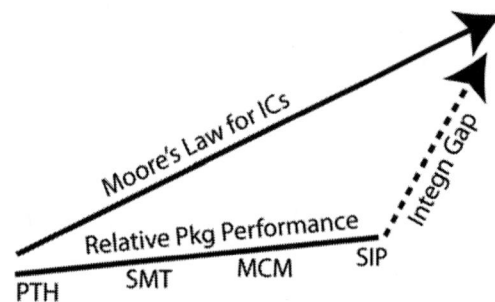

Figure 1. Semiconductor integration gap

Over the last 40 years transistor density in silicon integrated circuit (IC) chips has increased by a factor greater than 100,000 according to the phenomenon known as Moore's Law. Meanwhile, the ability to integrate silicon chips into systems has progressed more slowly. Package development can be traced from printed circuit boards (PCBs) having plated through holes (PTHs) around 1970. Surface mount technology (SMT) followed, multi-chip modules (MCMs), and systems in package (SiPs). The slow development of integration methods has resulted in an integration gap; this gap has dimensions of cost, performance, cooling, and scalability.

Technology for manufacturing flat panel displays (FPDs) has progressed steadily, almost exclusively in Asia. Currently the panels are as large as 7 by 8 feet, using 6 mask layers to provide sub-micron features at a cost of around $2,000 per panel (8th generation AMLCD[1] process). If adapted to the fabrication of copper substrates, this technology could enable a revolution in high density interconnect (HDI) and provide a major step toward a solution to the Integration Gap depicted in Figure 1. The adaptations can include known technical solutions, several of which are outlined in this paper. If in turn, this new substrate technology is used with Copper-Pillar-Well technology for integrating IC chips with substrates, advanced products like the 16 GByte memory module described herein become possible. They would perform better than conventional modules, and have potentially lower prices. Availability of such a commanding solution in a mainstream product like a DRAM module could propel the technology into solutions for other products. The other products may take advantage of the DIMM standard, thereby projecting the standard into non-memory applications, like supercomputers or data centers for example.

A new technology platform will not take hold unless all aspects of the solution are adequately covered. It is suggested that FPD technology offers a solution for high-performance and low-cost substrates, with 300 mm square copper panels also offering an interim solution. CPW technology offers a solution for assembly and rework of IC chips using flip chip attachments. The use of copper substrates offers a solution for cooling, and new test methods outlined herein offer a solution for testing. The memory module described in this paper is a vehicle for focusing all aspects of the solution on a specific application.

CONCEPTUAL DESIGN OF MEMORY MODULE

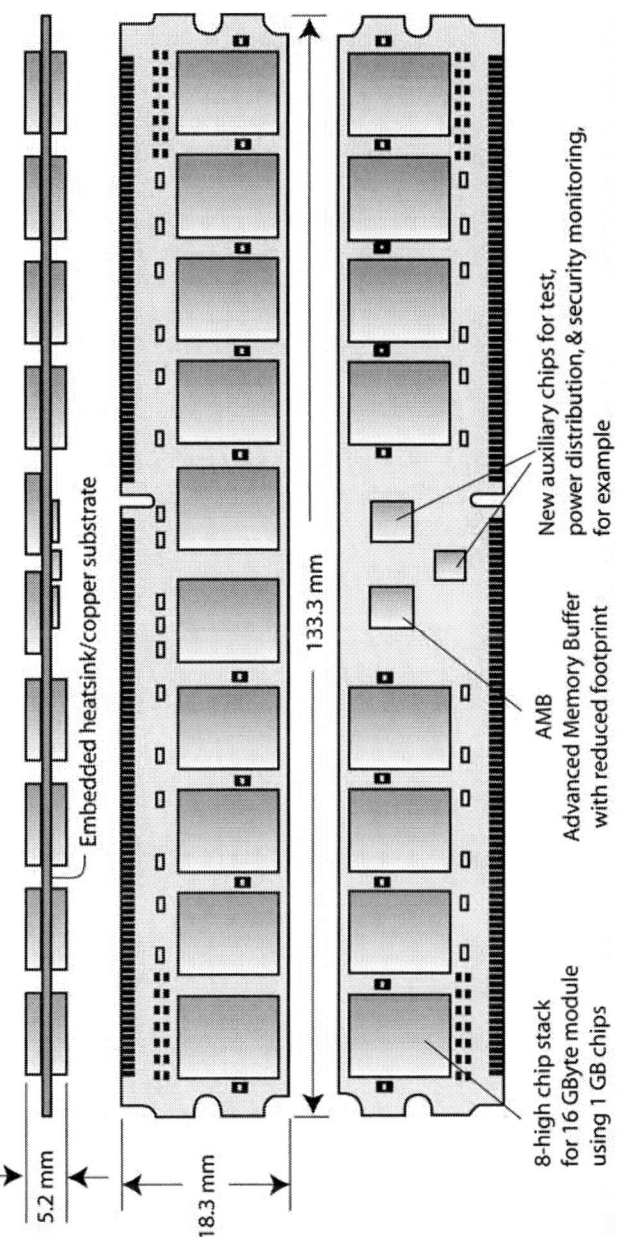

Figure 2. Proposed memory module employing a copper substrate.

In early VLP FB DIMM designs, fitting the AMB on the VLP height of 18.3 mm was a problem[2]. Although the chip size was much smaller, the footprint of the BGA attachment exceeded the 18.3 mm dimension. This was due to the pitch of the solder balls and the space required for routing traces around the solder balls using conventional printed circuit board (PCB) technology. As will be further described, the interconnection circuits on the copper substrates have a finer pitch than regular PCBs, corresponding to a High Density Interconnect (HDI) solution. Also flip chip connections in the form of CPW connectors have a pitch as small as 80 μm. Applying these improved capabilities allows the AMB footprint to be around 6-8 mm square (Chip Scale Package,

CSP), providing room for additional chips in the center region as shown. As described relative to the new test method, one of these chips should be a test chip. Other options for new capabilities may include a power distribution chip or a security monitoring chip, as examples.

THE CPW CONNECTOR

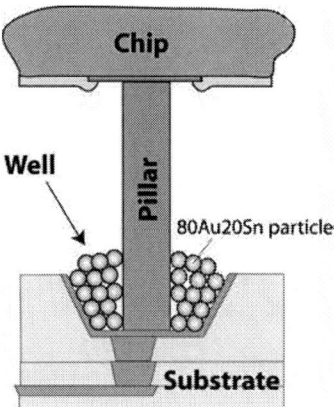

Figure 3. Detail of the Copper-Pillar-Well connector.

The CPW connector is designed for fine-pitch flip chip attachments. By using slender pillars, stress relief is provided between chip and substrate: the pillars can bend to relieve stresses in both x and y directions. Flexibility in the x-direction is required to relieve thermally induced stresses that occur as the chip and the substrate expand and contract at different rates. Additionally, flexibility in the y-direction is required in order for the IC chips to remain flat during temperature cycles. The requirement for flexibility of the chip attachment is more demanding if the chips are thinned, or if delicate high-k dielectric materials are used. Typical pillar dimensions include a 10 μm diameter and a 50-100 μm height. By providing this compliance an epoxy under layer can be avoided, and this leads to a simple and straightforward technique for re-working any chips found to be defective. In addition to this re-work capability, chips can be assembled on substrates and tested without melting the 80Au20Sn particles to form solder. This is made possible by the clean and conductive surface of gold and gold alloy particles. The 80Au20Sn alloy has been used as a flux-less solder for many years; the particles are dipped in dilute hydrochloric acid prior to use, to remove any tarnish (oxide) on their surfaces. Combining the slender copper pillar with a well at the substrate surface provides improved mechanical and thermal properties. The pillar-in-well connection is mechanically rugged, able to withstand thermal cycling and shock, and also provides a good thermal connection to the substrate. Filling the wells with particles like 80Au20Sn rather than with solder paste provides a new convenience in assembly and test. An IC chip can be aligned and placed with each pillar centered on and partially penetrating a well filled with particles, just through the force of gravity. Either a gentle insertion force or ultrasonic urging can be employed to insert the pillars fully into the wells, without damaging fragile chips. Finally, the wells provide a vertical tolerance of about ± 10 μm due to their depth of around 30 μm. This is useful to overcome planarity problems, such as uneven heights of copper pillars due to non-uniform plating rates during electro-forming, as well as dimples observed at the ends of the pillars.

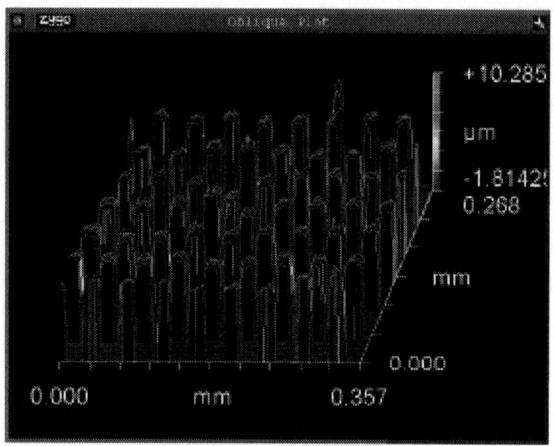

Figure 4. 10 μm diameter copper pillars.

Figure 4 shows 10 μm diameter copper pillars produced by Igor S. Zavarine et al.[3] using an electroplating process developed by Enthone-Cookson Electronics and NEXX Systems, Inc.

To improve the thermal connection between the die and the copper substrate, heat bumps may be employed alongside signal bumps, as shown in Figure 5. Since they are formed at the same time as the signal bumps, the heat bumps add almost no additional cost.

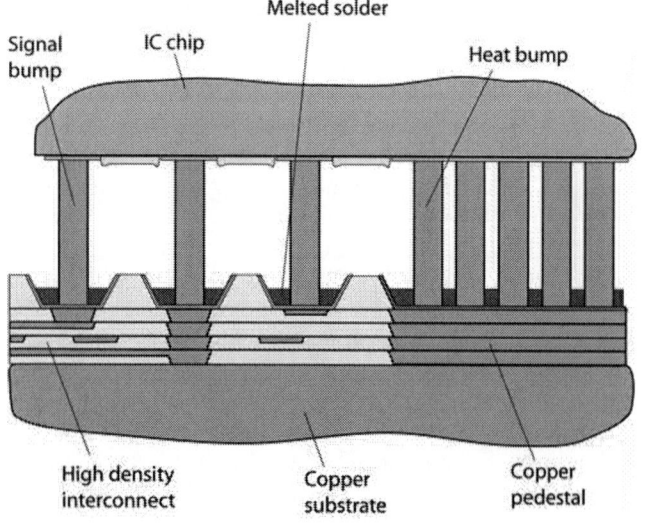

Figure 5. Use of additional copper pillars as heat bumps.

While the ability to add heat bumps has general utility, it is not required for the current VLP FB-DIMM application. This is because heat is more effectively removed from the back side of the die, as will be shown in reference to Figure 9. Heat removal from the back side is preferred over removal from the front side because of the high thermal resistance of the metal and dielectric interconnection layers underlying the input/output pads on the semiconductor chip.

FABRICATION OF THE COPPER SUBSTRATE

Figure 6 shows a sequence of steps for producing copper feed-throughs in a base copper substrate. A commercially available milling tool produced by LPKF Laser and Electronics[4] can be used to mill the cylindrical cavities shown. A combination of a porous backing layer and a thermal release layer is used. The porous backing layer enables vacuum-aided filling of the cavities with a glass frit paste. The thermal release layer provides a mechanism for separating the copper substrate from the backing layers prior to firing the glass frit. Firing is performed in a reducing atmosphere to minimize oxidation of the copper. The copper substrate and the feed-throughs so formed will be hermetic because both the copper and the glass have low moisture permeability.

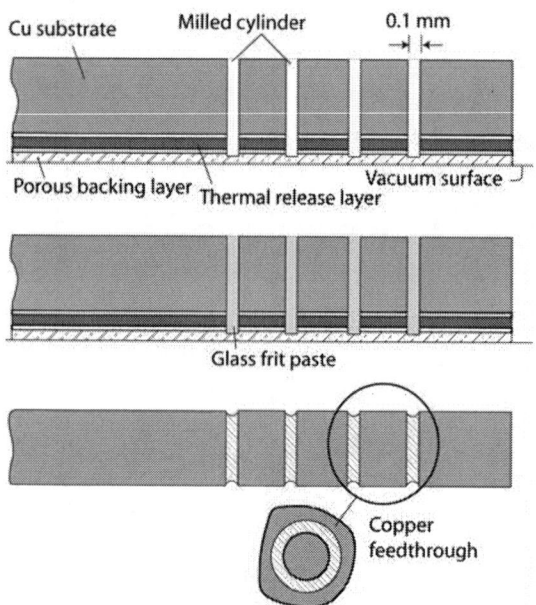

Figure 6. Fabricating copper feed-throughs in a copper substrate.

Figure 7 summarizes a process sequence for fabricating a dual damascene layer on the copper substrate.

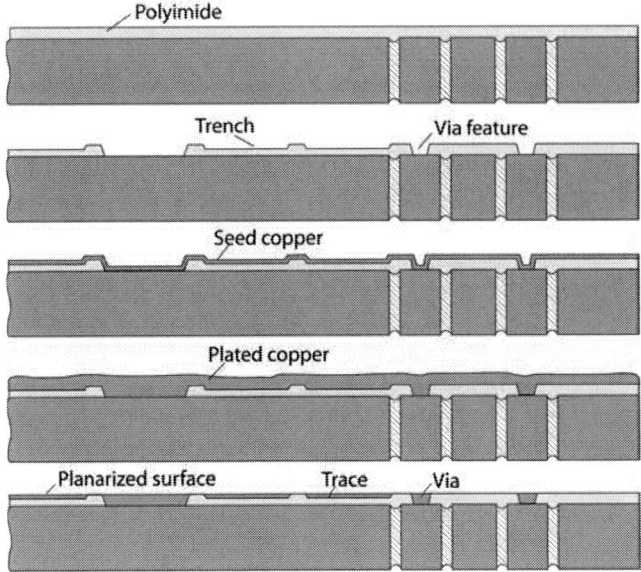

Figure 7. Fabricating the first dual damascene layer.

The method shown in Figure 7 is suitable if 300 x 300 mm copper panels are used, for example. Different methods will be required for the larger FPD panels, and these are not known at this time.

Figure 8 shows the result of fabricating 6 layers on each side of the copper substrate, including a well layer corresponding to the CPW well shown in Figure 3.

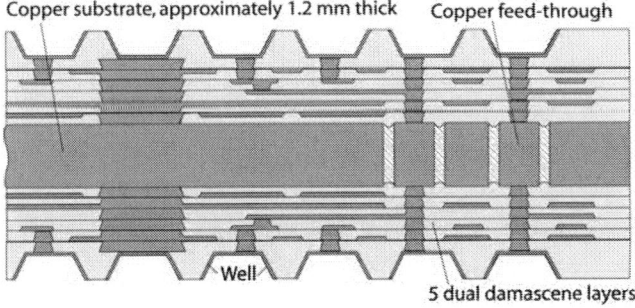

Figure 8. Built-up copper substrate with interconnection layers and wells.

ASSEMBLY TEST AND REWORK

The 16 GByte memory modules will include 144 memory chips at 1 Gbit each (they employ 9 bit words for error correction). The as-assembled compound yield will be too low to produce good modules without rework. Even the 8-high die stacks treated as separate components will have only 43% compound yield, assuming 90% chip yield. Consequently, effective rework methods will be required for producing the modules cost-effectively. The methods must also allow for automation, in order to process such a large die count reliably and cost-effectively.

Figure 9 shows a completed 8-high die stack for use in the module. The vertical axis is expanded to show the construction more clearly. To meet the module thickness of 5.2 mm shown in Figure 2, the chips and the copper substrates are each approximately 0.1 mm thick.

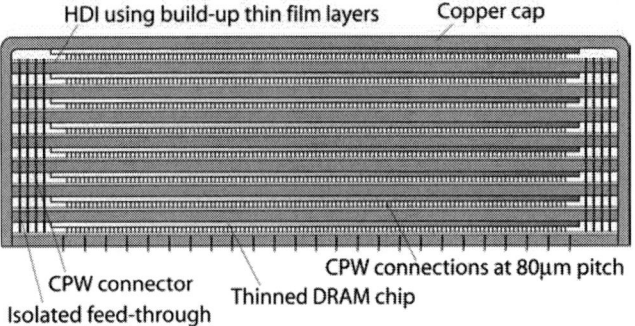

Figure 9. 8-high die stack module employing CPW.

The test concept includes a test chip embedded on the substrate (board) to be tested, as shown in Figure 10.

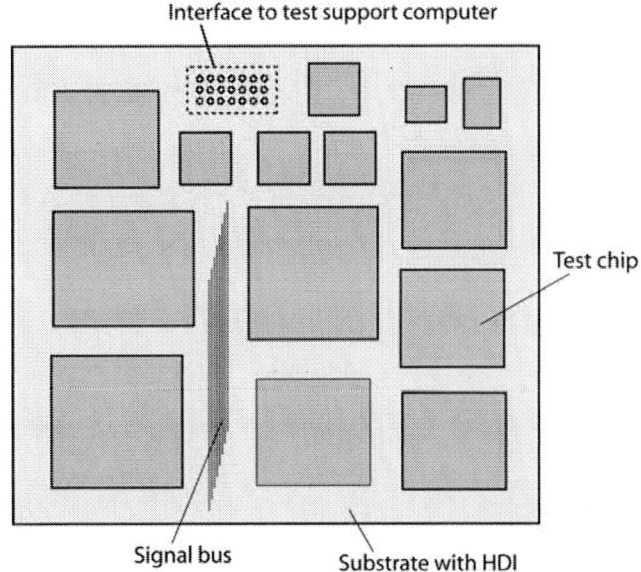

Figure 10. Generalized test concept.

The chips shown in gray in Figure 10 communicate with one another at full system speed over signal buses. If this were not the case the module would not function correctly. Adding the test chip perturbs the electronic system only slightly, adding a small amount of capacitive loading. Therefore, the system will still operate at full system speed. A system processor executing a test program can be used to stimulate the system which will respond by asserting data on the signal buses. Sampling circuits and comparators are provided on the test chip. The sampling circuits can capture the signals on the buses at full speed. The compare circuits can also operate at full speed, and the results of the comparisons can be communicated to a test support computer through a relatively slow speed connector interface as shown. Details of this testing technique are described in a published patent application[5].

The specific test solution for a VLP FB-DIMM is shown in Figure 11. In this case the test chip will also contain a processor capable of generating stimulus vectors; it will assert them on the data bus lines, in accordance with a test program.

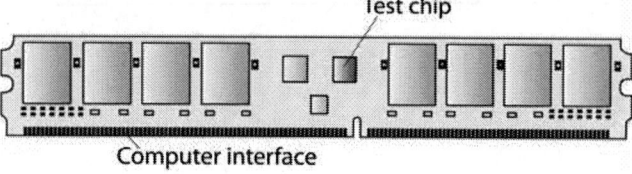

Figure 11. Test implementation for VLP FB-DIMM

We will examine the process steps for assembly test and rework for the case of 300 x 300 mm copper panels, having 8-high stacked die modules, as depicted in Figure 9.

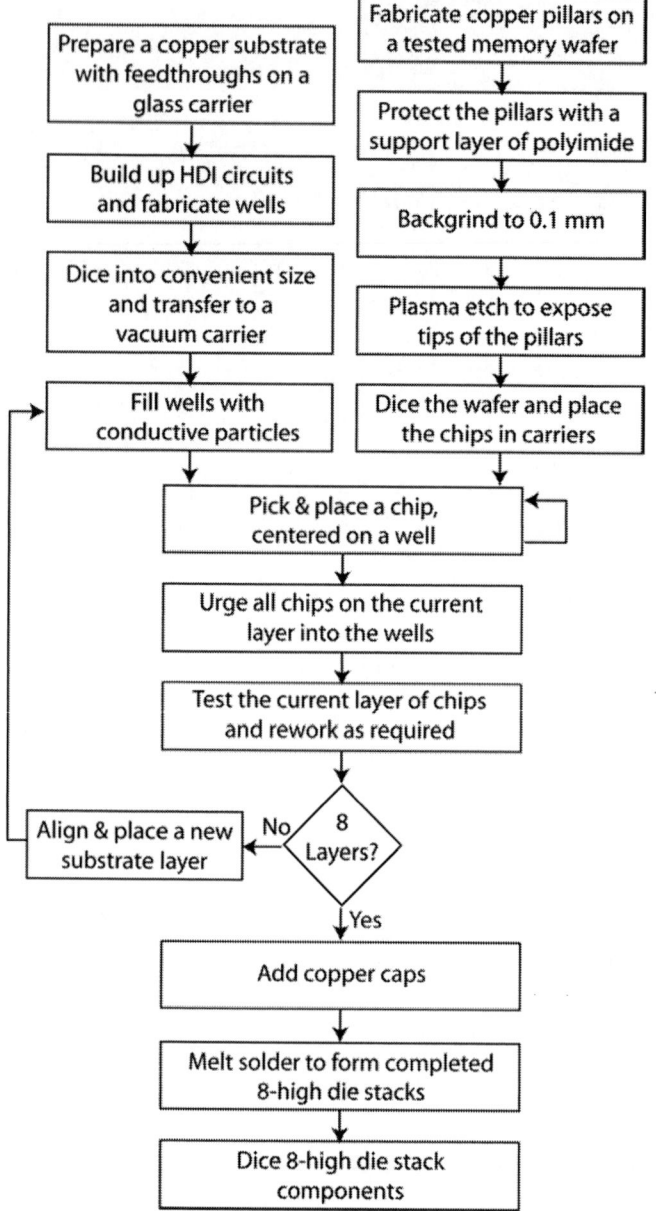

Figure 12. Flow chart for fabricating 8-high die stacks

A separate process is employed for integrating the 8-high die stacks into the memory modules, as shown in Figure 13.

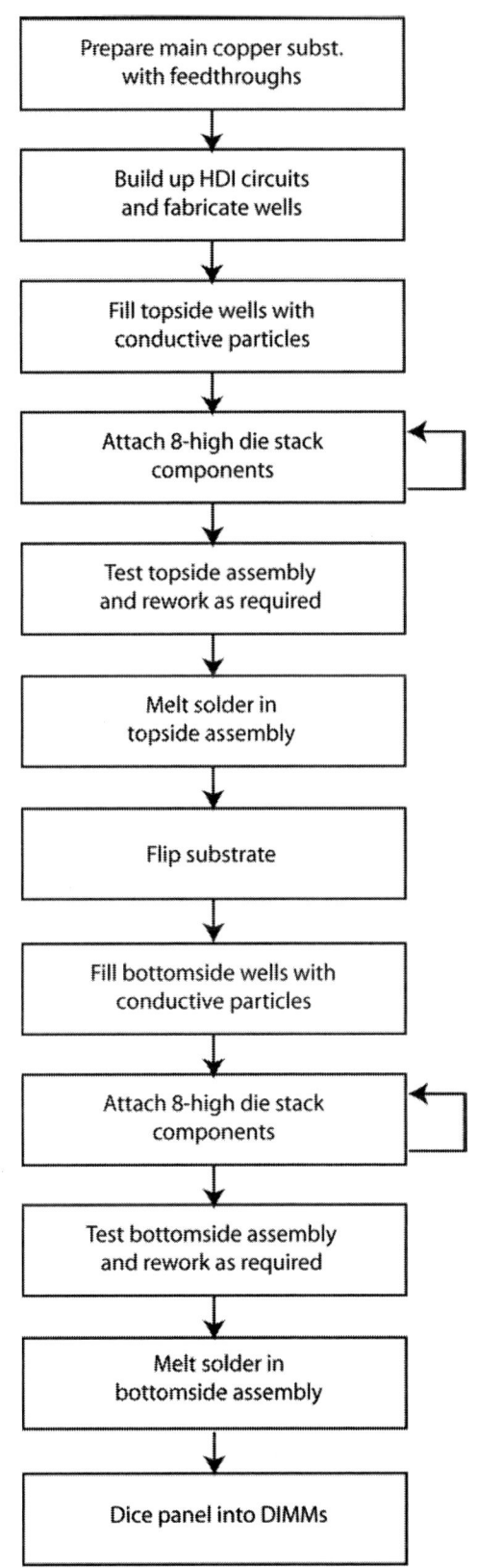

Figure 13. Flow chart for fabricating the DIMMs

PANEL FORM FACTORS

Figure 14 shows the 300 x 300 mm copper panel format that is compatible with Figures 6-9 and 11-13.

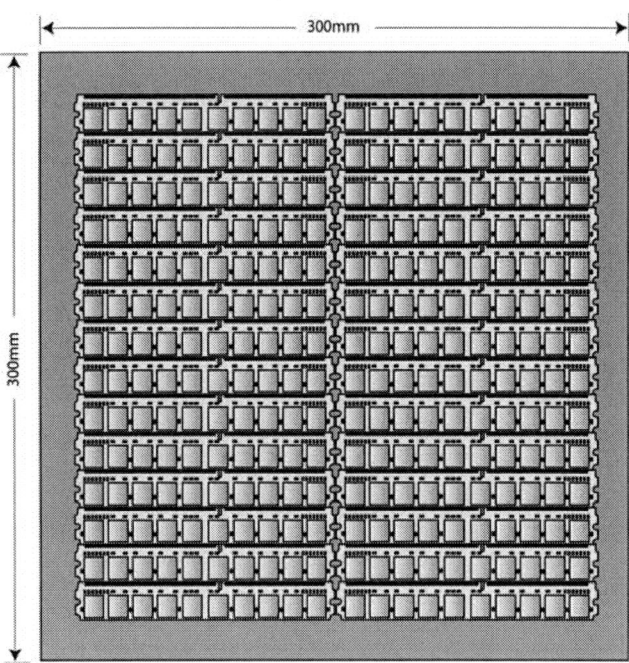

Figure 14. 300 X 300 mm copper panel with 28 modules

The 16 GByte modules have 144 memory die in each module, so the panel with 28 modules has 4,032 memory die. It is suggested that the processes described herein can support 100% assembly yield for this large number of die, and in addition, the rework process for replacing defective die can be automated. Automated rework is enabled by CPW technology, and in-module testing.

The 300 x 300 mm square panels can be processed by mainstream semiconductor equipment. For example, Suss Microtec offers the lithography cluster Lithopack 300 for coating, baking, exposing and developing 300 mm substrates. Usually the substrates are round wafers, but the equipment can be adapted to 300 mm square panels. The cluster includes the MA300 full-field proximity mask aligner, capable of exposing over 100 wafers per hour.

A more aggressive panel form factor takes advantage of FPD technology and is shown in Figure 15.

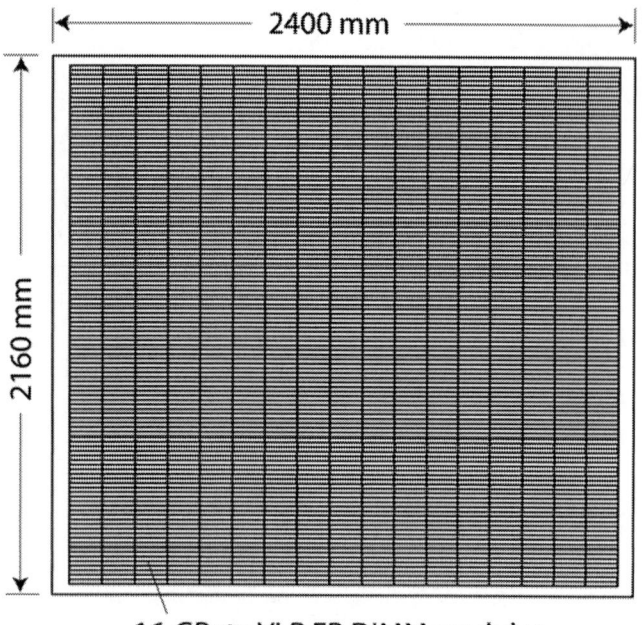

Figure 15. Copper panel employing FPD layout

Note that Generation 8 flat panels support 6 or more mask layers with 0.8 µm features, more than adequate for the HDI circuits described herein. A 6-layer circuit has a panel cost of around $2,000. The main copper substrate has 6 layers on each side. Since there are 1,955 VLP FB-DIMM modules per panel, the cost of the main substrate would be around $2. The stacking copper substrates supporting the 8-high die stacks on each side of the main substrate would raise the total substrate cost per 16 GByte DIMM to around $20, an attractive proposition.

The large panel shown in Figure 15 would include 281,000 memory die when fully assembled. Again, it is suggested that 100% assembly yield is achievable in this panel format using the rework processes described herein. Furthermore, these processes can be automated, at least prior to the point where the solder particles have been melted. If a faulty chip is detected after the solder has been melted, it can still be reworked manually. This can be achieved by applying hot inert gas through a shroud placed over the defective chip, removing the defective chip or assembly, sucking out the solder remaining in the wells, refilling the wells, and attaching a replacement chip or chip assembly.

A potential method for adapting the FPD process for fabricating thin copper sheets is to attach the copper sheet to the glass panel using a UV-activated release layer. After processing is complete, UV can be directed through the glass to the adhesive interface, reducing the adhesion forces to around 1% for detaching the copper sheet from the glass panel.

USING DIMMS IN NON-MEMORY APPLICATIONS

The copper-based DIMM described herein could be applied to a supercomputer or data center application as shown conceptually in Figure 16.

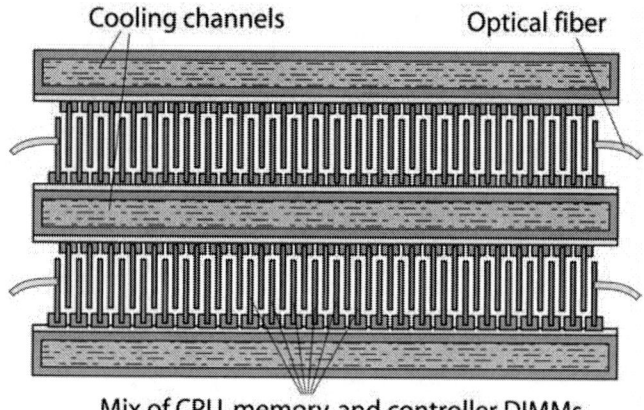

Figure 16. A non-memory application for DIMMs

As shown in the figure, end modules can be adapted for fiber optic communications. Internal modules can be mixed and matched, according to a customer's requirements. The DIMM standardization would lead to lower module costs, and a-la-carte configurations. For server applications, this type of architecture can support a miniaturization factor of 100X, according to a study conducted by the author. Such systems can be well-tested, repairable, and adequately cooled.

SYSTEM SCALABILITY

Modern electronic systems typically include such a diversity of component types and interconnection schemes that scalability becomes a far-off dream. However, Jedec physical and electrical standards for DIMM modules could be applied to a wide range of solutions, including servers, data centers, and supercomputers; the physical formats could be applied even more broadly. Such standardization could provide many benefits including lower development costs, lower production costs, faster time to market, and improved tailoring of systems to customer specifications. Another factor is that tool vendors are currently applying their efforts across a fractured industry, with poor return on investment. Increased standardization could improve their business and lead to improved software tools. It is suggested that the concepts contained herein could provide a foundation for such a standardized design platform.

CONCLUSIONS

- Advanced fabrication techniques can be applied to VLP FB-DIMMs, to produce prototype 16GByte modules in 2007. These modules can have improved thermal and electrical performance as well as higher density.
- To close the Integration Gap between IC chips and packaged systems, four parallel innovations are desired:

a) substrates having improved wiring density;
b) improved substrate materials for cooling denser modules;
c) new methods for functionally testing integrated and partially assembled modules in an automated fashion at full speed; and,
d) automated rework methods to improve rates of production, increase the level of integration, and dramatically lower reject costs by enabling 100% assembly yield.

- Scalability at the system level can be enhanced by using standard DIMM formats for memory and non-memory applications.
- The cost of HDI substrates could be dramatically reduced by adapting proven FPD technology to copper panel fabrication.

REFERENCES

[1] Active Matrix Liquid Crystal Display.

[2] Bill Gervasi, "Very Low Profile Fully Buffered DIMM', SimpleTech/Jedex presentation.

[3] Igor S. Zavarine, Xuan Lin, Chonglun Fan, Yun Zhang, Bill Wu, and Zhenqiu Liu, "Electroplated Copper and Copper Alloys for Wafer Level Packaging," Proceedings of the 2nd International Wafer Level Packaging Conference, San Jose, California, 2005, pp. 97-102.

[4] LPKF Laser & Electronics, Wilsonville, Oregon, USA.

[5] Peter C. Salmon, "Apparatus and Method for Testing Electronic Systems", Published U.S. Patent application, 20040176924.

SINGLE WAFER BUMPING

Yixiang Xie, Qiang Fu, and Solomon Basame
Surfect Technologies, Inc.
Albuquerque, NM, USA
yxie@surfect.com

ABSTRACT

The rapid growth in demand for bumped wafers is forecast to grow exponentially over the next ten years at a growth rate of 28.1% a year according to a research firm[1]. The key to win this growing market is to lower the cost of bumping wafers through the use of lower cost plating equipment and methods. The combination of lower cost equipment with novel process development will enable new applications and emerging markets. Low cost single wafer plating units have been designed and tested for 200 and 300 mm wafers, board plating manufacturing, and R&D facilities. The units not only apply to electrolytic plating but also are used in electroless plating processes. Single metal, multi-metal plating, and nano-particle plating processes have been developed in the units with smallest line feature of 20 μm and open via diameter of 50 μm. In this paper, single layer, multi-layer metals, metal alloy, and metal capsulation plating process, resulted images, and uniformity will be discussed.

Key words: Bumping, Plating, Interconnect, innovation

BACKGROUND

Since IBM initiated C4 bumping technology as a replacement for wire bonding, only a few large companies which occupy 10% microelectronic market use flip chip packaging technologies. The remaining 90% of the market will convert to bumping technology and its derivatives as it evolutes with cost-effective equipment, processes development in next 20 years in different ways[2]. The electroplating process has significantly more cost reduction potential than those to be replaced. Single wafer plating units have a significant difference from main frame conventional equipment. The latter has multiple units to accommodate pre-treatment, plating, rinsing, drying, and often have open unit, or expose wafer or parts to air between units. It is not difficult to see the direct disadvantages of a main frame equipment operation, which have multiple wafer or parts handling, metal surface oxidation due to exposure, and more maintenance efforts. And indirectly, large footprint for clean room occupation, longer operation time needed in wafer handling, limit to single metal or alloy plating, and lower through put. All the above add to high equipment and operation cost. A single wafer plating unit combines all plating associated process operations in one closed unit to eliminate all above disadvantages of main frame, and plug in in-line metrology and novel agitation mechanism[3] that is a key to successful plating. A single wafer unit is not necessarily a single unit equipment; it is a single or multi-unit plating computer.

PROCESS

Process development has been focused on tight pitch bumping with down to 50 μm diameter via and resolution line width down to 20 μm to cover wider applications. Figure 1 demonstrates a dry film pattern on a base metal deposited Si wafer.

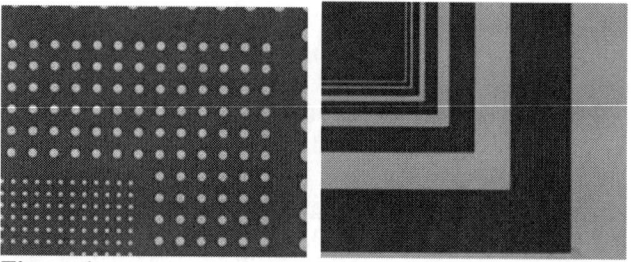

Figure 1
Bumps down to 50 μm diameter and traces to 20 μm width

The plating equipment used to develop these plating processes is Surfect Technologies, Inc.'s 200 mm and 300 mm plating units. Plating chemicals are from major vendors for the semiconductor industry. Surfect has developed capability for: Sn, Cu, Ni, SnAg, PbSn, and the extension of Au/Sn layer process[4], Cu/Sn layers, Ni/Cu/Sn stack, Ni nano-particle and Ag nano-particle in Sn plating bumps. This development follows Surfect's micro particle plating process[5].

Copper Plating

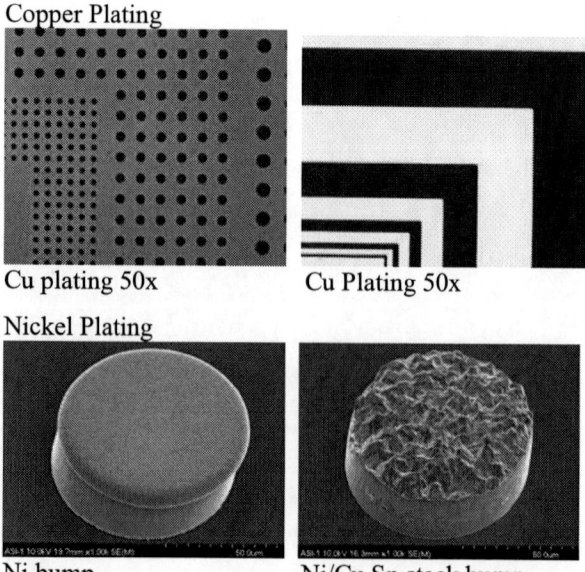

Cu plating 50x   Cu Plating 50x

Nickel Plating

Ni bump   Ni/Cu Sn stack bump

Tin Lead Plating – Layer and Alloy

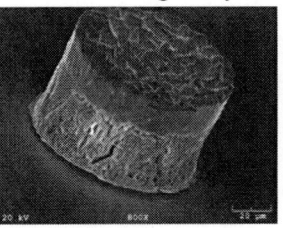
60/40 Pb/Sn layer

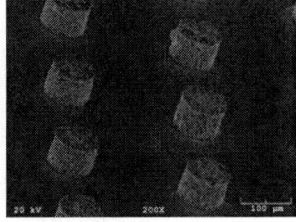
95/5 PbSn alloy

Silver Nano Particle Plating

Nano- Ag in Sn bump

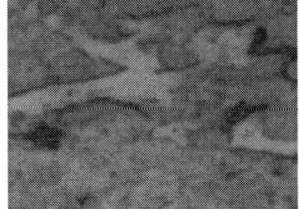
Bump cross-section Ag black

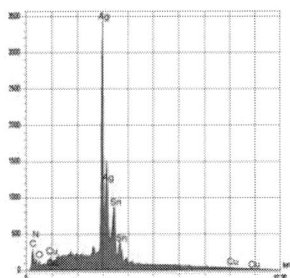
EDS showing Ag

Nickel Nano Particle Plating

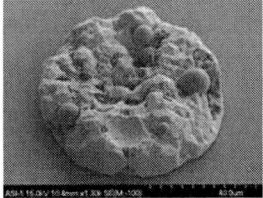
Nano- Ni in Sn bump

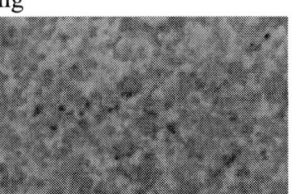
Bump cross-section Ni in dark

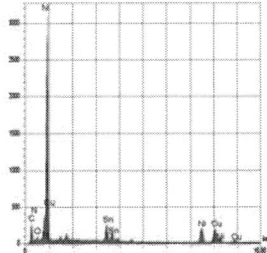
EDS showing Ni

Resolution data
50, 80, 120 µm diameters Cu bump WID

Distributions - Uniformity ±6%

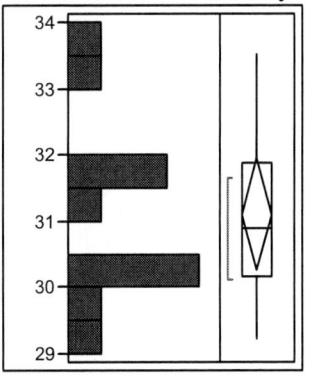

Quartiles

100.0%	maximum	33.520
99.5%		33.520
97.5%		33.520
90.0%		33.379
75.0%	quartile	31.883
50.0%	median	30.905
25.0%	quartile	30.160
10.0%		29.392
2.5%		29.230
0.5%		29.230
0.0%	minimum	29.230

Moments

Mean	31.105
Std Dev	1.3188942
Std Err Mean	0.380732
upper 95% Mean	31.942985
lower 95% Mean	30.267015
N	12

20, 50, 100 µm Cu trenches line WID
Distributions- Uniformity ±11%
Quartiles

100.0%	maximum	27.250
99.5%		27.250
97.5%		27.250
90.0%		27.186
75.0%	quartile	26.385
50.0%	median	25.970
25.0%	quartile	24.803
10.0%		23.244
2.5%		23.130
0.5%		23.130
0.0%	minimum	23.130

Moments

Mean	25.652
Std Dev	1.2132399
Std Err Mean	0.3836602
upper 95% Mean	26.5199
lower 95% Mean	24.7841
N	10

RESULTS AND DISCUSSION

Using Surfect Technologies, Inc. 200 and 300 mm wafer single unit plating computers, the above pictures and data demonstrate the capability of plating Sn, Cu, Ni, SnAg, PbSn,

Cu/Sn layers, Ni/Cu/Sn stack, Ni nano-particle and Ag nano-particle in Sn plating bumps and other patterns. Surfect Technologies, Inc. developed processes provide higher plating rates and improved morphology, uniformity of thickness, and bump height distribution when compared to industry standards. The layering deposition, particle deposition provides a new path of cost-effective, and quality in bumping applications. The multi-feature size plating uniformity development gives designers a wider choice of dimensions in design rules.

REFERENCES
[1] Mark LaPedus, EE Times: Semi News, July 12, 2006
[2] Steve Anderson, Surfect Technologies, Inc., Key Note, Semicon West, San Francisco, July 11, 2006
[3] Surfect Patent pending, 2006
[4] G. Minogue, Surfect Technologies Inc., A Novel Au/Sn Seal Ring Fabrication Technology, MEPTEC, May 11, 2004
[5] Paul Vianco, New Mexico Small Business Assistance Program: Surfect Technologies, SAND Report 2003-2991, September 2003

SiP – IDENTIFYING ISSUES FOR STACKED (3D) MULTICHIP PACKAGING ADOPTION

Larry Gilg
Die Products Consortium
Austin, TX, USA
gilg@dieproduct.com

MULTICHIP PACKAGING TRENDS AND DRIVERS

The term "packaging" as applied to microelectronics, delineates the materials, processes and methods for providing a satisfactory electrical, mechanical and thermal operating environment for an integrated circuit (IC). The resulting "packaged" device is termed a module. The term multi-chip module (MCM) was introduced in the mid-1990s as a technology that incorporated several unpackaged ICs on an interconnecting substrate. There was a flurry of activity to develop high-performance interconnecting substrates that could wring the utmost in performance from the bare, bumped or TAB'ed IC device. Advanced packaging came to be viewed as a value added process that offered the utmost performance with the smallest form factor (size and weight) for critical applications where cost was not the prime consideration. The resulting MCM developments gained a reputation as overpriced solutions good for solving niche problems only.

However, as makers of portable and handheld systems analyzed the opportunities for building smaller, lighter and higher performing devices with greater functionality, the benefits of the multi chip packaging approach using die products became clear. The challenge then became to develop designs and assembly technologies that result in a lower cost than traditional surface mount devices, allowing faster time to market than the system on chip solutions. The enabling factor in these advanced packaging technologies is reliance on die products.

DIE PRODUCTS

"Die products" is a term that describes IC devices that are true die-size packages, that is, when a die shrink occurs, the device size as seen by the customer will shrink as well. This means that the new wafer level packaging technologies in use today are included in the definition of die products, as well as the more traditional wire-bonded and flip-chip devices. Die products are available to the market through similar channels as packaged ICs.

Technology Imperatives

The continuing drive for enhanced digital content in consumer based mobile electronics products has created significant market demand for devices with smaller size while sustaining the historic trends of higher levels of integration, performance, manufacturing, and reliability at lower cost. These sometimes competing requirements are best fitted to solutions incorporating die products for stacked die modules.

SMALLER SIZE

There is no question that unpackaged or wafer level packaged die impose the smallest footprint on the interconnecting substrate. The chip to board area ratio is the metric used to measure the packaging effectiveness of the final product. For single chip packages, a chip to board area ratio of 0.5 is considered a small footprint. Stacked die modules can attain a chip to board area of close to 2 - a four-fold decrease in chip to board footprint over the single die solution.

Space savings in the z-direction is also highly leveraged by stacking bare die. Thinning the die maintains the overall package height to acceptable levels, usually below 1.4mm above the mounting substrate.

PERFORMANCE

Die products exhibit shorter interconnections to lower the propagation speed of signals through a high speed system. The line resistance of the interconnecting lines is lower because the packaging density increases, decreasing the length of the interconnecting wiring layers. Parasitic electrical elements are all reduced because of the reduced distances of the electrical connections to the wiring plane. Passive electrical elements can be included immediately adjacent to the chip I/O to promote dramatic improvements in the electrical performance of the device.

MANUFACTURING

Die products have been adopted in high-volume, hand-held and wireless applications, resulting in a low-cost infrastructure emerging through the utilization of proven packaging technologies. This enables new configurations, including not only stacked die modules, but also package stacking and folded flexible circuit modules which are being developed for a broad range of new applications.

COST

The total system solution cost using die products is favorable due to the higher levels of integration possible. The savings is attributable to reduced substrate size, simplified assembly manufacturing processes and most important of all, faster time to market with a differentiated product. Solutions featuring die products exhibit increased features per package area and for 3D modules, number of levels.

ISSUES FOR STACKED DIE MODULES

The success of stacked die modules in the marketplace is due to manufacturers taking advantage of ubiquitous technologies currently in use in the packaging industry that have matured for single chip packaged devices. However, many of these technologies require a unique "twist" to address specific issues for stacked die modules. In some cases, stacked die technology is pushing equipment and related process capabilities to their limits.

Wafer Thinning

A critical issue for stacked die modules is the total height of the stack, the requirement for which is usually less than 1.4 mm. This includes the height from the level of the attachment on the motherboard to the height of the mold cap.

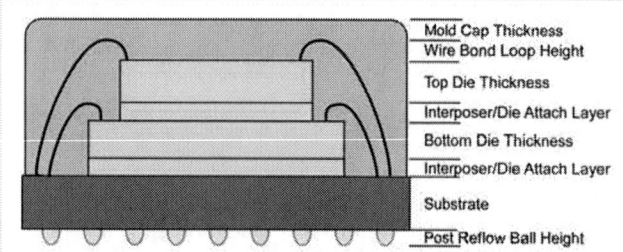

Figure 1: Typical stacked die arrangement. The total height of the stack, from the bottom of the mounting balls to the top of the mold cap is targeted to be less than 1.4 mm for most applications.

Thinning wafers is accomplished by backgrinding the entire wafer, to thicknesses less than 100µm (for 200mm) wafers being done in production today. Handling wafers this thin demands extreme care subsequent to the thinning process. At a 100-µm thickness, a 200-mm wafer can no longer support itself (150-µm for 300-mm wafers). Thin wafers will usually require stress relief to remove micro cracks from the backgrinding process. Otherwise the wafers will be extremely fragile since breaking strength is correlated with backside roughness.

Die Preparation

Die preparation is the process by which the wafer is singulated into individual dice in preparation for assembly. Die preparation consists of two major steps, namely, wafer mounting and wafer saw.

Die preparation is common to all packaging methods using silicon die. Optical visual inspection is usually conducted to inspect for defects before the wafers are released for production. Wafers are then mounted on a backing tape that adheres to the back of the wafer. The backing/mounting tape provides support for handling during wafer saw and the die attach process. The wafer saw process cuts the individual die from the wafer leaving the die on the backing tape.

Thinned wafers for stacked die modules require special backing tapes.

UV-GRINDING-TAPE

The tape applied on the active wafer side for device-protection during thinning loses 90% of its adhesive-strength through UV-irradiation, making it easy to remove.

UV-DICING-TAPE

UV tape is also used on the back side of the wafer after the backgrind process. The chips are then more readily picked by the die-bonder as the dicing-tape loses its adhesive-strength through UV irradiation

DETAPING AFTER MOUNTING

The grinding tape is detaped, or removed from the front side of the wafer after the wafer back side has been mounted to the dicing-frame.

SPECIAL DICING-TAPE PROVIDING GLUE FOR DIE-BONDING

The adhesive layer of the dicing tape goes with the die when picked from this type of tape. This same adhesive is then used as the attach mechanism in the die-bond-process for a traditionally active side up for wire bonding.

SPECIAL GRINDING-TAPE PROVIDING GLUE FOR FLIP-CHIP DIE-BONDING

Here the adhesive layer of the grinding tape remains at the wafer after removing the grinding tape. This adhesive layer on the active side of the chip is now used in the flip-chip bonding process.

DIE-ATTACH-FILM

The adhesive film is put on a wafer, diced with the chips and then used as adhesive layer during the stacking of these chips.

Wafer Singulation

The wafer is singulated into individual dice in preparation for assembly after the thinning process. The issues with singulation of thin wafers have prompted several different processes to be developed:

- Dicing through in one go
- Various step-cut systems
- Multiple blade dicers
- Laser-Dicing

Die-to-die Bonding

For the stacked die process, high die placement accuracy is essential to mount the upper chip accurately onto the lower one. Inaccurate die placement can lead to electrical failure (wire shorts) and impacts the epoxy bleed out. Specific alignment tolerances for the die placement must include theta alignment. See figure 2. Each die in the stack must be aligned with the substrate, and tolerance stack-up must not allow bond pads on a top layer die to mis-align with lower levels. Using backside-coated wafers results in evenly spread adhesive on the backside of the top die and no epoxy/die offset can occur in the die stack.

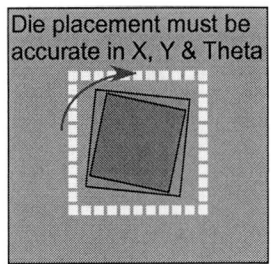

Figure 2: Die placement must be precise to maintain visibility and correct order of bonding pads on all die in the stack.

Care must be taken so that the bondline is the correct thickness and completely covers the surface of the supporting die.. Die that are cantilevered over a spacer are subjected to high stress along the bondline, and an incomplete bondline can cause cracking to occur. On the other hand, a thick bondline can cause wirebond loop to be exposed at the top of the stack.

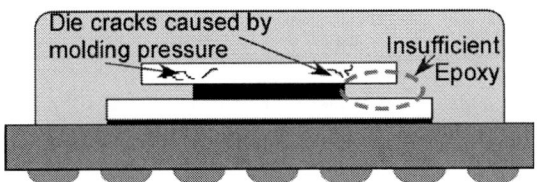

Figure 3: Incomplete epoxy dispense can result in die cracks in the finished module.

Low Loop Height Wire Bonding

Stacked die modules are characterized by several novel wirebonding practices. In order to maintain the target finished height of 1.2mm or less, low loop height control is a requirement. 125μ loop heigh with 25μ wire is achievable today. If lower profiles are needed, reverse bonding can offer loop heights of less than 75μ.

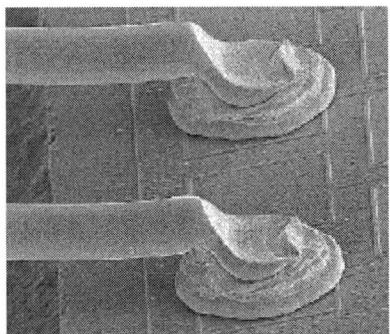

Figure 4: Reverse bonding is formed by putting a ball on the IC pad, much like a stud bump, then performing a stitch bond on the ball, rather than a normal ball bond attach to the IC pad.
Stack Die CSP Interconnect Challenges, Flynn Carson, Glenn Narvaez, HC Choi, and DW Son –ChipPAC, Inc., IEEE/CPMT Seminar

A wire bonder for stacked die modules must have sufficient optical depth of field to focus the image on a field that can have as much as 750μ height differential from the bottom die to the top of the stack. If the bonder does not have the resolution required, multiple passes may be needed for bonding individual die.

Thin and Dense Substrate

Much of the advantage of the stacked die module is due to the simplified I/O structure on the outside of the package. In other words, much of the wiring complexity is contained within the module, and there is no need to bring out all I/O to the motherboard. However, this requires a very dense interconnect for the stack, and to maintain the height limits, it must also be very thin. To further increase integration density and improve performance, substrate materials and fabrication technologies need to be improved. While onchip interconnect density continues to improve at 30% every two years, substrate interconnect density is improving at a much slower rate - enlarging the gap. Low cost processing and materials to enable sub 25 micron line and space interconnect are required to close this gap.

Most stacked die applications are targeted at lead free electronics and as a result must meet more demanding lead free reflow requirements. This includes the ability to withstand multiple 260 C reflow cycles. Many HDI laminate substrate structures which use lower Tg (glass transition temperature) materials exhibit excess warpage and degrade at these reflow temperatures. New materials are required to extend the maximum temperature limits of low cost HDI substrates.

High Yield Assembly Process

The assembly process for stacked die modules is similar to single chip modules, except that multiple passes through a particular step may be needed.

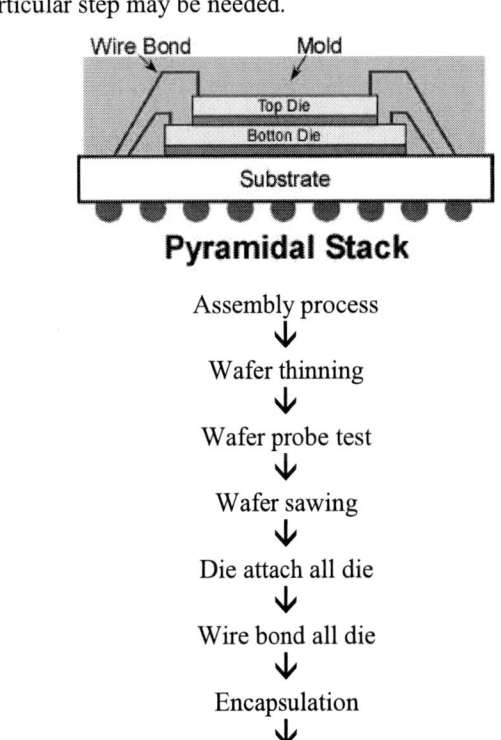

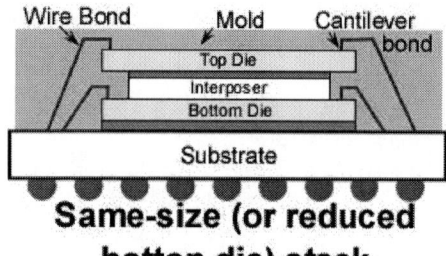

Same-size (or reduced botton die) stack

Assembly process

↓

Wafer thinning

↓

Probe

↓

Dice

↓

Bottom die attach

↓

Bottom die wire bond

↓

Top die attach

↓

Top die wire bond

↓

Encapsulation

↓

Module test

Testing of Stacked Die Modules

A stacked die module is physically identical to a single chip module. It may be socketed and tested much like a single chip package. However, since not all I/O on each chip is available on the outside of the package, there are certain important distinctions to single chip package final test. Initially, for adequate module yield, each IC to be used in a stack should be fully tested at the wafer level. This includes performance testing and reliability stress tests.

In the ideal case, if test methods that rely on KGD methods at wafer test are employed for all of the stack components, the final test may simply verify the stack assembly process. Test methods applied at the module level that enable defect analysis, diagnostics, fault localization and data distribution and analysis will lead to effective yield learning and reduced reliance on physical failure analysis on the failing modules themselves.

Failure Modes of Stacked Die Modules

The following failure modes need to be considered when designing stacked die modules.

- Thermo-mechanical Stress
- Package/Matrix Warpage
- Hygorswelling stress
- Vibration and Modal Analysis
- CFD/Heat transfer

- Popcorn & delamination
- Moisture diffusion & vapor pressure
- Fracture mechanic
- Solder joint fatigue
- Drop impact
- Failure modes (solder joint creep, fatigue damage, delamination, die and microvia cracking)

These same failure modes are generally applicable to all IC package formats, but stacked die modules require special consideration in several cases. Thermo-mechanical related failures account for about 65% of total failures in stacked packages. The root cause generally originates in product/process design phase. This means that the qualification focus must shift from physical prototyping to modeling and simulation. Similar to the issue of stacked module co-design, qualification of these packages requires an integrated approach to the electronic, thermal mechanical and material properties of the product using computer simulations and analyses.

SUMMARY

The hype surrounding the multichip module (MCM) packaging technology of the 1990s was based on the assumption that high reliability aerospace and computing applications would be the predominant users. Consequently, the expensive die products approaches developed in that era never lived up to the billing. However, a quick look inside today's portable products aimed at the consumer markets reveals that multichip packaging in the form of stacked die modules has survived quite nicely, and is in fact, becoming the packaging technology of choice for those applications.

The benefits of stacked die modules include:

Different die technologies and/or functions can be assembled in the same module.

Time to market can be enhanced by using integrating standard components into a new system in a package, rather than undertake a high risk system on a chip development.

Other technologies, such as MEMS, optical or vision components may be included in the module.

Differing assembly technologies can be incorporated at the subassembly level, simplifying the final assembly, bill of materials and parts inventory.

Savings on interconnecting substrate area, die assembly materials and simplification of testing can result in significant cost savings.

CHALLENGES IN FLIP CHIP DIE SORTING, HANDLING AND INSPECTION

Gerald Steinwasser
Mühlbauer, Inc.
Newport News, VA, USA
gerald.steinwasser@muhlbauer.com

ABSTRACT

The growing number of bare die applications naturally require faster and more efficient die sorting systems; however, at the same time, the quality requirements have become increasingly stringent, with more and more end users calling for zero defects! The capabilities of such systems have to combine speed (in excess of 10k UPH) and extreme vision inspection capabilities. These capabilities are challenged by the constantly growing variety of wafer materials and types, subsequent lighting influences, inspection criteria and various packaging media.

Keywords: Flip Chip, Known Good Die, Die Sorting, Wafer Map, Reel-to-reel Sorting, Carrier Tape, Bare Die Tape

BACKGROUND

Die Sorting operations are performed in many different ways. Usually, a wafer, which has been inspected and tested, bumped and sawn, is provided with the test results in form of a wafer map to the die sorting machine. Because at this point the status of the dice is known, the term "Known Good Die sorting" is also used. Instead of wafer maps, in a few remaining applications, the dice on the wafer are distinguished by ink dot markings. Only the Known Good Dice are then picked from the wafer based on the wafer map (or ink dot markings) and then placed into packaging media. During that pick-and-place process, the dice can be flipped and are also visually inspected. A variety of standardized materials is available as packaging media (carrier tape, waffle pack and Gel-Pak™, surftape and wafers). For most high volume applications, the preferred packaging media is still carrier tape, which is commonly accepted and standardized throughout the industry. The focus of this paper is on these high volume reel-to-reel applications.

Typical pick process from a bumped wafer

WAFER RECOGNITION

In the first step, a wafer is inserted into the machine and stretched on the wafer table. The stretching helps to open and widen the saw streets between the individual rows of dice. With the introduction of the 300mm (12in) wafers the capability and flexibility of the systems also had to be increased in that aspect. With the loading of the wafer, a barcode on the wafer frame is usually read. This barcode allows the system to recognize the wafer and is used for the subsequent operation(s) (e.g. lot handling- and SPC reports). It also allows the system to pull the corresponding wafer map file from a computer or network. The system reads the wafer map and looks for specific reference dice. These reference dice are also specifically marked on the wafer and by the combination of vision and wafer map, the wafer is aligned and brought into position to start the sorting process. Besides the wafer map, the decision of the machine to pick a die from the wafer is also influenced by the optical inspection of the die. The majority of the wafers at this stage are bumped and with the bumps facing upwards, the bump integrity is an important selection criteria.

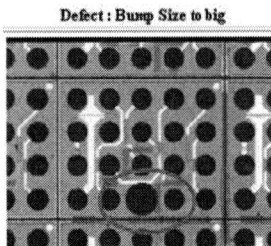
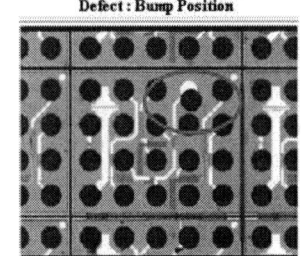

Other defects to look for at this point are broken dice, contamination, cracked dice, material voids and/or broken edges.

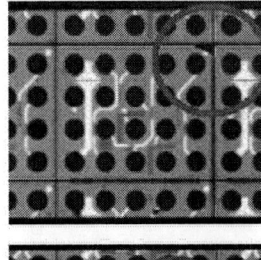
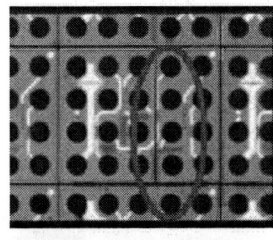
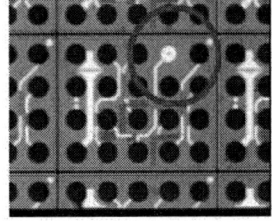

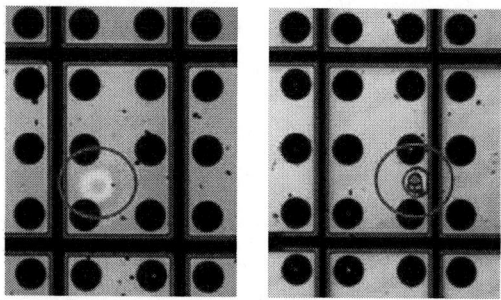

Examples of detectable defects

The selection of appropriate lighting in combination with the right camera resolution plays an important role regarding detectable defect details.

	Standard CCD Camera (640x480pix)			HighRes CCD Cam (1024x1024pix)		
Die-Size	Field Of View	Resolution	Detectable defect size	Field Of View	Resolution	Detectable defect size
[mm]	[mm]	[µm / pix]	[µm]	[mm]	[µm / pix]	[µm]
0,5	4,0 x 3,0	6,1	30 x 30	2,3 x 2,3	2,2	11 x 11
1,0	4,0 x 3,0	6,1	30 x 30	2,3 x 2,3	2,2	11 x 11
1,5	4,0 x 3,0	6,1	30 x 30	3,0 x 3,0	2,9	15 x 15
2,0	5,3 x 4,0	8,1	40 x 40	4,0 x 4,0	3,9	20 x 20
2,5	6,7 x 5,0	10,1	51 x 51	5,0 x 5,0	4,9	25 x 25
3,0	8,0 x 6,0	12,1	61 x 61	6,0 x 6,0	5,9	30 x 30
3,5	9,4 x 7,0	14,2	71 x 71	7,0 x 7,0	6,8	34 x 34
4,0	10,7 x 8,0	16,2	81 x 81	8,0 x 8,0	7,8	39 x 39
4,5	12,0 x 9,0	18,2	91 x 91	9,0 x 9,0	8,8	44 x 44
5,0	13,4 x 10,0	20,2	101 x 101	10,0 x 10,0	9,8	49 x 49
5,5	14,3 x 10,7	21,7	108 x 108	10,7 x 10,7	10,5	53 x 53
6,0	14,3 x 10,7	21,7	108 x 108	10,7 x 10,7	10,5	53 x 53
6,5	14,3 x 10,7	21,7	108 x 108	10,7 x 10,7	10,5	53 x 53
7,0	14,3 x 10,7	21,7	108 x 108	10,7 x 10,7	10,5	53 x 53
10,0	16,0 x 12,0	25,0	125 x 125	12,0 x 12,0	11,7	59 x 59
15,0	22,7 x 17,0	35,4	177 x 177	17,0 x 17,0	16,6	83 x 83
20,0	29,3 x 22,0	45,8	229 x 229	22,0 x 22,0	21,5	107 x 107
25,0	36,0 x 27,0	56,3	281 x 281	27,0 x 27,0	26,4	132 x 132

Table with camera comparison

DIE EJECTION, DIE PICK AND FLIP ACTION

The dice are ejected from the wafer with the help of a needle or needle array (depending on die size). The shape of the ejector needle(s) is important to avoid any penetration of the wafer foil, damage to the die (needle marks) or tape residue on the dice. After being ejected (released from the wafer tape), the die is usually turned 180 degrees (flip chip) and is then taken over by a pick & place tool. After the die is flipped, the exact position on the flip tool is recorded by a camera and the offset for the place position is calculated. During the handover to the pick & place tool, a die presence check is performed. This is usually done with an integrated sensor, which checks for the die on the pick & place tool. New materials, e.g. glass or other translucent materials, made it necessary to work with different type of sensing techniques (e.g. vacuum, infrared etc.).

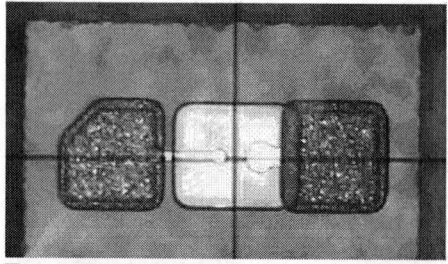

Example of a translucent die substrate

The vast majority of (flip chip) bare die applications to date have been in a size range of approx. 0.8x0.8mm up to 7x7mm square. With the expansion in the size range, new challenges also arrived. On the low side, the handling of extremely small dice provides mainly new challenges regarding the packaging media. On the high side, the limitations in field of view and camera resolution are setting limits to the detectable defect size.

HighRes CCD Camera (2048x2048pix)		
Field Of View	Resolution	Detectable defect size
[mm]	[µm / pix]	[µm]
4,0 x 4,0	2,0	10 x 10
4,0 x 4,0	2,0	10 x 10
4,0 x 4,0	2,0	10 x 10
4,0 x 4,0	2,0	10 x 10
8,0 x 8,0	4,0	20 x 20
8,0 x 8,0	4,0	20 x 20
8,0 x 8,0	4,0	20 x 20
8,0 x 8,0	4,0	20 x 20
8,0 x 8,0	4,0	20 x 20
8,0 x 8,0	4,0	20 x 20
8,0 x 8,0	4,0	20 x 20
12,0 x 12,0	5,9	29 x 29
12,0 x 12,0	5,9	29 x 29
12,0 x 12,0	5,9	29 x 29
27,0 x 27,0	13,2	66 x 66
27,0 x 27,0	13,2	66 x 66
27,0 x 27,0	13,2	66 x 66
27,0 x 27,0	13,2	66 x 66

Table with capability of high resolution camera

Thinned Dice ($\approx$ 140 - 40 µm)
The main challenge in handling these dice comes from their brittleness (e.g. GaAs). To avoid any damage to the dice, a perfect synchronization between die ejector, pick- and pick & place tool is very important. Minimum forces to hold the dice during the handover are also critical.

Ultra-thin Dice ($\leq$ 30 µm)
With the development of ultra-thin dice, which are flexible and make the use of die ejector needles useless, a new type of release mechanism had to be developed. A special wafer release method is used in these applications. The release of the die is achieved by a non-mechanical die ejection system in which the wafer release is activated by a hot-air blast. The hot air blast causes the tape to lose its adhesive function and the die can then be picked up with a special vacuum pick-up tool. Other methods are using ultrasonic wave to release the dice.

Depending on the bumped wafer surface, different pick-up tools must be used. Most common are vacuum pick-up tips, which pick the dice between the bumps on the die surface. Another type of pick-up tips is made out of soft rubber. These soft tips 'lay' themselves between the bumps without damaging them. In case the bump array doesn't leave enough room to pick the dice or the danger of damage is too great, a so-called 'collet' tool is used, which grips the dice on the out side of the die. Special pick-up tips that take the flexibility of the substrate material into account have also been already developed and are available on the market for ultra-thin die applications.

DIE PLACEMENT AND INSPECTION

Die Placement

For an accurate placement of the dice it is essential that the camera over the place position be able to see the cavity in the carrier tape where the die is to be placed. Because bare dice (flip chip) are usually very shallow components, the corners inside the cavity need to be very well defined – for good visual recognition and for a good fit of the die into the cavity. There are still only a limited number of specifically made bare die tapes in the market. While traditional carrier tapes are manufactured with a negative thermoforming process, the bare die tapes are manufactured with a positive forming process.

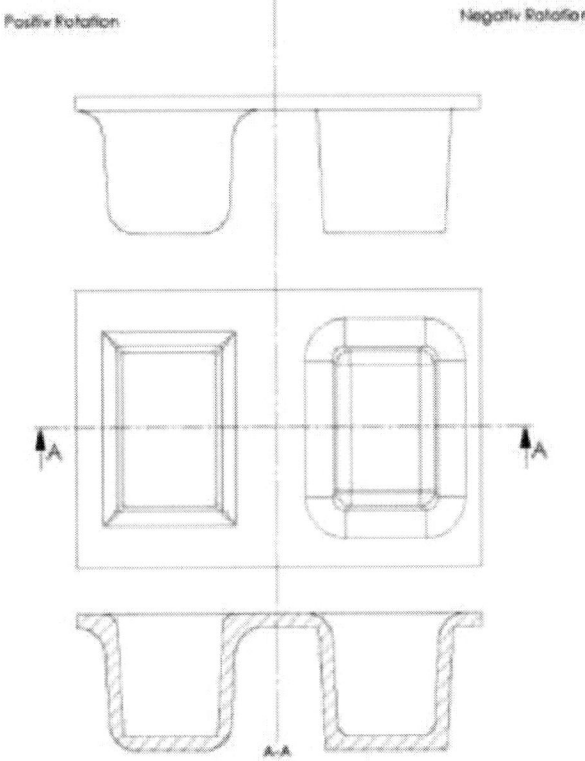

Comparison of positive and negative formed cavities

A good cavity recognition process is needed to calculate the placement offset when the die is handed over from the pick-up tool.

Examples of a good and an incompletely formed cavity

A non-matching cavity can create problems for the correct positioning and later removal of the die in the assembly machine. Because of the low mass of smaller dice, a cavity with a center hole is preferred. This way, vacuum can be applied to the bottom of the cavity, and thus create a

positive force to hold the die in the cavity. Static protection is also very important. With die sizes becoming smaller and smaller, the limits for the use of carrier tape are soon to be reached (e.g. the limit of punching a center hole into a cavity is approx. at 0.5mm Ø). At this point, other media like surf tape, which has an adhesive surface, comes into play.

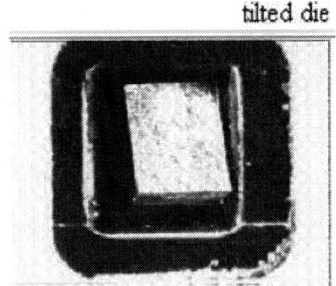

Example of a tilted die caused by an improper cavity

Backside Inspection

One of the most important steps during the die sorting process is the inspection of the backside once the die is placed into the cavity. Here, several quality aspects must be verified within a fraction of a second. Usually the camera, which also checks for the correct positioning of the die in the cavity, is used for this purpose.

One main aspect of the backside inspection is to check for any possible chipouts, which might have been caused by the sawing process (die separation). The limits for chipouts are usually defined by a length from the outside borders of the die towards the center of the die.

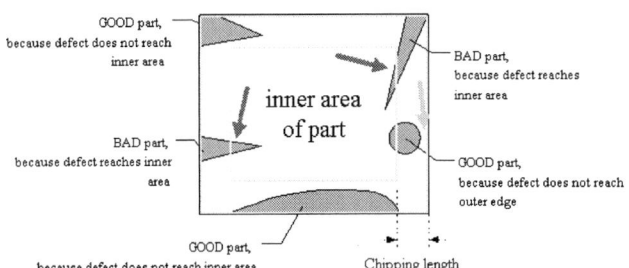

Schematic to describe chipout criteria

Another aspect of the backside inspection is the orientation of the die, also known as pin one location and the integrity of the laser mark. To recognize all these features, the inspection system must be able to 'align' the die within the field of view almost instantly, which can be difficult depending on the tolerances of the dice within the cavity.

Examples of dice in cavity with laser mark

Also, adequate lighting is absolutely essential. Color variations in different areas of the wafer and the use of backside coatings make the development of newer and more sophisticated lighting system an on-going and continuous process.

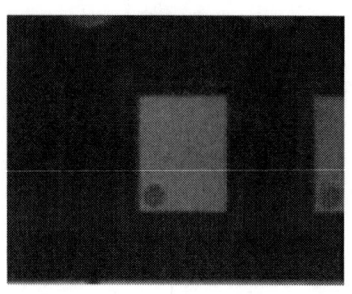 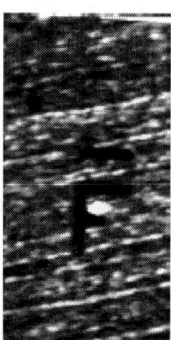

Examples of dice with pin one and laser mark

Recognized rejects are removed from the cavity and replaced by the next die to ensure 100% good product in the carrier tape. Another challenge to consider is the tombstoning effect. 'Tombstones' are dice that are taller than they are long and wide and, therefore, have the tendency to 'fall over' in the cavity. The index movement can potentially cause displacement out of the cavity from the position where the die is placed and the position where the cover tape is applied.

There are two methods currently in use for the sealing of carrier tape. The hot-seal process 'welds' the cover tape onto the carrier tape with a heated sealing shoe. In the cold sealing process, a pressure sensitive tape is applied in a similar manner but without heating.

To ensure a good output as close as possible to 100%, another post seal camera can check once again after the cover tape is applied. Depending of the clarity of the cover tape, a repetition of the inspections performed by the placement camera is in certain cases possible.

Examples of post seal inspection (through cover tape):
Top image – turned die
Bottom images – filled and empty pockets

Again, the correct illumination is essential to achieve the correct contrast for the vision camera to see.

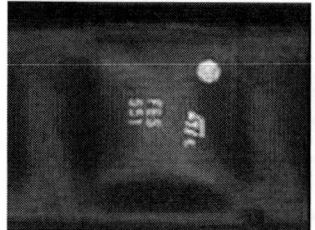 

Images of post seal view with different lighting solutions
Left image – external coaxial illumination
Right image – dark field illumination

NEW DEVELOPMENTS

The use of barcodes is seen as a possibility to satisfy the increased demand for traceability. In turn, this creates another new challenge to process these images and interpret them correctly.

Images with first (unsuccessful) tests of a 2D barcode on an approx. 1.5x2mm die

With the use of more 300mm (12in) wafers, die sizes are also growing. However, the defect details that need to be detected on larger dice are still down into the 10-20 micron level. This brings us to the current possible limits of available camera technology (see previous: Table with capability of high resolution camera). A new generation of die sorting equipment for larger die sizes and highest resolution cameras needs to be developed.

To achieve better results when inspecting for various defects, multiple images with different, defect-specific illuminations have also shown promising results. When only one image is taken, a compromise for the illumination used usually needs to be made to find as many defects as possible. However, the nature of the defects is not always detectable with just one type of lighting. By making multiple images in short succession, the illumination can be optimized for each type of defect, making the inspection much more accurate. Increased computer speed and more advanced software algorithms allow for processing multiple images in parallel at the highest speeds and thus still achieving high throughput numbers on the die sorting systems.

Another promising approach under development to inspect larger dice is to split the inspection area into different sections, wherein the individual sections are processed in parallel.

OVERVIEW OF MEMS WAFER LEVEL PROCESSES AND PATENTS

Ken Gilleo, Ph.D.
ET-Trends LLC
Warwick, RI, USA
Ken@ET-Trends.com

ABSTRACT

MEMS, the technology that adds *mechanical elements* to electronic chips is one of the most important fields of the 21st century. The MEMS industry continues to grow at a steady pace that will begin to accelerate as more advanced devices move from lab to fab. But many unique challenges remain for both fabrication and packaging. Micro-Electro-Mechanical Systems (MEMS) manufacturing methods tap into the semiconductor industry so that the "MEMS factories" already exists. Electronic circuitry can be fabricated on the same wafer as the mechanical elements to build "smart" MEMS chips. But MEMS wafers, unlike "pure" electronic chips, are typically fragile and extremely sensitive to micro-contamination as might be expected. One solution to mechanical shock and micro-particle contamination sensitivity is to exploit wafer-level processing. Today, MEMS wafers are capped and hermetically sealed at wafer-level, but most commercial processes are far from optimization. Commercial methods typically use multi-step capping processes followed by double-singulation to separately release the cap and MEMS chip. But the resulting capped MEMS chip is only a "pre-package" that must still go through a "final packaging" step. The capped chip can often be overmolded once the mechanical elements are protected from direct contact with molding compound. But some capped MEMS devices must be packaged in cavity enclosures to avoid the stress of encapsulant shrinkage. New wafer-level processes, especially those that produce a complete package, have been proposed and may achieve commercial success. This paper will provide an overview of existing methods and new processes, including a survey of the most recent patents and pending applications.

Key Words: caps, capping, MEMS, micro-mechanical, packaging, wafer-level

BACKGROUND

MEMS represents the synergistic and strategic fusion of mechanical, electrical and electronic subcomponents into a single, self-supporting integrated system. Optics can also be added to create the subset MOEMS (micro-opto-electro-mechanical systems), also called optical-MEMS. The most advanced MEMS designs form electronic logic elements on the same chip with the mechanical features. But other strategies produce components on wafers that must be fusion-bonded to produce the finished product. The resulting products have extraordinary functionality. The logic circuitry can be viewed as the *brain*, and the mechanical components add the *arms, hands, fingers, eyes, nose, ears, mouth* and other life-like input/output functions. MEMS also brings control -

electrical, mechanical, optical and electromagnetic. MEMS micro-actuators thus add the *muscle* to move mechanical parts and manipulate materials. *Merging of motion, sensing, control and computation* on a single chip or wafer-assembled device is a major leap in technology.

We can build minute, nearly invisible machines that perform the same mechanical, optical, electrical and electronic functions found in the complex and often-massive equipment of the macro-world. We are essentially shrinking the every-day equipment of our world onto tiny chips of the macro/nano-worlds. MEMS comes close to, "shrink the world onto a chip", and it is the ultimate System-on-Chip (SoC). MEMS has exceptional potential because the technology can integrate just about everything – motors, mechanical mechanisms, optical systems, sound, biochemical analysis, materials manufacturing, wireless, and computation. The industry is already well established. Electronic giants including Intel, IBM, Texas Instruments, Bosch, and Motorola, are some who have been in this field for a considerable time. National governments have been sponsoring MEMS R&D for a long time, and Sandia National Laboratories in the US, continues to be one of the foremost developers. More recently, several start-ups have launched commercial products. Many of the emerging devices have fabrication and packaging issues that are device-specific, making challenges ahead more numerous. But there appears to be an increasing focus on wafer-level processes for cost-effective solutions.

MEMS FABRICATION OVERVIEW

There are several established MEMS manufacturing processes, but those based on silicon semiconductor methods are dominant since they achieve massively-parallel processing and extreme miniaturization for exceptional economics. The most popular MEMS fabrication process is *surface machining* where sacrificial materials, like SiO_2, are selectively formed and then etched away to leave 3D structures typically made of silicon. Photolithography is used to produce patterned "structural" and "sacrificial" sections on a wafer. Surface micromachining can be used in existing fabs with minimal modification. 3D mechanical structures are produced when the sacrificial sections are etched away leaving the permanent structure parts. Multiple layers of permanent and sacrificial patterns are used to generate 3-dimensional elements. The most advanced MEMS process today SUMMiT V, comes from Sandia National Laboratories and employs up to 5-layers. The process has been licensed by semiconductor fabricators.

Integrated electronics can also be fabricated within a MEMS line using CMOS processes. But some strategies fabricate the logic on separate wafers that must be bonded to the mechanical wafers. Also, some of the 3D MEMS mechanical structures, like pumps, are best produced by bonding wafers together with the appropriate sub-structures. The industry has therefore developed several wafer-bonding processes.

MEMS wafers are often very sensitive to mechanical shock, especially in unpackaged form, and are especially vulnerable to particulate contamination like residue from wafer sawing. The mechanical motion zone of a MEMS chip must be protected during singulation with a temporary mask or by wafer-level packaging techniques. Even a signal micro-particle can render a MEMS chip useless by impeding motion - a *sand in the gears* problem. But the most significant MEMS challenge is still packaging since solutions must provide environmental protection without restricting chip mechanical motion.

MEMS PACKAGING REQUIREMENTS
The most common electronic packaging methods for electronic devices utilize plastic overmolding where resin contacts the active face of the die. Unfortunately, this nearly 60-year old plastic packaging encapsulation approach cannot be used for MEMS [1]. Direct contact of the MEMS mechanically-active surface by molding compound or other encapsulant, would "freeze" the moving parts. There is another MEMS packaging issue still in debate. The hermeticity level requirement has not been determined for most MEMS devices. The default position is to go with full-hermetic packaging. While this may be the prudent strategy, hermetic packaging has been costly to a degree where packaging becomes many times more expensive than the device. Some devices can require specific atmospheres, such as high vacuum, or the presence of lubricants, or anti-stiction materials. Small MEMS movable elements can become permanently stuck together when smooth surfaces make contact (stiction) and hydrophobic, low surface energy additives are helpful. The application of wafer-level processes not only can reduce cost, but also provides the best logistics for introducing coatings, additives, and atmosphere control agents.

MEMS WAFER BONDING
Wafer bonding began in the MEMS field, probably with automotive applications, but is now being applied to non-MEMS areas such as chip stacking [2]. Accelerometers and gyroscopes are commonly sealed at wafer-level to produce a micro-cavity. Pumps, detectors, resonators, and the general area of RF-MEMS, uses wafer bonding. While it is possible to form cavities using single wafer techniques, wafer-bonding strategies are more suitable for complex structures. A MEMS pump, chemical reactor, fluidic controller, and many analyzers, are designed using two or more wafers that must be bonded in a final step. Common bonding methods are listed; the first two don't require addition of bonding medium.

1. Silicon Direct Bonding (SDB); considered a high-temperature process suitable for both Si and its compounds, notably, SiO_2 and Si_3N_4. Typical bonding temperatures range from 800°C to 1200°C where the bond can reach the intrinsic strength of bulk silicon. Thermally-sensitive devices can be bonded, although with reduced strength, at temperatures between 200°C and 400°C by using chemical surface activation methods.

2. Anodic; Silicon-to-Glass; heat/current; the silicon wafer and glass substrate are brought together and heated to a high temperature. A large electric field is applied across the join, which causes an extremely strong bond to form between the two materials.

3. Ziptronix Process; this method avoids elevated temperatures. A series of chemical treatments are applied to silicon, the chemistry is removed, and the wafers for mating are brought into contact to form covalent bonds. The wet chemistry is apparently specific to the substrates. The wafers are completely dried before bonding. The primary process, that uses $LiTaO_3$ and quartz, is said to produce high bond strength without heating.

4. Adhesive, inorganic; glass frit, lower temperatures, popular approach for capping.

5. Adhesive, metallurgical; solder; eutectic bonding utilizes a metal that forms an alloy with the substrate; e.g.; gold-silicon eutectic composition.

6. Adhesive, polymer; thermoset, thermoplastic, radiation-cured materials can be used. This is a low-cost method that is quite versatile. The major drawback is that the seal is not hermetic. However, high moisture-barrier plastics, like LCP (Liquid Crystal Polymer) may be suitable for some applications. It's worth noting that many MEMS devices may not require a hermetic seal for packaging. Polymers should also be suitable for bonding fluidic chamber components and any systems that will sample liquids and gases.

Several companies, including EV Group and SUSS MicroTec, offer specialized wafer bonding equipment with very high precision and controlled atmosphere environments, in addition to heating capability. While wafer bonding can be used to construct MEMS devices, these applications will not be covered since, the focus here is wafer-level packaging.

MEMS WAFER-LEVEL PACKAGING
Accelerometers are an important MEMS product and find wide application in the automotive field as motion sensors. Virtually all vehicles use accelerometers, or newer gyroscopes, as air bag sensors. These MEMS inertial sensors detect rapid deceleration occurring in a crash and react in milliseconds to fire airbags. Common designs use

two MEMS chips per bag to insure high reliability and all but eliminate false firing that plagued the earliest systems. MEMS inertial sensors are also beginning to find use for anti-roll-over, load leveling, suspension adjustment, and for GPS backup (in tunnels and other signal-blocking zones).

Early packages were hermetic ceramic enclosures that have been well-known for many decades in the high-reliability and military electronics areas. The MEMS device was placed in such a cavity package, anti-stiction agents were added as required, and the lid was sealed. Such packages worked well, but were more expensive then the MEMS chips. As volumes grew, and cost-constraints increased, a low-cost, but high-reliability option was sought. Even though ceramic packaging costs were reduced with leadless devices, like the LCCC (Leadless Ceramic Chip Carrier), a completely different approach continued to be developed. (See Figure 1) The idea of capping the mechanically-active MEMS area finally reached fruition.

Figure 1A - Ceramic MEMS package (ADI)

Figure 1B - Latest MEMS Ceramic Packages

Analog Devices, Inc. (ADI) pioneered MEMS accelerometers and many innovations in packaging, including the addition of volatile anti-stiction agents into the cavity of the package. ADI, Motorola, and others, eventually developed and commercialized wafer-level capping methods that are in use today. The ADI process has been well described in the literature and will be used as an example of the present state of MEMS WLP. The first steps involve etching alignment keys on the top of the wafer and cavities on the bottom. But a cut

capture pad, or "pre-cut", is also etched that can be seen in Figure 3. The 3rd step is to apply a glass frit slurry by screen printing and then dry the "adhesive ink".

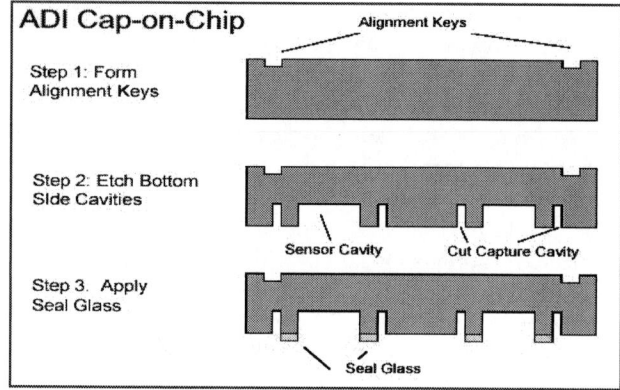

Figure 2 - ADI WL-Capping Steps 1-3

The next step (4) aligns the cap array to the MEMS wafer followed by thermally fusing the glass frit. The step requires fairly precise alignment that can be done using the topside alignment keys or by infrared optical methods. Step #5 involves precision sawing through the cap wafer without damaging the MEMS wafer. The alignment key is designed to serve as the saw blade aligner and it is only slightly offset from the pre-cut below as can be seen in Figure 3. The precut may be temporarily filled with a protective material as shown in Figure 4. The cap is now singulated, but the MEMS wafer is still in one piece. The reason for the separate singulation step is to make the MEMS bonding pads accessible for elemental wire bonding.

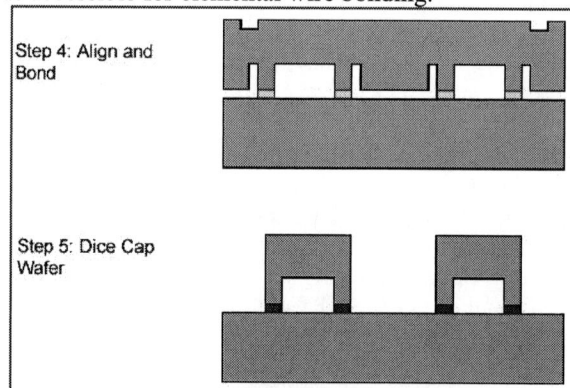

Figure 3 - ADI Capping Steps 4-5

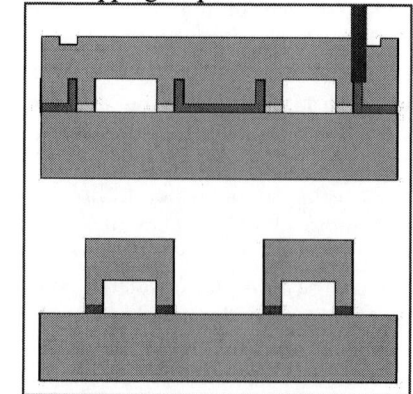

Figure 4 - ADI Capping Separation Detail

The final "pre-packaging" step is to singulate the MEMS wafer just prior to final packaging. But once the MEMS devices are protected by capping, the contamination-sensitivity issue is eliminated since moving parts are now protected. Without the cap, even a tiny saw residue particle could completely disable the mechanics of the device. The singulated capped chip still needs to be packaged. All of the capping steps have only provided some level of protection to the mechanically-active section. The chip must be electrically-connected and placed, or formed, into a package. However, the capped devices can now be handled like a conventional electronic chip - almost. While the capped MEMS chip is protected from the environment, the device is still sensitive to external forces, including stress created by transfer molding. However, many capped MEMS inertial chips can be overmolded by conventional methods. Figure 5 shows a cross-section of a capped and overmolded MEMS accelerometer. The capped chips are also being packaged in molded plastic cavity packages that eliminate stress from EMC.

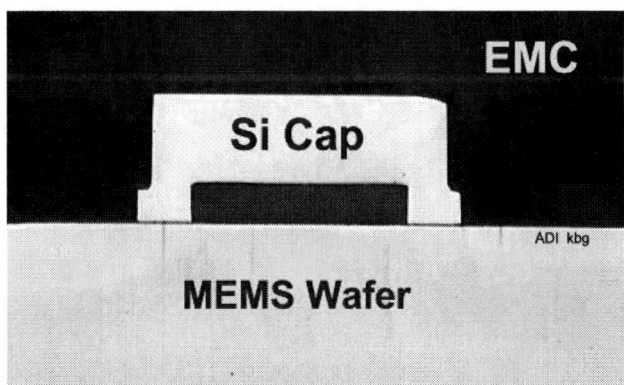

Figure 5 - Packaged MEMS Cross-section

Other Capping Concepts
While these hermetic caps work quite well, the multi-step, double-singulation process begs to be simplified. The desire is to saw both the capping wafer and device wafer in a single step. But the challenge of one-step singulation is how to provide bonding pad access. There are at least four basic strategies; (1) vias through chip, (2) vias through cap, and (3) wafer-bonding with pre-cut caps, and (4) singulate bonded caps by non-mechanical means. Through-chip via processes have been developed over the past decade and this same concept is being used for 3D stacked chips. A MEMS chip with pads routed to the backside by means of through vias could be capped by one of the wafer-bonding methods listed earlier, and then singulated in one step by simultaneously cutting through the cap and device wafers.

Vias-through-cap designs have been an active area of development and several patents have issued with many more pending. One of the earlier access-through-cap concepts, from Motorola, is shown in Figure 6 where the cap has openings to access bonding pads on the MEMS, or other device. Selectively-applied glass frit is used as the wafer bonding agent, but it also forms the vertical walls. The cap wafer is therefore relatively simple but the frit-application process would seem to be a challenge as well as wire bonding through

a small opening. However, the design does allow the device and cap wafers, once joined, to be singulated in a single step.

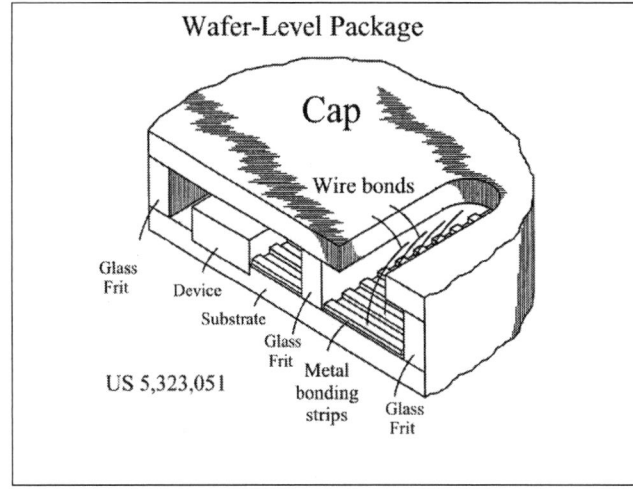

Figure 6 - MEMS cap (reference 3)

An even earlier process describes a wafer-level capping process where individual caps are held in a fixture that is actually an etched wafer. The wafer-fixture is pressed against a MEMS, or other active device wafer and the caps are bonded, followed by lifting the fixture. The process has several problems but the most significant is that the caps must be loaded and extra space must be provided around each active device to accommodate the caps as seen in Figure 7 (reference 4).

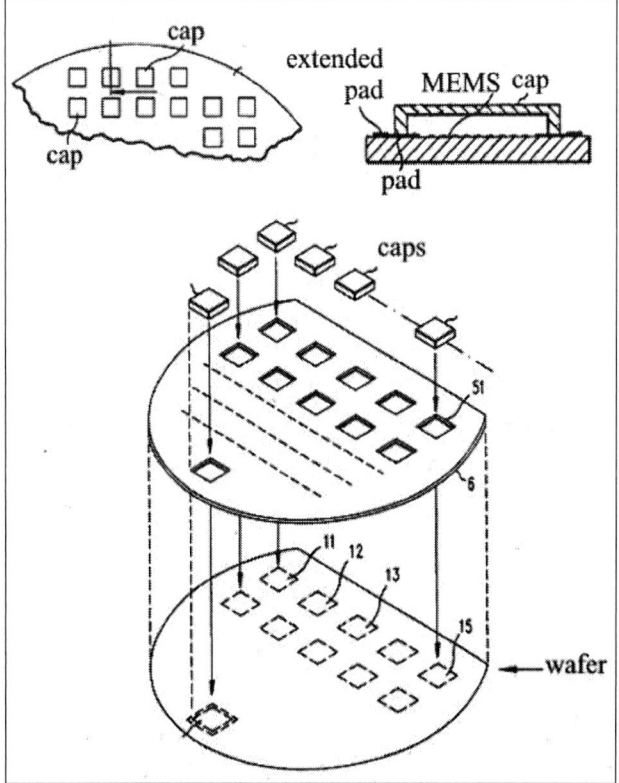

Figure 7 - Capping Using a Fixture

143

Recent Wafer-Level Capping Patents

A number of patents describe wafer-level caps that have through-via interconnects. For example, a Rockwell patent [5] describes a MEMS structure with a protective cap having electrical traces embedded in a nonconductive substrate. The electrical trace includes a first terminal end that is exposed to the peripheral region of the device, and a second end that is connected to the MEMS structure to facilitate operation of the device. The processes are rather complicated, however.

Several patents describe caps with conductive vias and methods for producing and applying. The cap with through vias for wire bonding or direct chip attach appears to be the most practical wafer-level packaging strategy. However, many of the via forming and filling processes are complicated. A recent patent issued to Sony, and shown in Figure 8, is typical of through-cap via inventions.

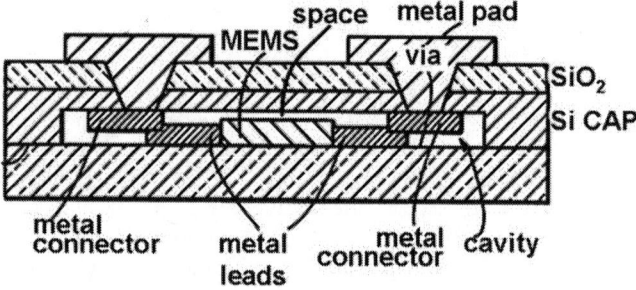

Figure 8 - Through Via Cap [reference 6]

A patent granted to ADI [7], a company that is very active in building a capping portfolio, describes a wafer-level process where the cap is selectively etched after bonding, to expose bond pads (see Figure 9). The cap is made with photoimageable glass, such as Foruran (Schott) or Fotoform (Corning). This seems to be an improvement over earlier ADI processes, but the capped chip still needs to be packaged.

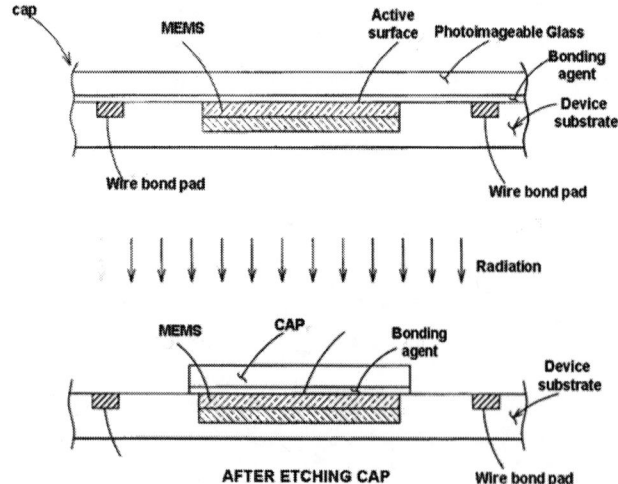

Figure 9 - Photoimageable Cap

A Motorola patent describes the use of a pre-etched wafer cap with cavities to accommodate each MEMS chip, but also a deeply etched trench around each cavity. The cap wafer is bonded to the device wafer, using glass frit, and the cap wafer is ground down to the trench level with cap singulation resulting [8]. This also seems to be an improvement over existing methods, although the capped singulated devices still need to be packaged.

The ultimate WL-capping process should produce a complete WLP. While many capping patents are really "pre-packaging" concepts, more recent IP describes methods of producing ready-to-bond WLPs. For example, National Semiconductor [9] describes a WLP where the conductors extend over the edge of the die and to the bottom as shown in Figure 10. But most recent patents describe through-vias such as the one issued to Sanyo and shown in Figure 11 [10].

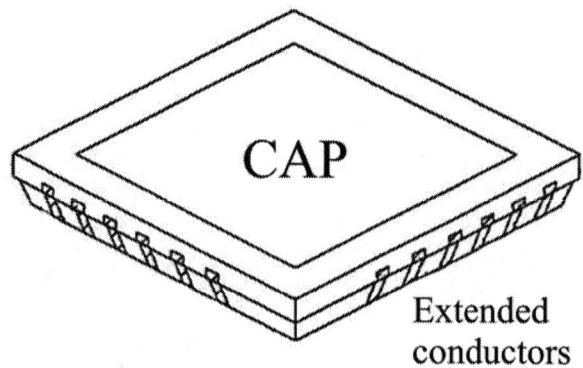

Figure 10 - MEMS WLP

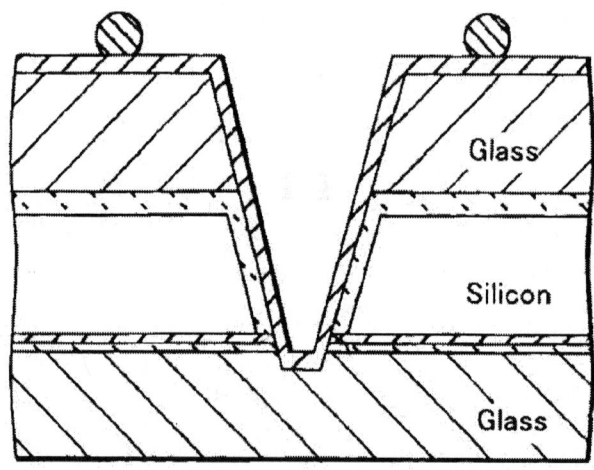

Figure 11 - Through-Via Cap

US patent class 438/113 describes many other capping methods that achieve partial and full WLP. Pending patent applications also tend toward full WLP with through-vias such as the one shown in Figure 12 [11]. The method can be used for optical devices by using glass as the cap. The WLP interface can be solder balls or a plug-in pin type that could be fabricated using MEMS processes. The idea of using MEMS fabrication techniques for package connection can be taken a step further. Figure 13 shows the plug-in MEMS

144

concept that would also allow fluids and gases to flow through the connections [12].

While most WLP methods start with a complete capping wafer, it is theoretically possible to apply caps that are completely singulated. While early concepts used fixtures, an expandable wafer tape can be used to eliminate handling.

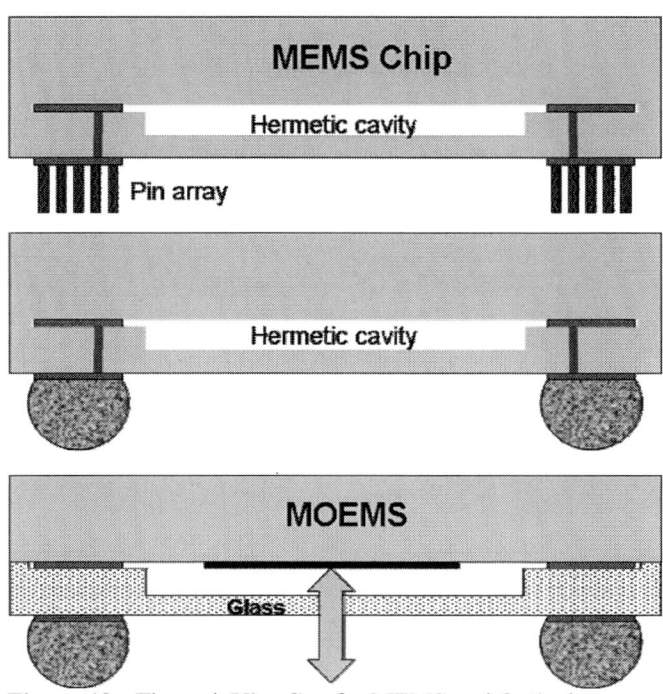

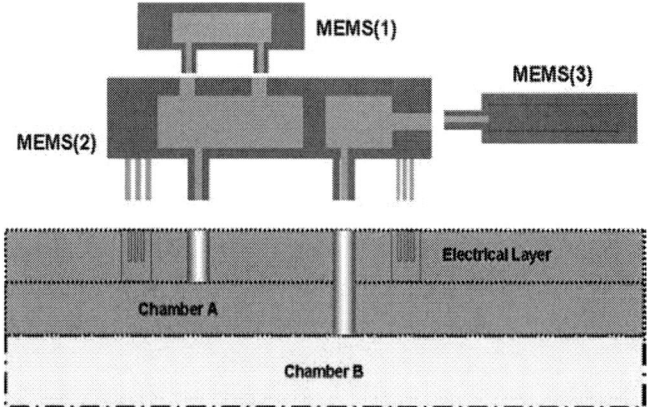

Figure 12 - Through Vias Cap for MEMS and Optical Devices

Figure 13 - Plug-In Modules

Figure 14 shows the caps on tape concept. Recesses are first etched into a wafer that is placed on wafer tape and then singulated into caps. The tape is expanded to separate the caps to the required spacing. The separation can be asymmetrical by using differential expansion shown in Figure 15. The wafer-bonding agent, such as glass frit or polymer adhesive, can be applied to the etched cap wafer before singulation.

Figure 16 shows how the smaller caps can be aligned and bonded to the MEMS wafer leaving bonding pads exposed. This is still a "pre-packing" concept and further package processing is required.

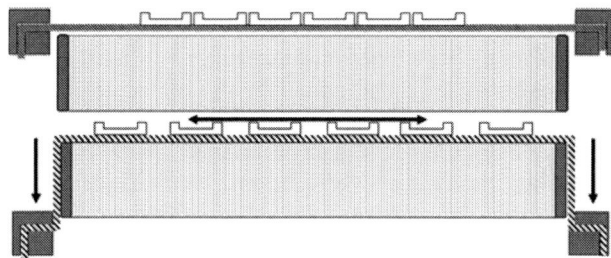

Figure 14 - Expanded Tape with Caps

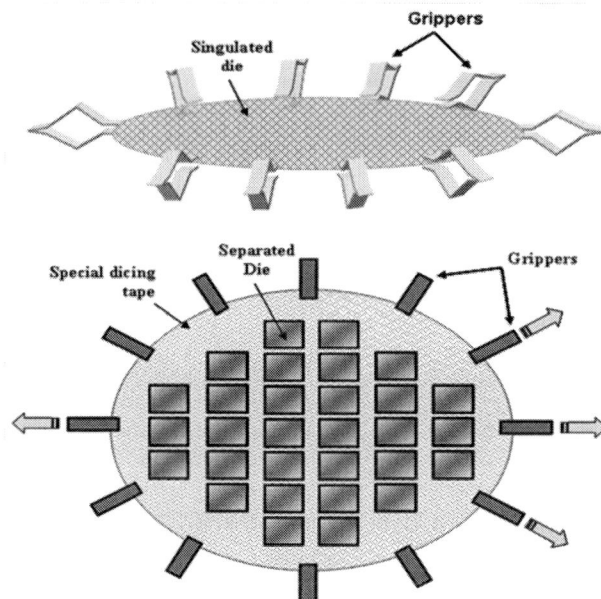

Figure 15 - Asymmetric Tape Expansion

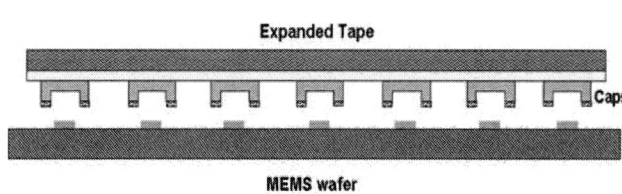

Figure 16 - Wafer-Bonding Caps onto MEMS Wafer

FUTURE PATHS

MEMS, MOEMS and many optical devices have special requirements that can be provided by wafer-level processes. MEMS especially benefits from wafer-capping and enclosure methods since the highly contamination-sensitive mechanical structures are protected before leaving the clean room environment. Capping/encapsulation prior to wafer singulation will become a standard method since it is a very reliable and efficient means of protection from saw residue. The trend, shown by reference to the patent literature, will be to form complete wafer-level packages to replace "pre-packaging" methods in use today.

MEMS is advancing from basic inertial sensors and ink jet chips to much more complex systems. Many MEMS devices, especially fluidic chips that will be used in medical applications, will require wafer bonding to form cavities and channels. Again, the trend will be to form complete packages. Some devices will require fluidic connectors, such as those shown in Figure 13. Several universities are developing fluidic coupling technology where fittings are fabricated by MEMS processes. The day may come when MEMS devices are plug-in - like Lego® blocks. This would enable the reconfiguration of micro-machines by simply swapping out various modules, much as we do today with macro-world equipment.

CONCLUSIONS

Wafer-level processing is the ideal solution for building complex MEMS devices with cavities and channels. MEMS devices such as rocket motors, pumps, fluidic controllers, chemical reactors, and analyzers can benefit from wafer bonding. WLP packaging is also the ideal approach to protecting fragile and contaminant-sensitive MEMS devices. MEMS capping has become the standard method for inertial sensors, such as accelerometers and gyroscopes, but present methods do not provide a complete package. The industry will move from "pre-packaging" to full-WLP in the future. The large number of issued patents and pending applications suggest that MEMS WLP is a very active field. Expect MEMS to adopt WLP as the best solution to a myriad of unique problems.

REFERENCES

[1] Rose Jr., George, M., United States Patent 2,538,593 January 16, 1951.

[2] Tong, Q.-Y, Gösele, U., Semiconductor Wafer Bonding: Science and Technology, Wiley, NY, December 1998.

[3] Adams, et al, "Semiconductor wafer level package", US Patent 5,323,051, June 21, 1994.

[4] Sahakian, V. K., "Integrated circuit chip sealing assembly", US Patent 4,907,065, March 6, 1990.

[5] Harris, et al. "Method for fabricating an isolated microelectromechanical system (MEMS) device incorporating a wafer level cap", US Patent 6,768,628, July 27, 2004.

[6] Brady; F. T., "Wafer-level through-wafer packaging process for MEMS...", US Patent 6,660,564, December 9, 2003.

[7] Karpman, et al, "Cover cap for semiconductor wafer devices", US Patent 6,534,340, March 18, 2003.

[8] Warfield, T. J., "Method for packaging semiconductor devices ", US Patent 5,604,160, February 18, 1997.

[9] Prabhu, A., "Electrical die contact structure and fabrication method", US Patent 7,067,354, June 27, 2006.

[10] Wakui, et al, "Manufacturing method of semiconductor device", US Patent 7,064,046, June 20, 2006

[11] Gilleo, K. "Reconnectable Chip Interface and Chip Package", US Patent Application 2005/0012191 A1, published Jan. 20, 2005.

[12] Gilleo, K. "MEMS Device Assembly", US Patent Application 2004/0163717 A1, published August 26, 2004.

[13] Gilleo, K., "Wafer-Level Assembly Method for Semiconductor Devices", US Patent Application 2006/0012020 A1, published January 19, 2006.

[General Reference] Gilleo, K., MEMS/MOEM Packaging: Concepts, designs, materials and processes, McGraw-Hill, New York, NY June 2005.

ACKNOWLEDGEMENT

ADI capping diagrams were provided by Dr. Larry Felton, Analog Devices, Inc., Cambridge, MA.

EFFECTS OF PLASMA PRETREATMENT ON FLIP CHIP AND CSP SUBSTRATE LEVEL ASSEMBLY YIELD AND RELIABILITY

Daniel Baldwin, Ph.D.
Advanced Assembly Process Technology - AdAPT Laboratory
The George W. Woodruff School of Mechanical Engineering
Georgia Institute of Technology
Atlanta, GA, USA

Paul Houston and Brian Lewis
Engent, Inc. – Enabling Next Generation Technologies
Norcross, GA, USA

ABSTRACT

A comprehensive study is performed investigating the influence of plasma pretreatment on assembly yield for solder flip chip in package assemblies and chip scale package (CSP) assembly to SIP modules. In addition, a detailed reliability screening is performed on the assemblies to assess the impact of pre assembly plasma treatment on long term reliability of the assemblies. Various plasma treatment techniques and two substrate surface finishes are included in the experimental analysis. The flip chip and CSP devices are underfilled with fast flow, snap cure underfill material. Baseline assemblies without plasma pretreatment are tested for comparison. The test vehicles are subjected to liquid to liquid thermal shock or air to air thermal cycle testing and analyzed using electrical test, CSAM, X-ray, and microsectioning.

Visual inspection of the plasma treated samples revealed higher more uniform fillet shapes especially in the corners of the components compared with the untreated devices. CSAM analysis indicated no significant difference between the plasma treated and non-plasma treated samples in terms of underfill uniformity and void formation during processing. The flow times for the plasma treated flip chip samples were 12 to 20 percent faster than the non-plasma treated samples. The flow times for the plasma treated CSPs were 55% faster than the flow times for the untreated samples. As for flip chip reliability differences, the plasma treated test vehicles with Au finish had the highest reliability with a 70 improvement over untreated samples. In contrast, the untreated flip chip assemblies with an OSP finish had 78% higher reliability compared with the plasma treated counterpoints. For the CSP components, the reliability of the components were generally the same as the non-Plasma treated samples.

Key Words: Plasma, Underfill, Flip Chip, Reliability, Yield

INTRODUCTION

The demand for miniaturized electronics continues to drive advances in system in package modules and high density board level assembly technologies. This in turn further miniaturizes the solder interconnects required for substrate level assembly. The associated micro-miniature solder interconnects are challenging to yield particularly in the low parts per million range required for many applications. Moreover, the thermomechanical reliability and mechanical reliability of the joints decreases rapidly with solder volume decrease typically requiring underfill or underfill like materials be applied to achieve adequate reliability.

A comprehensive study is performed investigating the influence of plasma pretreatment on assembly yield for solder flip chip in package assemblies and CSP assembly to SIP modules. In addition, a detailed reliability screening was performed on the assemblies to assess the impact of plasma pretreatment on long term reliability of the assemblies. Various plasma pretreatment techniques and two substrate surface finishes (OSP and ENIG) were included in the experimental analysis. Test vehicles consisted of 208 micron pitch solder bumped flip chip assemblies with 88 IO and an 84 IO, 7 mm size, 0.5 mm pitch CSP assemblies.

Flip chip and CSP devices were underfilled with fast flow, snap cure underfill material. Baseline assemblies without plasma pretreatment were tested for comparison. The test vehicles were then subjected to liquid to liquid thermal shock (LLTS) between -40°C and 125°C with 5 minute dwells or AATS between -40°C and 125°C with 10 minute dwells. The components were tested every 100 cycles for continuity in LLTS and every 200 cycles in AATS. The flip chip components were analyzed using scanning acoustic microscopy every 400 cycles and after C-SAM were baked at 80°C for four and a half hours so that moisture from the scanning would not induce delamination or increase delamination growth. Micro-section analysis confirmed the delamination and solder joint cracking as the root causes of failure.

EXPERIMENTAL PLAN

No clean fluxes provide a cost effective assembly solution for flip chip and fine pitch chip scale package (CSP) assembly particularly for CSP components of 0.4 mm or lower pitch and solder flip chip applications. However, the flux residues remaining are often identified as a concern for higher reliability requirements which generally demand underfill for these components. Flux residues can compromise underfill adhesion causing a notable decrease in reliability. This study investigates if plasma pretreatment of the components and substrates improves these assemblies in terms of process yield and thermo-mechanical reliability.

The analysis is flip chip and CSP assemblies consists of assembly of flip chip and CSP test vehicle assemblies, analysis of the assembly yields based on electrical continuity tests and X-ray inspection, thermal shock reliability testing, thermal cycle reliability testing, and failure mode analysis.

For the flip chip assemblies, a daisy chain PB8 solder bumped flip chip component was used. The flip chip test vehicles consisted of 12 components per board using 5 mm (edge dimension) flip chip devices having a 203 μm pitch between joints and 120 μm diameter eutectic lead-tin solder bumps. Two substrate surface finishes were used for the flip chip test vehicles including a Cu-Ni-Au or ENIG finish and a Cu metallization with an organic solderability preservative (Cu-OSP). For the chip scale package assemblies, a CABGA84 component was used. The CSP test vehicles consisted of 10 components per board using 7 mm (edge dimension) packages having a 0.5 mm pitch between joints, 84 IO, 356 μm eutectic lead tin solder balls. The substrate surface finish used for the CSP test vehicles was Cu-Ni-Au or an ENIG finish. Figure 1 shows assembled flip chip and CSP test vehicles.

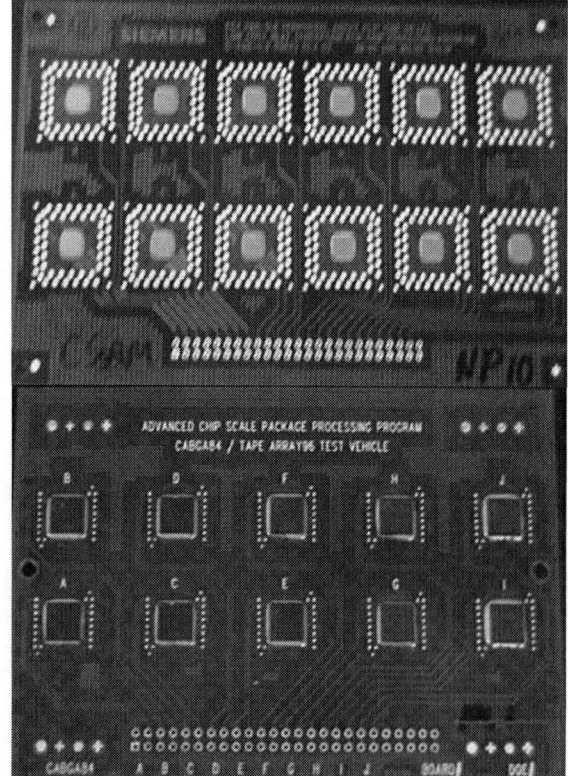

Figure 1: Flip Chip and CSP Test Vehicle Assemblies

Half the test vehicles including the chips, packages, and substrates were plasma treated prior to assembly in order to assess its impact on yield and reliability. In all of the results that follow, references to plasma treated and non-plasma treated refer to pre-reflow treatment of the substrates and components. No cleaning or plasma treatment was performed on the assemblies post reflow. The plasma treatment for the chips, packages, and substrates consisted of the following batch process using a commercial batch plasma system.

- Power: 400 W
- Gas: O_2
- Gas flow: 100 sccm
- Pressure: 200 mtorr
- Plasma treatment time: 3 min.
- Sample location: Ground shelf
- Shelf location: Powered and Ground 2 shelf spacing

All of the flip chip test vehicles were assembled using a conventional flip chip assembly process with capillary flow underfills. The bumped flip chip devices were dip fluxed with a commercial no-clean, low residue tacky flux. The chips were then vision aligned and assembled onto the test vehicle substrate. Convection reflow was used to attach the eutectic lead-tin solder joints to the substrate. Post reflow the flip chips were underfilled with a commercial fast flow, snap cure underfill chemistry. Underfill flow times were tracked and high speed camera videography was used to characterize the flow. The assemblies were analyzed visually and using scanning acoustic microscopy to determine differences in the underfill flow and fillet quality between the plasma treated and non-plasma treated samples.

The CSP test vehicles were assembled using a conventional SMT process with the addition of underfill application to the CSPs. The boards were first printed with a no-clean eutectic lead-tin solder paste material. Next, the CSPs were picked, vision aligned and placed on the boards. Convection reflow was used to attach the packages to the substrate. Post reflow the CSPs were underfilled with a commercial fast flow, snap cure underfill chemistry. Underfill flow times were tracked and high speed camera videography was used to characterize the flow. The assemblies were analyzed visually and using scanning acoustic microscopy to determine differences in the underfill flow and fillet quality between the plasma treated and non-plasma treated samples.

To evaluate the thermo-mechanical reliability of the plasma treated and baseline non-plasma treated assemblies, thermal cycle and thermal shock environmental stress testing was performed. The test vehicles were subject to liquid to liquid thermal shock (LLTS) between -40°C and 125°C with 5 minute dwells or air to air thermal shock (AATS) between -40°C and 125°C with 10 minute dwells. The flip chip devices and CSPs were tested every 100 cycles for continuity in LLTS and every 200 cycles in AATS. The flip chip components were analyzed using a scanning acoustic microscope every 400 cycles and post C-SAM were baked at 80°C for four and a half hours so that moisture from the scanning would not induce delamination or increase delamination growth.

Failure mode analysis was performed using CSAM analysis microsection techniques, and SEM analysis confirming solder joint cracking as the causes of failure.

RESULTS AND DISCUSSION

Analyzing the yield of the flip chip test vehicle assemblies it was found that the plasma treated OSP surface finish boards had a lower solder interconnect yield relative to the OSP surface finish boards not subject to plasma treatment. This is likely caused by a negative interaction between the Argon plasma and the organic surface protectant over the Cu pads suggesting that plasma treatment of OSP finished boards should be avoided. No statistical difference in interconnect yield was found between the plasma treated and non-plasma treated flip chip and CSP assemblies mounted on the ENIG surface finished boards.

Figure 2 shows a typical X-ray micrograph of a non-plasma treated CSP assembly post reflow. Figure 3 shows a typical X-ray micrograph of a plasma treated CSP assembly post reflow. No quantifiable voiding differences were seen in the X-ray micrographs between the plasma treated and non-plasma treated components.

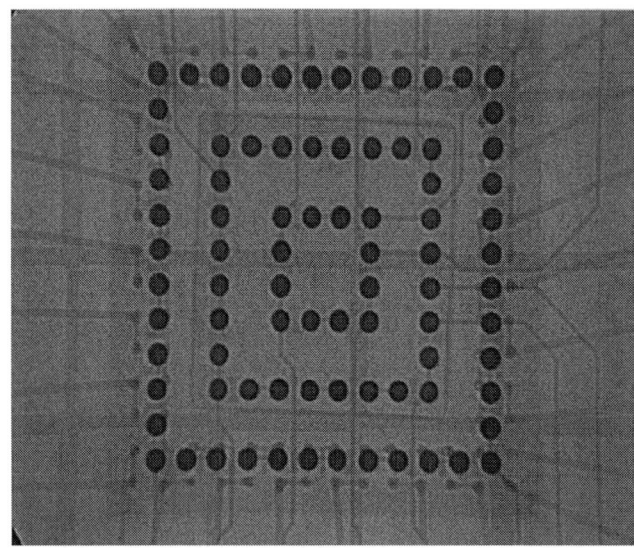

Figure 3: X-ray Micrograph of a Plasma Treated CSP Assembly Post Reflow

The underfill flow times for the test vehicle assemblies are given in Figure 4. Both flip chip and CSP test vehicles plasma treated prior to assembly had faster underfill flow times relative to their untreated counterparts. Flow times for slower flowing underfill materials would have shown a much larger difference. In addition, flow times were longer for components with the higher standoff heights.

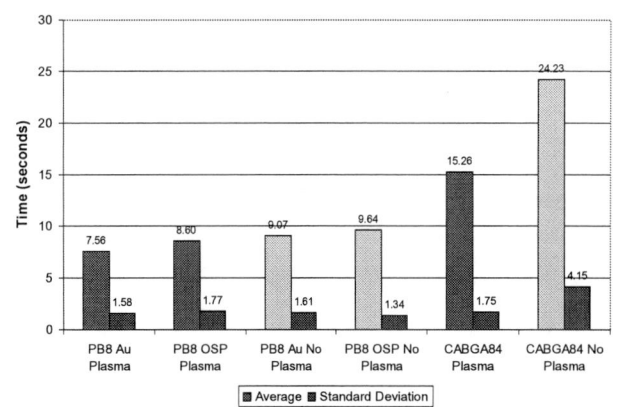

Figure 4: Underfill Flow Times for the Flip Chip and CSP Test Vehicles With and Without Pre-Assembly Plasma Treatment.

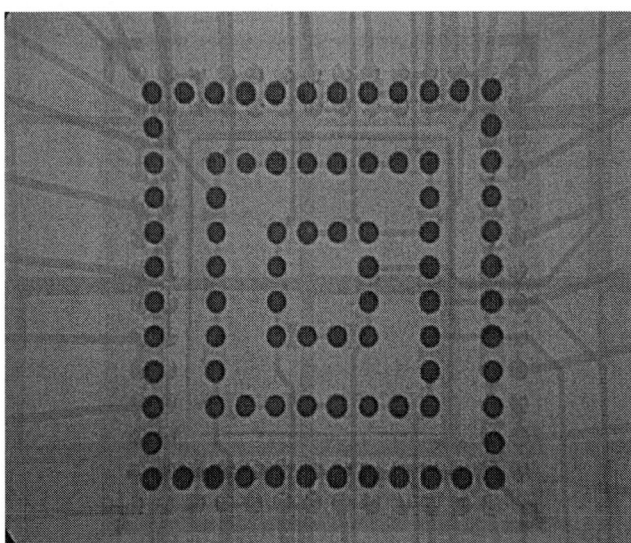

Figure 2: X-ray Micrograph of a Non-Plasma Treated CSP Assembly Post Reflow

Post underfill cure, the plasma treated test vehicle assemblies had more uniform underfill fillets on all sides of the devices/components. This was especially evident at the corners of the component. Figures 5 and 6 show micrographs of post underfill cure flip chip test vehicle assemblies. Notice for the non-plasma treated device, the enlarged underfill fillet along the upper edge of the chip corresponding to the dispense edge of the device. The plasma treated device had nearly equal fillet widths on all four edges of the device. Notice the even fillets on the plasma treated samples.

Figure 5: Typical Micrograph of a Flip Chip Test Vehicle Without Pre-Assembly Plasma Treatment Illustrating Non-Uniform Underfill Fillet Formation.

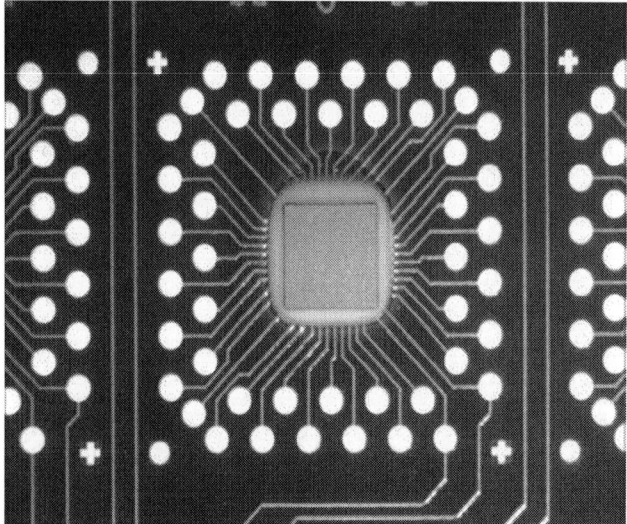

Figure 6: Typical Micrograph of a Flip Chip Test Vehicle With Pre-Assembly Plasma Treatment Illustrating Uniform Underfill Fillet Formation.

Side view micrographs of the flip chip test vehicles post underfill cure are shown in Figures 7 and 8 illustrating non-plasma treated and plasma treated, respectively. Notice the difference in the fillets in the corners of the device. The fillets for the plasma treated components flows up the edge of the component higher compared with the non-plasma treated assemblies.

Figure 7: Typical Edge Micrograph of a Flip Chip Test Vehicle Without Pre-Assembly Plasma Treatment Illustrating Lower Underfill Fillet Formation in Corner Region.

Figure 8: Typical Edge Micrograph of a Flip Chip Test Vehicle With Pre-Assembly Plasma Treatment Illustrating Underfill Fillet Formation in Corner Region.

Figures 9 and 10 show micrographs of post underfill cure CSP test vehicle assemblies. Notice for the non-plasma treated device, the enlarged underfill fillet along the upper edge of the package corresponding to the dispense edge of the device. The plasma treated assemblies had nearly equal fillet widths on all four edges of the package. Notice the even fillets on the plasma treated samples.

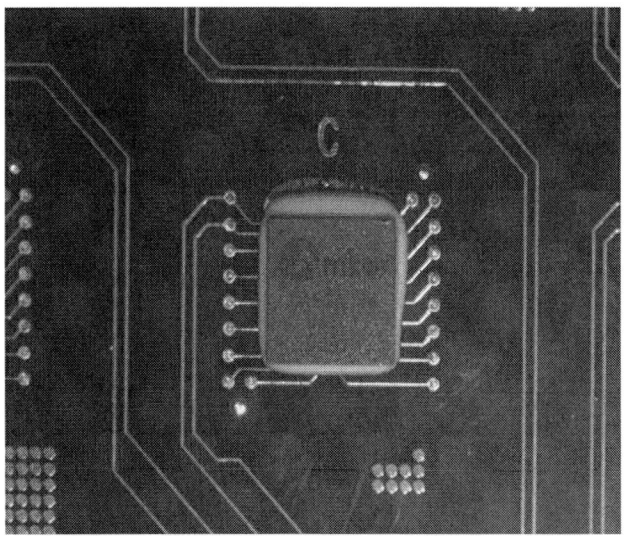

Figure 9: Typical Micrograph of a CSP Test Vehicle Without Pre-Assembly Plasma Treatment Illustrating Non-Uniform Underfill Fillet Formation.

Figure 10: Typical Micrograph of a CSP Test Vehicle With Pre-Assembly Plasma Treatment Illustrating Uniform Underfill Fillet Formation.

Side view micrographs of the CSP test vehicles post underfill cure are shown in Figures 11 and 12 illustrating non-plasma treated and plasma treated, respectively. Notice the difference in the fillets in the corners of the device. The fillets for the plasma treated components flows up the edge of the component higher compared with the non-plasma treated assemblies.

Figure 11: Typical Edge Micrograph of a CSP Test Vehicle Without Pre-Assembly Plasma Treatment Illustrating Low Underfill Fillet Formation in Corner Region.

Figure 12: Typical Edge Micrograph of a CSP Test Vehicle With Pre-Assembly Plasma Treatment Illustrating Improved Underfill Fillet Formation in Corner Region.

Notice the difference in the underfill fillets in the corners of the device and the fillet formation up the side of the component. The fillets for the plasma treated components more completely covers the component corners relative to the non-plasma treated. Parts were underfilled from the opposite side from the side shown.

Scanning acoustic microscopy (CSAM) of the underfilled flip chip devices showed no discernable difference between the plasma treated and non-plasma treated assemblies. Figures 13 and 14 show typical CSAM micrographs at the chip to underfill interface of plasma treated and non-treated samples, respectively.

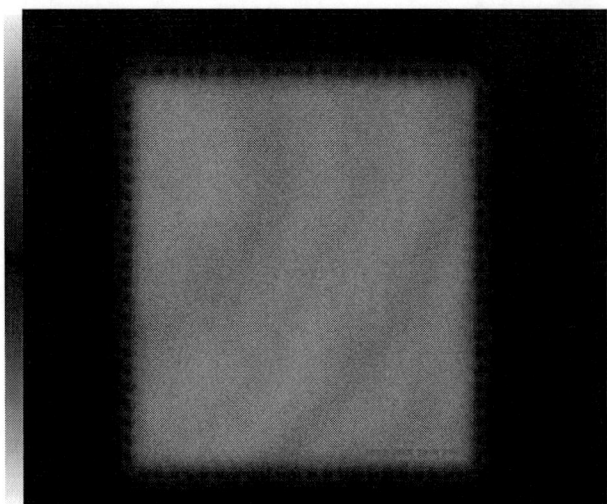

Figure 13: Micrograph of a Flip Chip Test Vehicle With Pre-Assembly Plasma Treatment Illustrating Uniform Underfill Flow.

Figure 14: Micrograph of a Flip Chip Test Vehicle Without Pre-Assembly Plasma Treatment Illustrating Uniform Underfill Flow.

Reliability Analysis

Thermal shock environmental stress testing was used to assess the impact of pre-assembly plasma treatment on the thermo-mechanical reliability of both the flip chip and CSP assemblies with underfill processed using no clean flux. The test vehicles were subject to LLTS or AATS between -40°C and 125°C. The flip chip devices and CSPs were tested every 100 cycles for continuity in LLTS and every 200 cycles in AATS.

Figure 15 plots the reliability of the flip chip assemblies subject to LLTS with the reliability statistics given in Table 1. In terms of mean time to failure (MTTF), the plasma treated flip chip assemblies mounted on ENIG had the highest reliability, and the plasma treated assemblies mounted on OSP had the lowest reliability. The flip chip assemblies without Plasma treatment mounted on OSP have similar reliability performance to the ENIG assemblies

without plasma treatment. In this case, the failure rates were similar in both instances as implied by the Weibull slopes. For the OSP surface finish assemblies, pre-assembly plasma treatment clearly has a negative impact on reliability decreasing the relative MTTF by over 23% and inducing a failure mechanism that increases the failure rate by nearly 40%.

Table 1: Weibull parameters for flip chip assembly reliability subject to LLTS

Surface Finish	Plasma Treated		Not Plasma Treated	
	Slope	MTTF	Slope	MTTF
ENIG	4.59	5273	5.48	4386
OSP	6.59	3350	4.71	4369

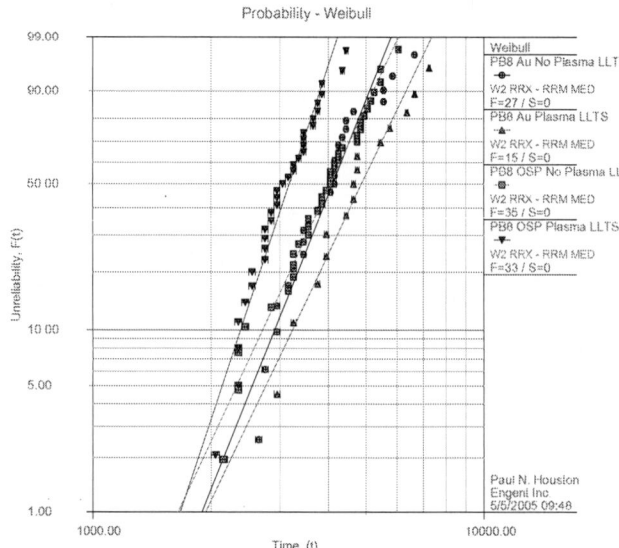

Figure 15: Flip Chip Test Vehicle Reliability Subject to LLTS Testing. Weibull Plot showing percent failed devices versus number of thermal cycles. Legend is shown on the upper right side of the figure for assemblies with and without plasma treatment on both ENIG (Au) and OSP surface finishes.

The LLTS reliability performance of the underfilled CSP assemblies is plotted in Figure 16. The reliability performance was nearly equivalent for CSP assemblies with and without plasma treatment prior to no clean assembly. The reliability statistics are given in Table 2. The data sets appear to be from nearly the same population, meaning that there is relatively little statistical difference between their reliability performance. Hence, plasma treatment prior to assembly does not appear to have beneficial or detrimental impact on the CSP reliability.

Table 2: Weibull parameters for CSP assembly reliability subject to LLTS

Surface Finish	Plasma Treated		Not Plasma Treated	
	Slope	MTTF	Slope	MTTF
ENIG	6.18	6299	4.69	6772

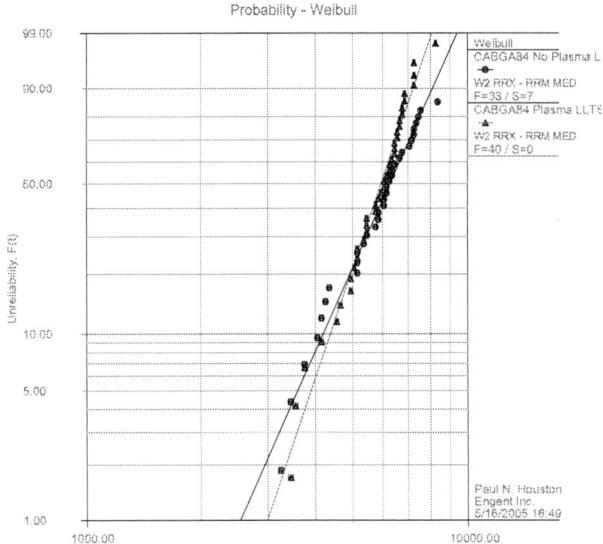

Figure 16: CSP Test Vehicle Reliability Subject to LLTS Testing. Weibull Plot showing percent failed devices versus number of thermal cycles. Legend is shown on the upper right side of the figure for assemblies with and without plasma treatment on an ENIG surface finish.

LLTS testing is particularly useful in driving interfacial delamination in underfilled assemblies. To characterize interfacial delamination, CSAM analysis was performed on the flip chip test vehicles every 400 cycles. Figures 17 and 18 show acoustic micrographs of ENIG substrate flip chip test vehicles at 800, 1600, 2400, and 3200 cycles with and without plasma treatment, respectively. The very light gray and white areas on the CSAM micrographs indicate regions of delamination (or voiding). Onset of underfill to chip delamination occurs at approximately the same number of cycles for the plasma treated and non-treated assemblies in Figures 17 and 18. Once the onset of delamination occurred, the assemblies without plasma treatment tended to delaminate faster relative to the plasma treated assemblies.

Figures 19 and 20 show acoustic micrographs of OSP substrate flip chip test vehicles at 800, 1600, 2400, and 3200 cycles with and without plasma treatment, respectively. The very light gray and white areas on the CSAM micrographs indicate regions of delamination (or voiding). Onset of underfill to chip delamination occurs earlier for the plasma treated assemblies relative to the non-treated assemblies. This gives further evidence of the negative interaction between the oxygen plasma and OSP coating on these flip chip assemblies. Similar to the ENIG assemblies, the delamination rate is faster for the assemblies without plasma treatment compared to the plasma treated assemblies for the OSP substrates.

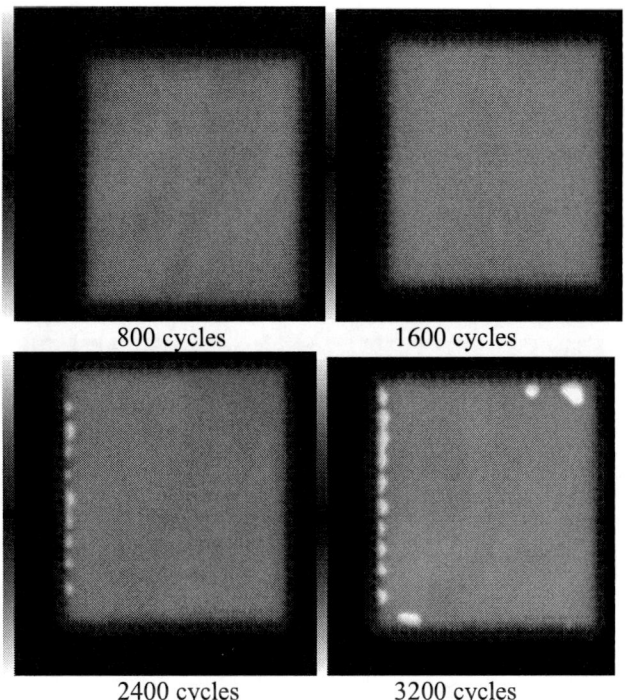

Figure 17: CSAM Micrographs of a Typical Plasma Treated Flip Chip Test Vehicle With an ENIG Substrate Under LLTS Testing.

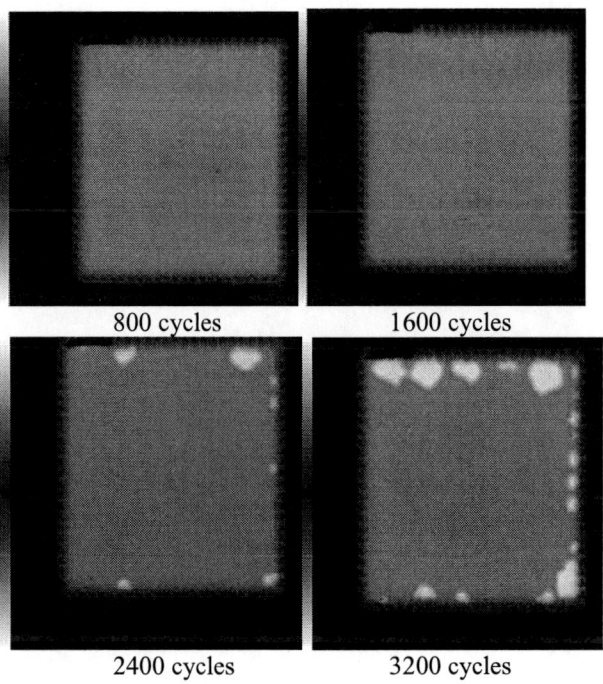

Figure 18: CSAM Micrographs of a typical Non-Plasma Treated Flip Chip Test Vehicle With an ENIG Substrate Under LLTS Testing.

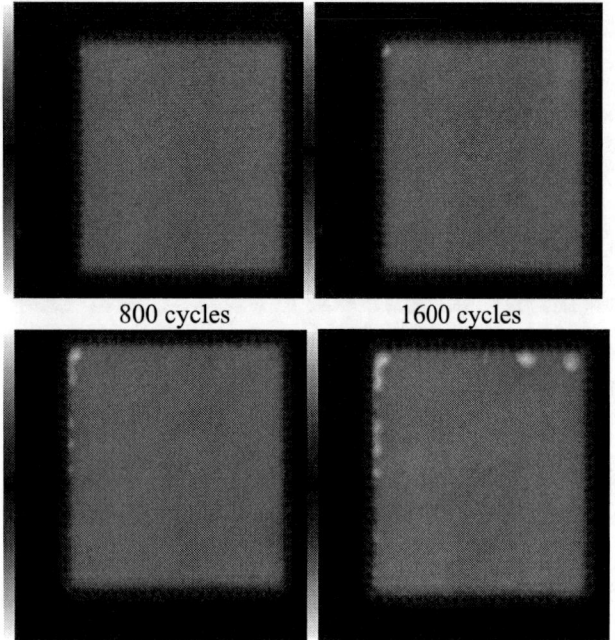

| 800 cycles | 1600 cycles |
| 2400 cycles | 3200 cycles |

Figure 19: CSAM Micrographs of a Typical Plasma Treated Flip Chip Test Vehicle With an OSP Substrate Under LLTS Testing.

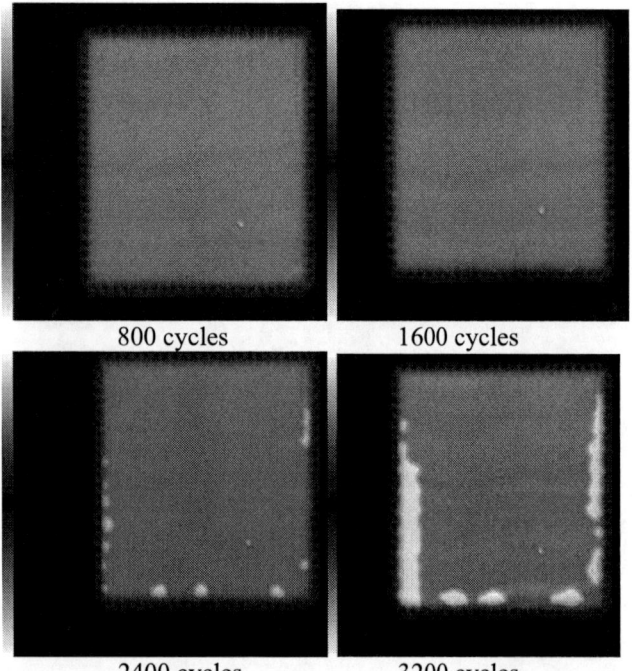

| 800 cycles | 1600 cycles |
| 2400 cycles | 3200 cycles |

Figure 20: CSAM Micrographs of a typical Non-Plasma Treated Flip Chip Test Vehicle With an OSP Substrate

A Weibull plot of flip chip assembly reliability is shown in Figure 21 for assemblies subject to AATS. The reliability failure table is given in Table 3, and the reliability statistics given in Table 4. The plasma treated flip chip assemblies mounted on ENIG tended to have higher reliability relative to the other assemblies. The plasma treated assemblies mounted on OSP had the lowest reliability. The reliability data sets from the flip chip assemblies mounted on OSP

substrates appear to be from the same population denoting no discernable difference in reliability when plasma treatment is used. None of the plasma treated gold samples had failed prior to 4000 cycles. For the non-plasma treated samples mounted on ENIG substrates, 1 sample failed before 3000 cycles and 6 failed by 4000 cycles. For the samples mounted on OSP substrates, neither plasma treated or non-plasma treated samples had any failures prior to 2000 cycles. Both OSP test vehicles with and without plasma treatment had one failure between 2000 and 3000 cycles. By 4000 cycles, 7 of the plasma treated samples and 5 of the Non-plasma treated samples had failed.

Table 3: Weibull parameters for flip chip assembly reliability subject to AATS

Surface Finish	Plasma Treated		Not Plasma Treated	
	Slope	MTTF	Slope	MTTF
ENIG	16.94	5634	7.02	5503
OSP	5.67	5905	6.63	6056

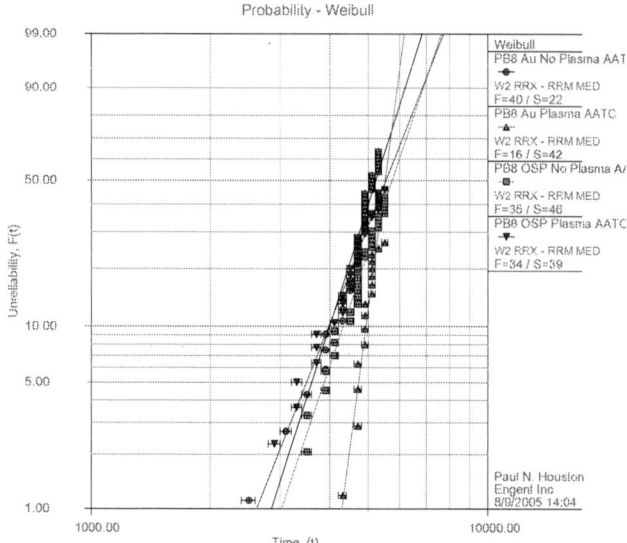

Figure 21: Flip Chip Test Vehicle Reliability Subject to AATS Testing. Weibull Plot showing percent failed devices versus number of thermal cycles. Legend is shown on the upper right side of the figure for assemblies with and without plasma treatment on both ENIG (Au) and OSP surface finishes.

Table 4: Failure table for the flip chip test vehicles showing at a specific number of cycles the number of failed units over the total number of unit tested.

		1000	2000	3000	4000
PB8 OSP	Plasma	0/73	0/73	1/73	7/73
PB8 OSP	Non-Plasma	0/81	0/81	1/81	5/81
PB8 Au	Plasma	0/58	0/58	0/58	0/58
PB8 Au	Non-Plasma	0/62	0/62	1/62	6/62

Figures 22 and 23 show acoustic micrographs of ENIG substrate flip chip test vehicles at 800, 1600, 2400, and 3200 AATS cycles with and without plasma treatment, respectively. Onset of underfill to chip delamination occurs earlier for the non-plasma treated assemblies compared with the plasma treated flip chip assemblies. Once the onset of delamination occurred, the assemblies without plasma treatment tended to delaminate faster relative to the plasma treated assemblies.

Figures 24 and 25 show acoustic micrographs of OSP substrate flip chip test vehicles at 800, 1600, 2400, and 3200 AATS cycles with and without plasma treatment, respectively. Onset of underfill to chip delamination occurs earlier for the plasma treated assemblies relative to the non-treated assemblies. This gives still further evidence of the negative interaction between the oxygen plasma and OSP coating on these flip chip assemblies. For the OSP substrate assembles, the delamination rate is faster for the plasma treatment assemblies compared to the non-plasma treated assemblies.

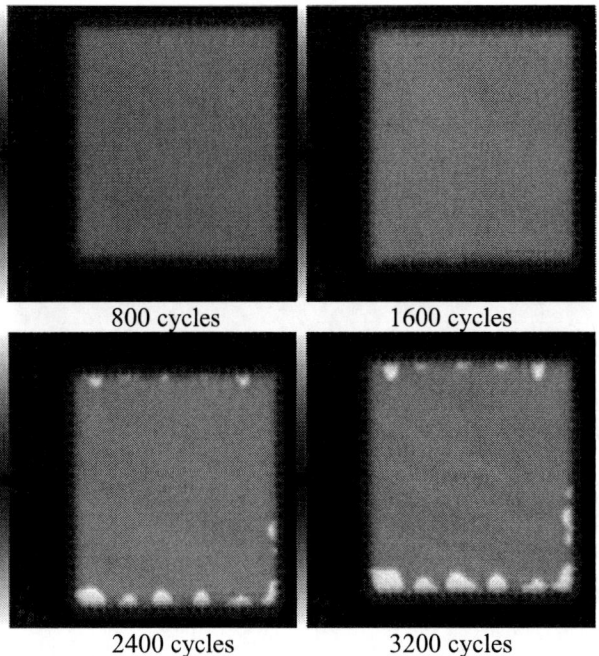

Figure 23: CSAM Micrographs of a typical Non-Plasma Treated Flip Chip Test Vehicle With an ENIG Substrate Under AATS Testing.

Figure 22: CSAM Micrographs of a Typical Plasma Treated Flip Chip Test Vehicle With an ENIG Substrate Under AATS Testing.

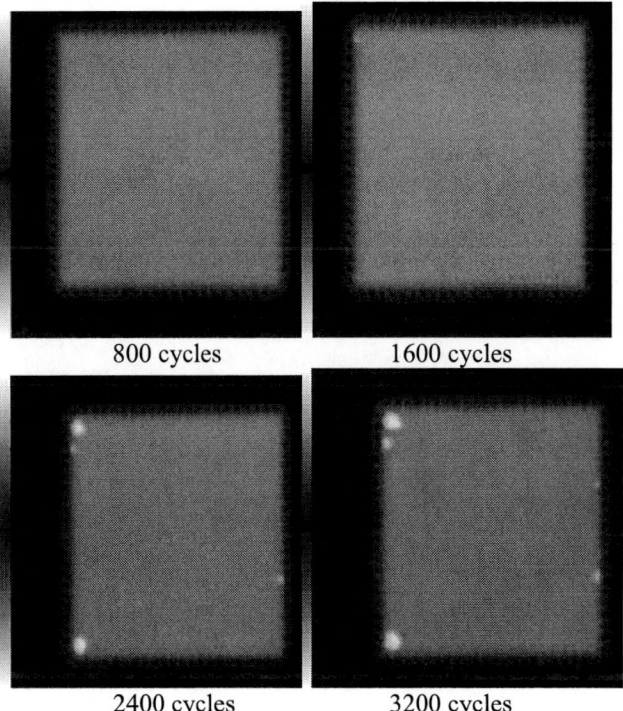

Figure 24: CSAM Micrographs of a Typical Plasma Treated Flip Chip Test Vehicle With an OSP Substrate Under AATS Testing.

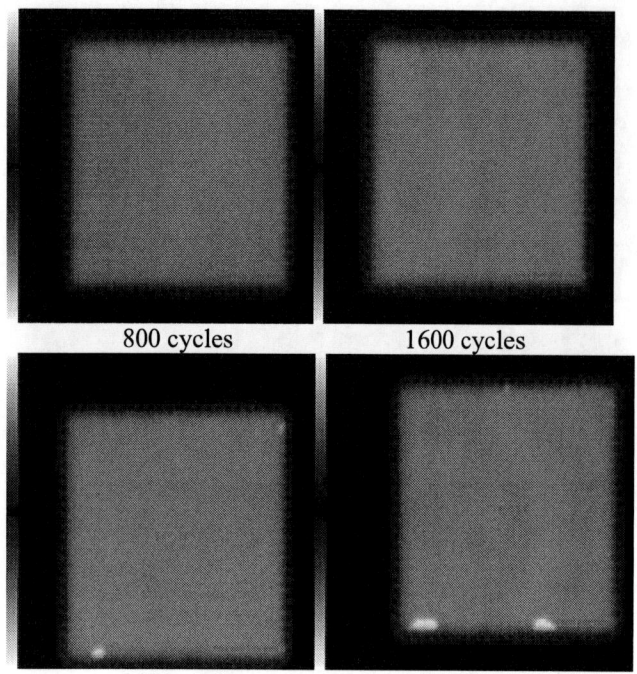

| 800 cycles | 1600 cycles |
| 2400 cycles | 3200 cycles |

Figure 25: CSAM Micrographs of a Typical Non-Plasma Treated Flip Chip Test Vehicle With an OSP Substrate Under AATS Testing.

A Weibull plot of CSP assembly reliability is shown in Figure 26 for assemblies subject to AATS. The reliability failure table is given in Table 5. None of the plasma treated samples had failed prior to 3000 cycles, but one component failed before 4000 cycles. For the Non-Plasma treated samples, 1 sample failed before 3000 cycles and 2 had failed by 4000 cycles. Through 6000 AATS cycles, the CSP assemblies appear to have similar reliability performance between and plasma treated and non-treated samples.

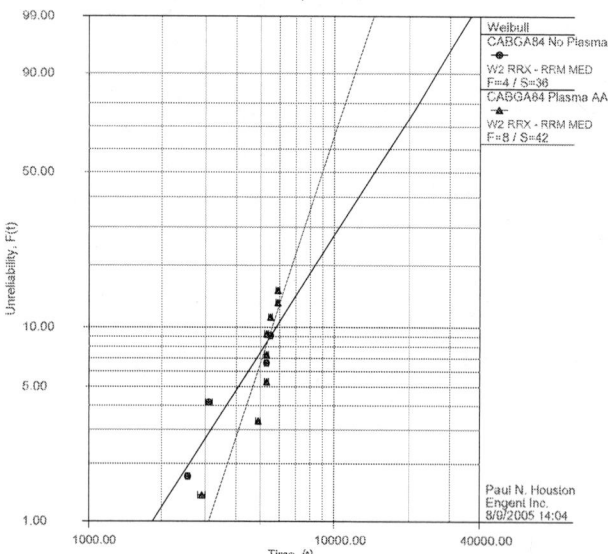

Figure 26: CSP Test Vehicle Reliability Subject to AATS Testing. This Weibull Plot is preliminary as the insufficient assemblies reached failure prior to test termination.

Table 5: Failure table for the CSP test vehicles showing at a specific number of cycles the number of failed units over the total number of unit tested.

		1000	2000	3000	4000
CABGA84	Plasma	0/50	0/50	0/50	1/50
CABGA84	Non-Plasma	0/40	0/40	1/40	2/40

Figure 27: SEM Micrographs of a typical Non-Plasma Treated Flip Chip Test Vehicle With an ENIG Substrate Under LLTS Testing.

CONCLUSIONS

This study investigated the influence of plasma pretreatment on assembly yield and thermo-mechanical reliability of solder flip chip in package assemblies and CSP assembly to SIP modules. Various plasma pretreatment techniques and two substrate surface finishes (OSP and ENIG) were analyzed. Visual inspection of the plasma treated samples revealed higher more uniform fillet shapes especially in the corners of the components. CSAM analysis indicated no significant difference between the plasma treated and non-plasma treated samples in terms of underfill uniformity and void formation during processing. The flow times for the plasma treated flip chip samples were 12 to 20 percent faster than the non-plasma treated samples. The flow times for the plasma treated CSPs were 55% faster. As for flip chip reliability differences, the plasma treated assemblies with Au finish had the highest reliability with a 70 improvement over untreated samples. In contrast, the untreated flip chip assemblies with an OSP finish had 78% higher reliability compared with the plasma treated counterpoints. For the CSP components, the reliability of the components were similar to the non-Plasma treated samples.

REFERENCES

[1] Petasch, W., et al., "Microwave Plasma for Superior Substrate Pre-treatment in Chip Packaging Technologies," Business Briefing: Global Semiconductor Manufacturing Technology, 2003.

[2] Haji, H., "Application for Stacked Package and Flip Chip Bonding," 8th Annual Known Good Die Workshop, Napa CA, September, 2001.

[3] Ryu, S. J., "Improvement of the Flip Chip BGA Defects using Atmospheric Pressure Plasma Surface Treatment Equipment," 39th International Symposium on Microelectronics, IMAPS, San Diego, CA, October, 2006.

EMBEDDED IC POLYIMIDE MULTI-LAYER SUBSTRATE

M. Okamoto, S. Ito, S. Okude, T. Suzuki, O. Nakao, T. Ito and R. Yamauchi, Ph.D.
Fujikura Ltd.
Sakura, Chiba, Japan
m_okamoto@fujikura.co.jp

ABSTRACT

We have developed the IC chip embedded polyimide multilayer substrate as a small semiconductor package of the next generation. This substrate was made by stacking the circuit layers based on the polyimide film. We have achieved the process of embedding chips and laminating films at the same time, utilizing the conductive paste for the connection between layers and the connection to the chip. The chip has the copper rewiring layer using wafer level process to make this connection possible by the paste. At the same time, the chip is thinned down to 85um and embedded in the substrate. We have confirmed this substrate endured the reflow test of JEDEC Level2 and HAST.

Key words: embedded, WLP, polyimide, conductive paste

INTRODUCTION

Recently, miniaturization and enhancement of performance in portable electronics such as cellular phones, digital still cameras and etc. are rapidly advanced. Along with it, miniaturizing and offering high performance and large capacity are also demanded for components installed in those portable electronics. In the field of semiconductor packaging, the System in Package (SiP) technology, that is to involve two or more bare ICs in one package and to systematize functions, attracts much attention due to the background above mentioned. The conventional SiP was to mount two or more ICs on a substrate side by side and was disadvantageous to the miniaturization of the package. In recent years, however, the package form that stacked ICs on an interposer substrate and connected by bonding wires has come to the forefront.[1] Thus, the technology that stacks ICs in three dimensions has great advantage to miniaturization and is becoming a mainstream in SiP field in which high densification of mounting area and wiring space is required. However, some new problems are pointed out to those stacked packages. One is the depreciation in signal integrity at high frequency because the loop length of bonding wires that connects each IC increases when the number of stacking ICs increases. There is also a problem that the increase of the number of pads on ICs disturbs the miniaturization of packages due to the increase of bonding wires. On the other hand, the technology to reduce the thickness of Si chip has been developed for the target of making packages thin. Because it has become possible to thin IC wafer down to 100um, embedded IC technology is more realistic. It is also able to reduce the package size by embedding ICs and passive devices in this substrate. Wafer level package manufacturers propose the multi chip package to which this embedding technology is applied, and the prototype of SiP with built-in bare chips and surface mounting devices such as resistors and capacitors besides IC chip has been reported. We have developed the embedded IC multilayer substrate by combining Wafer Level Package[2] (WLP) with the polyimide multilayer substrate that uses conductive paste for interlayer connection.[34]

STRUCTURE AND FEATURES OF THE EMBEDDED IC SUBSTRATE

Structure and features

Figure 1 shows the cross sectional structure of the IC embedded substrate that we have developed. WLP chip is embedded in adhesive layer of the multilayer substrate and is surrounded by spacers of polyimide films. The rewiring layer of WLP chip and circuit layer of substrate are connected by Interstitial Via Hole (IVH) that are filled with conductive paste. Because the embedded IC is thinned down to 85um and a thin PI film is employed as insulating layer and base of substrate, the thickness of the substrate is able to be less than general surface mounting chip devices.

Process

The making process of this substrate consists of two-stage wiring formation process, which is printed wiring board level and wafer level. Because each process is independent, it is possible to produce them separately. First, wiring formation at wafer level is shown in Figure 2. The process starts with the formation of insulating layer on the substrate. This insulating layer becomes the base of wiring connected with I/O pads of IC and also plays the role to relief stress from surroundings and protects circuit on IC surface. After the aperture to connect with IC is patterned by using photosensitive resin, the copper wiring is formed by using semi-additive process on it. This wiring aims to realign the pad pitch to fit the design rule of the printed wiring board to connect with IC. Specifically, the pad pitch of 100um or less on bare IC has been expanded to the land pitch of 300um or more on the insulating layer. Finally IC is diced after grinding silicon substrate mechanically. We tried to decrease the thickness of substrate by thinning IC down to 85um in this study. Next, the fabrication process of substrate is explained along with Figure 3. Copper Clad Laminated sheet (copper foil and polyimide film are laminated) is applied as a starting material, and circuit is formed to it by etching. Thermosetting bonding sheet in B stage is laminated to substrate, then blind via hole is opened by using laser irradiation. IVH is formed with filling the conductive paste into blind via hole by using screen printing method. To keep coplanarity of the substrate when IC is built in, the spacer layer is provided. The spacer consists of polyimide film and adhesive layer, and is made the same thickness of the IC. Then substrate in which circuit and IVH are formed, IC, spacer and base film are stacked in order, and are pressed in single batch press by vacuum hot press. In this process, substrate and IC adhere together with a thermosetting adhesive, at the same time, each copper pad of substrate and IC is connected with conductive paste electrically. Figure 4 shows the cross sectional photograph of joint portion of substrate and embedded IC. The conductive paste includes metal composition that diffuses and forms alloys while heating and pressing process, therefore connection of high reliability is realized.[5]

EVALUATION
We made the IC embedded substrate using ICs that had two wiring layer on it. Figure 5 shows the external photograph. The size of embedded IC was 1.2mm x 1.2mm, and the thickness including resin wiring layer was 85um. Table 1 shows the design rule concerning wiring for wafer level and substrate. Total thickness of the IC embedded substrate was 0.25mm, and the thinness that fell below the size of surface mounting chip of 0603 size (0.3mm thickness) was achieved. We made substrate that had daisy chain circuit (Figure 6), and evaluated its reliability. Test condition and results are shown in Table 2. No error of disconnection and significant external change were confirmed. In addition, results of tests done solely for the multilayer substrate are collectively shown in Table 3. This substrate satisfies electric resistance and

insulating performance required for the package substrate. We plan to do further examination and evaluation according to the specification of application for this substrate in future.

CONCLUSION
We have developed the IC embedded polyimide multilayer substrate by combining WLP technology with polyimide multilayer substrate technology using single laminating process. These technologies realize an extremely thin structure and simple fabrication process. This substrate is to add on above-mentioned features to the IC built-in substrate, which originally has features of connecting between components by the most direct way, and thus, is suitable for high speed transmission. It is expected that this substrate will advance to the practical stage as next generation package in various field.

REFERENCES
[1] T. Akazawa, "All of SiP Technology," Kogyo Chosakai Publishing, Inc., 2005

[2] Dr. N. Sadakata, "Wafer-level Chip Scale Package," Fujikura Technical Review, No.99, pp.77-80, 2000

[3] O. Nakao, "IVH Multi-layer Printed Circuit Board Laminated with Polyimide Films," Fujikura Technical Review, No.103 , pp.48-52 , 2002

[4] S. Ito, "All Polyimide Multi-layer IC Substrate," Fujikura Technical Review, No.107 , pp.37-41 , 2004

[5] M. Okamoto, "Polyimide Multi-layer Board using Alloying Paste," Proceedings of the 19th JIEP Annual Meeting, pp.125-126, 2005

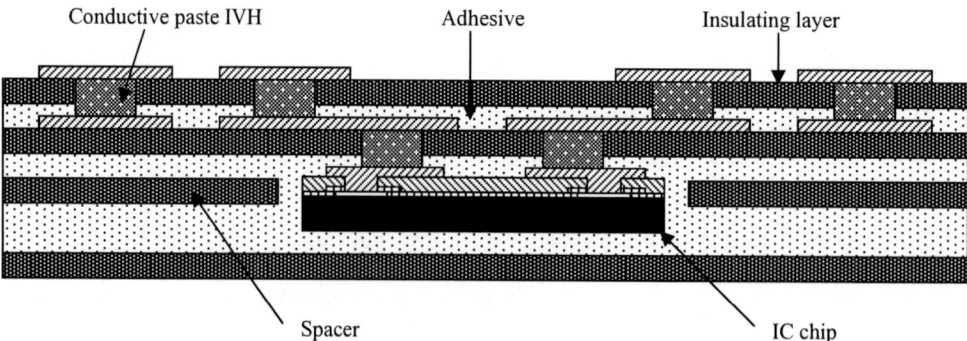

Figure 1. Structure of IC embedded substrate

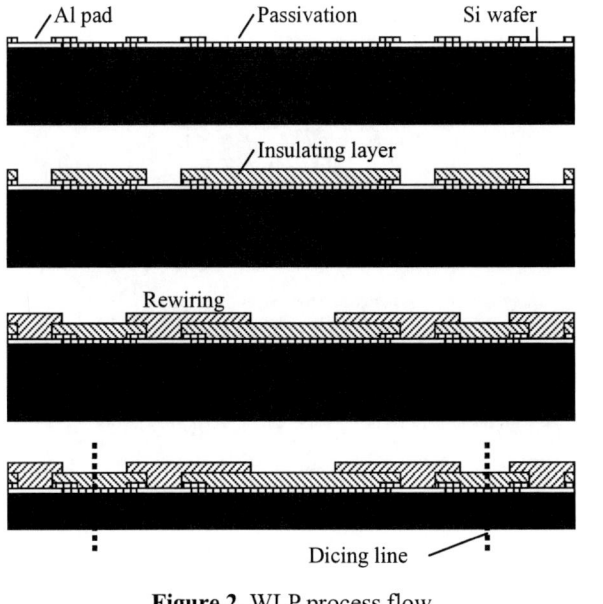

Figure 2. WLP process flow

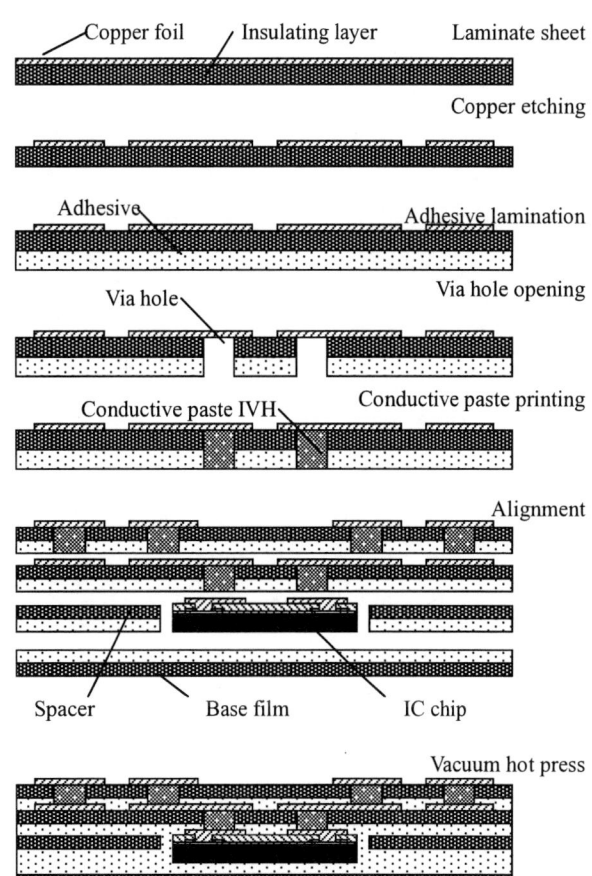

Figure 3. IC embedded substrate process flow

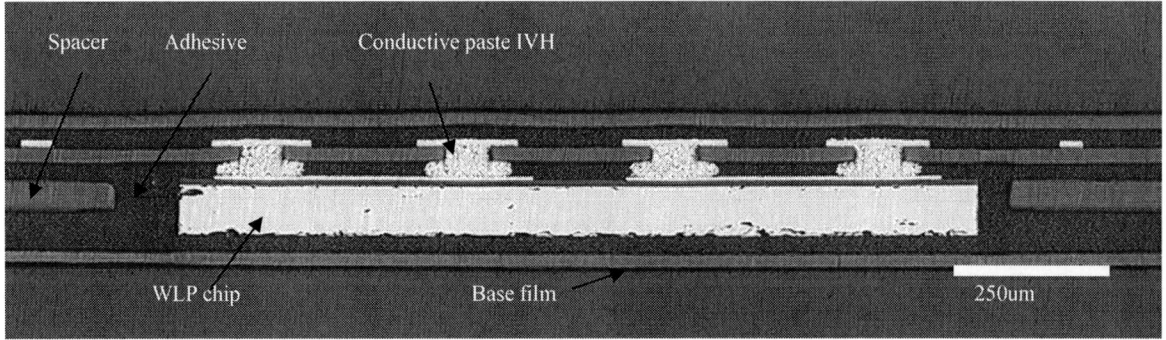

Figure 4. Cross-section of IC embedded substrate

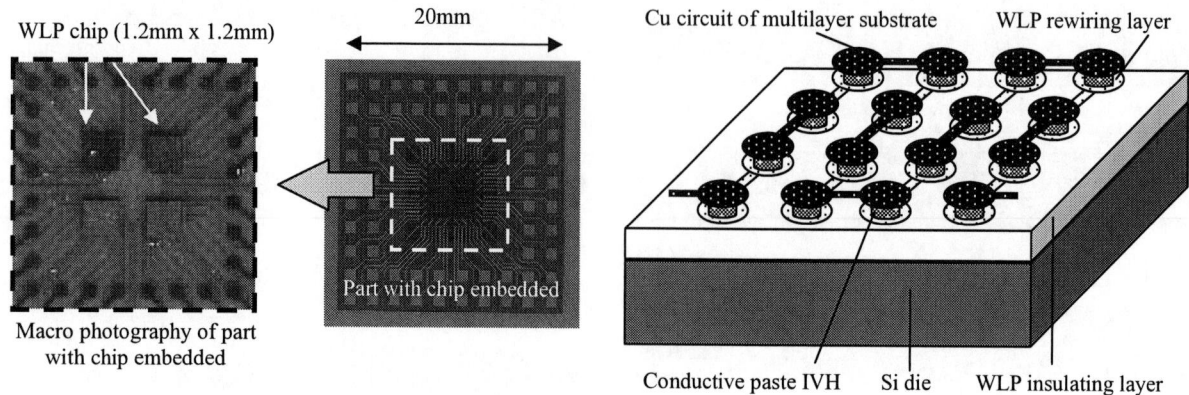

Figure 5. Appearance of IC embedded substrate (4 chips embedded)

Figure 6. Schematic illustration of daisy-chain test element

Table 1. Design rule

Wiring of wafer level Minimum line/space	10um/10um
Wiring of printed wiring board level Minimum line/space	50um/50um
Wiring of printed wiring board level Diameter land/via	250um/100um

Table 2. Results of reliability test

Item	Condition	Result
Moisutre reflow test as JEDEC Level2	85 degrees centigrade 60%RH 168hours Hot air reflow (max 260 degrees centigrade) 3times after moisture absorption	pass
Thermal cycle test (Air)	(Pretreatment: Moisture reflow JEDEC Level2) -25 degrees centigrade 9min/125 degrees centigrade 9min 1000cycles	pass
Highly Accelerated temperature and humidity Stress Test	(Pretreatment: Moisture reflow JEDEC Level2) 130 degrees centigrade 85%RH 192hours	pass

* Specification: No error of externals, void and delamination resistance variance <20%

Table 3. Results of reliability test solely on multilayer substrate

Item	Condition	Result
Thermal cycle test (Air)	-65 degrees centigrade 30min/125 degrees centigrade 30min 1000cycles	pass
Thermal cycle test (Hot oil)	20 degrees centigrade 20s/260 degrees centigrade 10s 200cycles	pass
Temperature-Humidity-Bias Test	85 degrees centigrade 85%RH 50V 1000hours	pass

* Specification: No error of externals, void and delamination resistance variance <20%, insulating resistance >500M-ohm

STAIR-STEP IC PACKAGES FOR LOW COST AND HIGH PERFORMANCE

Joseph Fjelstad
SiliconPipe, Inc.
San Jose, CA, USA
jfjelstad@siliconpipe.com

INTRODUCTION

The electronics industry's birth is marked by many with the invention of the transistor at Bell Labs, however others place the date much earlier with Lee De Forrest's invention of the vacuum tube. An intermediate birth date was suggested in the National Geographic in an issue published late in 1945. The article contained a statement that 1945 marked the beginning of the new "Era of Electronics". However, going back in time again, one will find that the term "electronic" was actually coined around 1910, substantiating the claims of those who give De Forrest credit. While determining a birth date is an entertaining exercise, a pragmatist might well argue that it was the invention of the IC that marked the birth of the electronics revolution we see about us today. As important and the IC is, it is useless without a supporting interconnection structure and experts are increasingly coming to agree that it is the IC package that holds the key to unleashing the performance potential of current and future integrated circuits. Today's multi-gigahertz processors operate at internal clock speeds that outstrip the electronic interconnection infrastructure's ability to support them. There is clear need to fill this performance gap. The potential of the processor becomes difficult to tap. While on-chip clock speeds keep pace with Moore's Law, short range transmission of those signals is being degraded by the IC packaging and circuit interconnection which have not been able to keep pace.

While optoelectronic interconnection solutions are being worked on as possible future solutions, electronic interconnection technologies are likely to the used far into the future. This is both due to the dominance of electronics and significant number of impediments to near term adoption of optical solutions. Optical data transmission technologies are stymied by such road block issues as high cost; high power requirements; limited infrastructure and lack of experienced engineers. Even so, for long distance signal transmission, optoelectronic solutions are unparalleled. However, it is cost effective transmission of signals over short distances that is needed. Lacking immediate cost effective and reliable solutions for optical interconnections, the near term challenge is to solve the problem of designing and constructing an optimal electrical signal channel. Digital signals in the gigabit-per-second range must overcome many of the same problems that have faced the designers of RF and microwave circuits in terms of managing signal loss. Designers must manage signal transitions along an often tortuous path from chip-to-chip

and in doing so overcome the parasitic effects of all interconnections, beginning with the IC package.

IC PACKAGING'S 3D CHALLENGE

After the semiconductor itself, the IC package is the critical next connection in the hierarchy of electronic interconnections. Design of packages has undergone several distinct eras, beginning with through hole technology (pin in hole) in the 1960s. Transition to surface mount packaging occurred in the late 1970s, with area array packaging taking place in the late 1980s. Chip scale and wafer leveling packaging arrived in the mid 1990s and 3D packaging, characterized by stacked chip and stacked package technologies, recently arrived. The effects of integration can perhaps be best characterized by the steady increase in the percentage of silicon relative to the size of the circuits that support them. For example a highly dense circuit of the early 1970s might have had silicon represent at most perhaps 5% of the total area of the circuit board. Surface mounted package assemblies with two side assembly could reach perhaps 25% active silicon. A densely packed circuit assembly using chip scale packages could theoretically reach beyond 100% and now with 3D packages employing stacked chips or package structures from 2 to 8 high, the upper limit is unlimited in theory. The practice of such technologies is moving into full swing. For example, Tessera (San Jose, CA) fabricated a demonstration SO-DIMM memory module which had 200% and 400% active silicon on the module. Understandably thermal considerations for such structures are not inconsequential but the capability has been demonstrated and is a harbinger of things soon to come. While performance, functionality and economy are improved by such structures, there are practical limits to what can be achieved using both traditional and the new 3D structures. To overcome these limitations, there is need to reconsider once again, how IC interconnections are accomplished especially if copper is going to serve as the interconnection medium.

In recent years, copper has been seen by some specialists as having a limited future as an interconnection medium for high speed devices. This limited expectation for copper has been largely based on the historical performance of copper interconnection technology, which has been advanced in fits and starts of improvement. Looking back over time, one can see that, for many years, copper interconnection exceeded semiconductor speed capability. However, since the mid 1990s, performance of copper has fallen behind, while integrated circuit performance has improved in geometric

161

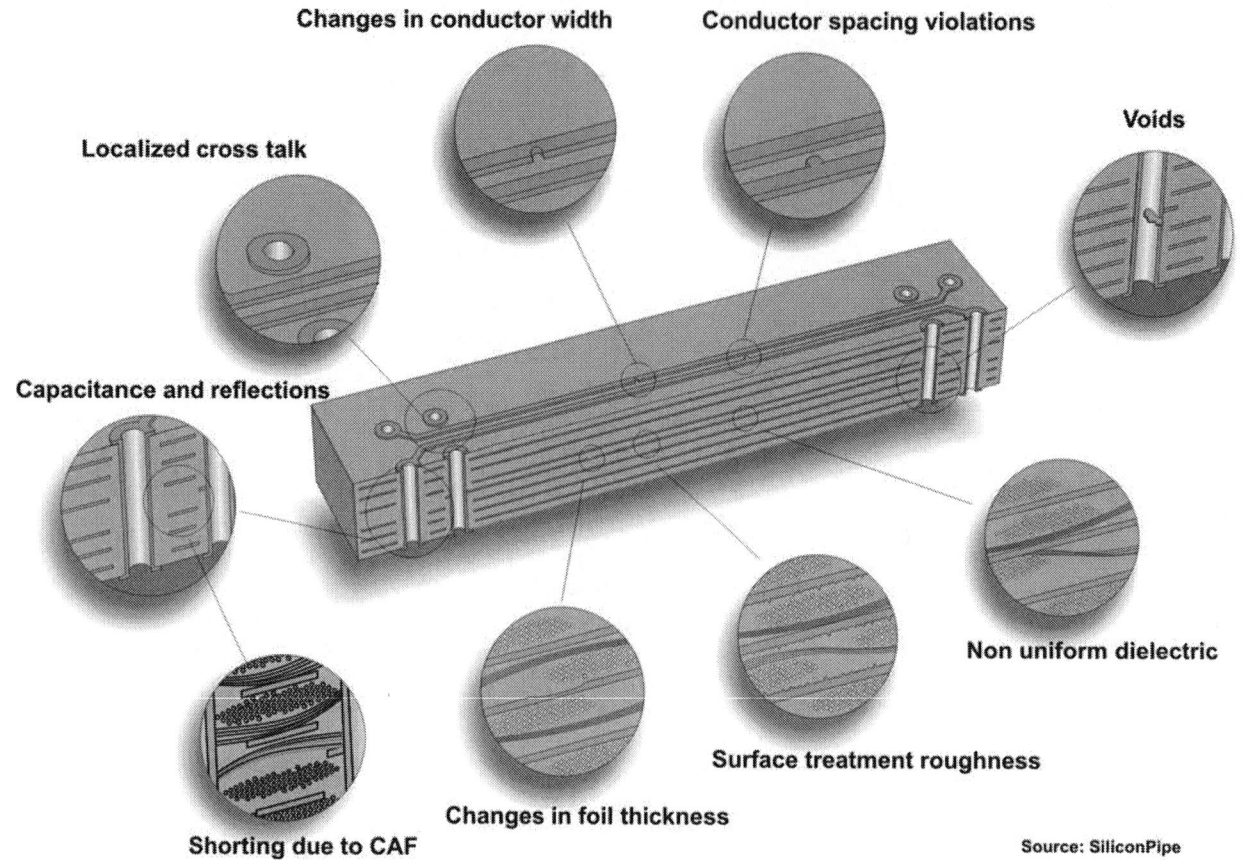

Figure 1 Printed circuits are prone to a number of design and manufacture artifacts that can impact signal integrity.

fashion in line with Moore's Law. A first impression might give rise to the thought that the gap between the technologies is unbridgeable, however a closer look reveals the presence of potential solutions using today's manufacturing technologies and infrastructure. In fact, with the proper choice of laminate materials, PCB architecture and system design, copper based IC packaging and interconnection technology it appears may actually be able to jump ahead of semiconductor technology.

SIGNAL PATH IMPEDIMENTS
The fundamental limits of copper based circuits and interconnection technology are already well known. Copper transmits signals at near the speed of light in air (or vacuum) but in practice the copper must be supported and surrounded by a dielectric. The dielectric and its electrical properties are key factors in speed, but so also are the physical characteristics of the channel. Fundamentally any disturbance in the signal path will degrade or reduces signal rise time, which is critical to signal integrity. In an electronic design, these impediments are defined by a number of different physical factors including the materials and interconnection path shapes that are a part of the design itself. Conductor loss, dielectric loss and impedance discontinuities manifest in a particular design (connectors,

through-hole vias, material changes, manufacturing defects and the like) are primary contributors to signal degradation. Now with the increase in processor speeds into the multi gigahertz range along with the speed of silicon in general, the electronics industry has reached a point where concerns once limited to the design of RF and microwave products have now crossed over into the realm of digital signal designs.

In the work of high speed PCB design, manufacture and assembly, there are a number of different elements impacting signal integrity. Common examples include inconsistencies in the physical properties and electrical properties of the dielectric, variations in signal trace width, changes in circuit spacing and uneven copper thickness. Even the type of foil adhesion treatment can affect the signal. (See Figure 1) All these common attributes must be taken into account as part of the signal integrity engineer's ability to accurately predict and assure maximum performance for a given design. When tied to the range of common electrical parameters such as resistance, dielectric loss, conductor loss, stray capacitance, signal skew and inductance (which can lead to cross talk) and the potential reflections due to electronic stubs from common circuit features such as vias, engineers quickly see a compounding and confounding confluence of design challenges.

Improvements in materials now having lower loss tangents and lower dielectric constants, coupled with better manufacturing materials and processes, have achieved some excellent performance gains. However, if one pays greater attention to the message offered by signal integrity experts relative to those performance limiting elements associated with printed circuit design and manufacture, some significant opportunities start to appear, beginning with the IC package.

REFINING SIGNAL PATHS

A reasonable path to addressing the problem is to simply avoid designing using the traditional approach of routing all signals within the PCB. Instead, a designer can segregate and separately handle the critical signal lines on a more easily controlled signal path. This is in keeping with the fundamental objective of high speed circuit design: get the signal to its target as directly and cleanly as possible. Knowing that the shortest distance between two points is a straight line, one can quickly envision multiple alternatives. But to achieve the objective, one will likely find they must depart from traditional design layout. As an example, it has been shown that a properly designed signal channel, in copper, is capable of carrying data without pre or post-emphasis well into the 20 gigahertz (40 gigabits per second) range [i], [ii] but not using standard methods.

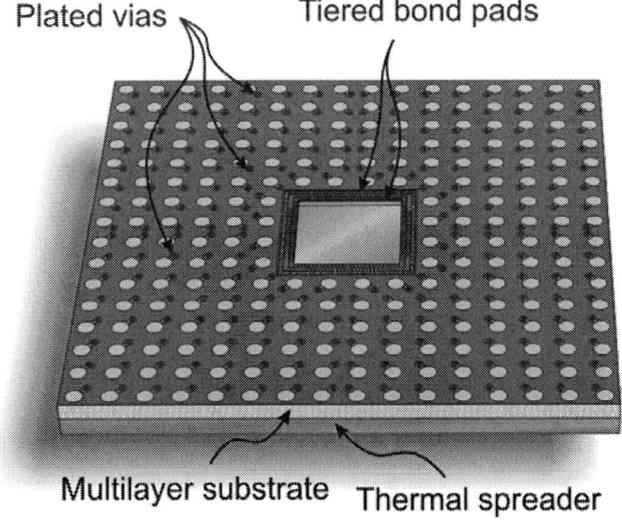

Figure 2 Traditional tiered wire bond layers for IC packages require plated vias which can sap performance.

When using traditional approaches to interconnection the ICs have signals (and power and ground) connected by means of wire bonds or flip chip solder balls. The signals traverse and exit the package often after tunneling their way through a number of layers of interconnection then travel down through larger solder connections into the PCB then through plated vias in the substrate. This path is repeated, in

reverse, up through a second set of solder connections into another chip package to the chip. It is evident even to one not skilled in circuit design, that the path is not optimum. There must be a cleaner path for signals. And there is. However, the new approach to IC package design and interconnection, necessarily breaks from traditional methods in order to yield the desired solution. Fortunately, the solution is simple. Simplicity, in thought, in design is a concept that has echoed throughout history from the ancient Greeks to modern times. Einstein warned that: "Things should be made as simple as possible but not simpler". So it is also that simplicity, when properly executed and applied to IC packaging, will yield the commonly sought objectives of higher speed, lower cost and greater reliability. It will also yield those objectives with greater facility.

STAIR STEP PACKAGING (SSP)

Using the precepts just counted a new approach to IC packaging was conceived. The effort was targeted to significantly simplify the design and manufacturing process to lower cost, while simultaneously offering the potential for much improved electronic performance. The concept builds on a idea first employed for improving the density of wire bonded IC packages such as pin grid arrays (PGA) in the mid 1980s. This technique is generally referred to as a tiered wire bond pad IC package structure. (See Figure 2) The method was used to successfully address the I/O density problems in applications where on chip pad pitch density exceeded the feature size limitation and routing capability of standard IC packaging technology.

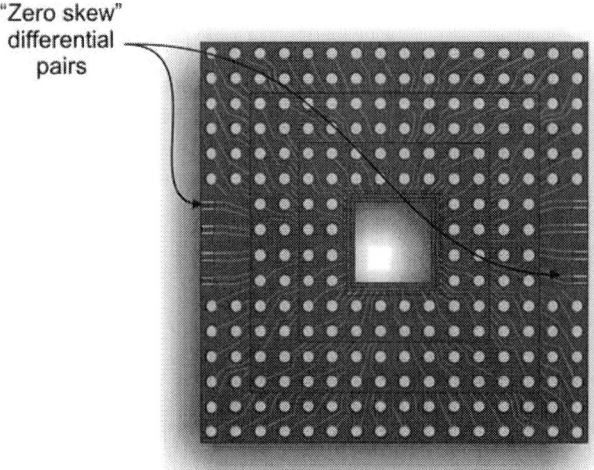

Figure 3 The stair step package (SSP) concept offers many advantages while obviating the need for plated through hole vias and saving materials.

In contrast, the new proprietary (patent pending) packaging technology [iii] takes the earlier tiered contact concept further extending it to the I/O terminations on the package surface. The result is a stair stepped or "wedding cake" like IC package structure. (See Figure 3) The new package structure has been dubbed a Stair Step Package

163

(SSP). The advantages of the simplified method are significant in all of the prescribed areas including, cost, performance and reliability. A point based evaluation of the concept relative to such commonly applied metrics relative to IC packaging will make clearer the benefits of the new approach.

Looking first to cost, it is noted that cost is reduced because of the reduction in the number of manufacturing steps and especially the elimination of plated vias. This concept follows earlier work in the creation of the "off the top" OTT™ package, which was designed for direct interconnection between one or more IC using the top surface of the IC package to bypass and obviate the need for plated vias in critical path signals[iv], [v] and can be used in combination with the earlier concept. These simple structural elements of the package also allow for improved manufacturing yields. Electrical testing cost should also be greatly reduced, if not eliminated. This is asserted because the circuits are only on one side (though a ground layer may be provided on the second side of each layer if desired) and they can thus be easily examined visually for shorts and opens and non-uniformities which could affect electrical performance. Moreover, because each layer only has as much material as is required for the circuitry used, there is less waste, especially in volume manufacture.

Qualified materials and equipment to accomplish assembly exist today and can be used with little, if any, modification. Finally, an added advantage is available in that individual layers can be inventoried for creating custom packages, including mixed pitch I/O if desired, on a moments notice. Turning to performance, one who attempts to determine the cost of standard circuit substrates will find that the same plated vias which add cost and consume routing space within the circuit can also be an impediment to electrical performance by creating unwanted capacitance. Thus by eliminating plated vias from the package, signal performance is significantly improved. In fact, complex field solver analysis, more frequently now required for characterizing critical signals as they pass through the package, is not needed. Moreover, differential pairs, common to most of today's high speed circuit designs, can be designed such that they have virtually zero skew and so that cross talk can be almost completely eliminated.

If this were not enough benefit, the clarity of the signal channel has yet another benefit that is not commonly recognized and that benefit is power savings. A clear channel means that much lower voltages are required for reliable signal transmission. For example, in one application of an embodiment of this general type of package, which was designed to demonstrate the fundamental capability of direct and via less signal transmission, the total power required was *less than two percent* of the anticipated requirement[vi]. Moreover, because of the versatility of the Stair Step package method, substrates of differing material types can be used when and where required so higher cost materials that are often desirable of high speed signals can be used sparingly and mixing of I/O pitch on a package is also possible. (See Figure 4) This approach also opens up routing channel opportunities to internal I/O terminations.

Reliability is the last item on the list. Once more, obviating the need for plated vias, which are often the "Achilles Heel" of interconnections, will help to improve overall package reliability. Presently under study, but not yet proven, is the expectation that the periphery I/O of the package having longer reach when making connections to the board will elongate, creating column-like terminations. Solder joints elongation has been identified as a condition that appears to provide improved solder joint reliability in some studies.[vii],[viii]. If concerns remain about assembly, because of the differing ball heights, ball height can be made uniform by simple using course pitches on outer (periphery) I/O rings. (See again Figure 4)

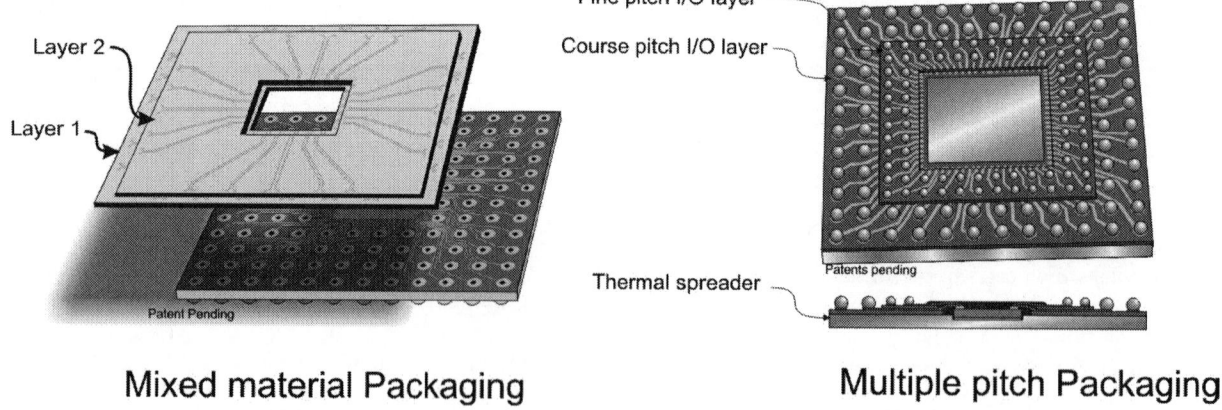

Mixed material Packaging

Multiple pitch Packaging

Figure 4 The stair step packaging concept has a some far reaching implications including the ability to fit (or retrofit) extant IC packages with OTT capability for multilevel interconnections and as well to create structures having different I/O pitches and either different solder ball or metal post heights to maintain planarity

Another benefit of the technology that finished IC packages can be fully tested and characterized at speed. Over all performance is improved greatly while power requirements actually drop. In demonstration system the output driver of the IC was able to operate at less than 2% of the projected power and yet was driven nearly 3X further than anticipated and through twice the number of connectors (i.e., two connectors rather than only one). While the single channel test was run at 10Gbps, the data indicated that the structure could achieve 25Gbps performance (See eye diagram in figure 5) with increased power and signal pre emphasis.[ix]

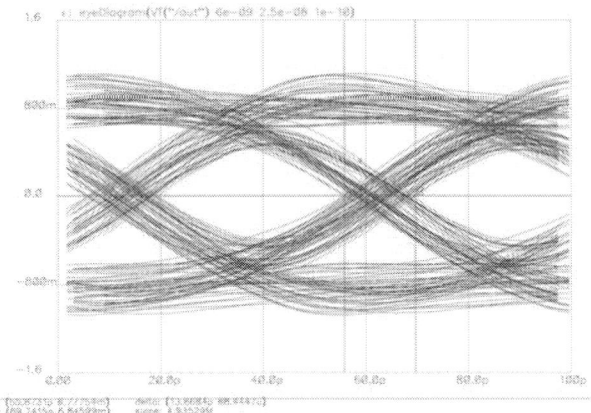

Figure 5 Eye diagram of OTT performance through two connections and 75mm channel at 20 Gbps

FUTURE DESIGN CHALLENGES
While these structures have clear benefits, they do require changes in EDA design concepts, the design tools and both manufacturing and assembly as well. However, all these areas are impacted but the challenges are relatively small and incremental. For example the OTT and SSP packaging concepts were vetted by the CTO of a major IC packaging foundry, and they indicate that these structures appear to be "business as usual" from an assembly perspective but that there would be a learning curve. Clearly both IC die packaging foundries and certain select customers desire the ability to do full parametric testing at speed. Board level assembly is a bit more complicated and is not presently a drop-in solution for most assemblers. Still, OTT is proving a compelling option to the end users requiring high speed and low cost. Moreover the wide application of OTT packaging technology in 10bps backplanes, high speed memory systems and in FPGA has shifted the current thinking of where architectural barriers exists. In combination, there potential appears to be imagination limited. Figure 6 illustrates and champions this notion.

CONCLUSION
Stair step packaging structures for IC die offer significant performance benefits. These seminal packaging and interconnection structures for ICs eliminate the need for vias on critical signal routes, while simultaneously offer the potential for materials savings of up to 30% because each successive layer uses less material. The physical structure of

the solution is simple and in terms of manufacturing techniques easier to build, with fewer steps (no drilling, no plating.) Design costs are reduced as signal layers are directly converted to micro-strip or strip-line through added metallization. Moreover, there is little need for signal integrity tools and experts since channels are highly predictable. From a performance perspective the plated via related impedance concerns are eliminated as is cross-talk. Skew is easily managed and a lower inductance on power feed paths can be provided. The technology also uses current materials and manufacturing techniques (flex or rigid) and optical inspection will predict electrical performance simplifying test. Cost economies of continuous (reel-to-reel) flow manufacturing is possible when flex substrates are used. Another unique potential advantage is that the technology can be used for both custom (design on demand) and COTS (assemble on demand from stored layers of standard configurations) package structures.

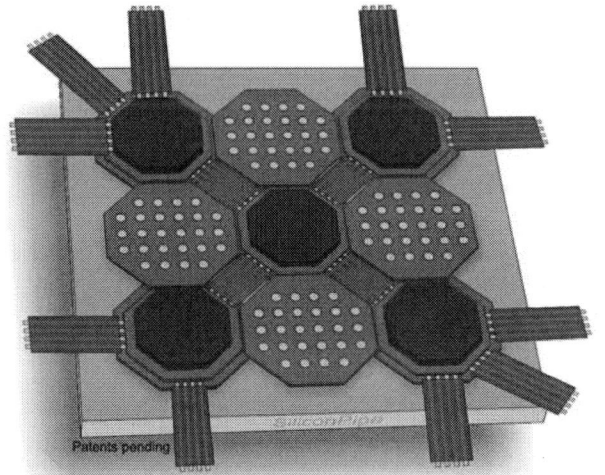

Flex for power, ground and high speed signals

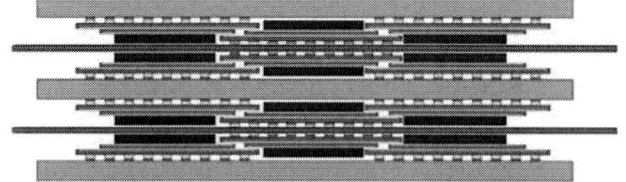

3D stacked circuit and package assemblies

Figure 6 Redefining interconnection using SSP and OTT concepts offers a wide range of potentially valuable architectural opportunities not possible using more traditional approaches. Top I/O as shown in the upper image can be used either for clean power & ground or signals.

In summary, a wide range of IC packaging technologies have been developed over the years. The newer methods described here, appear well suited to helping to meet the challenges. Flexible circuit based solutions specifically are offering some unique opportunity to designers to design complex systems, without adding complexity.

The package and interconnection structures and methods described facilitate the creation of shorter, cleaner signal paths. The methods will extend the performance limits of copper interconnections well into the future and offer the benefits of higher performance at lower I/O power and lower overall cost, while reducing layer count and complexity of both the IC package and the substrate to which it is mounted.

REFERENCES:

[i] Allan, R "Coax Backplane Connector Handles Rates To 40 Gbits/s" *Electronic Design Magazine*, Oct 13, 2003, pp 39-40

[ii] Merritt, R. "Board Design Revamped" *Electronic Engineering Times*, October 13, 2003, Pg 45

[iii] United States Patent Application 20050093152

[iv] Fjelstad, J., "Getting from Chip to Chip Fast, Novel High Speed Chip to Chip Signal Transmission Structures" *Global SMT and Packaging,* December 2003

[v] Fjelstad, J, Yasumura, G and Grundy, K., "3-D Partitioning of Printed Circuit Design – 'Elevated Highway Bypass' Packaging Design", *Advanced Packaging Magazine* February 2005 pp: 12-19

[vi] http://www.eetimes.com/press_releases/prnewswire/showPressRelease.jhtml?articleID=X310227&CompanyId=1

[vii] Lau, J. et. al. "HDPUG's Failure Analysis of High-Density Packages' Lead-Free Solder Joints" presented at APEX 2003 to the IPC SMEMA Council

[viii] Heinrich, S.M, et. al., " Improved Yield and Performance Of Ball-Grid Array Packages: Design and Processing Guidelines for Uniform and Non-uniform Arrays" *Components, Packaging, and Manufacturing Technology, Part B: IEEE Transactions on Advanced Packaging*, Volume: 19 , Issue: 2 , May 1996 Pages:310 – 319

[ix] Grundy, K. et al., "Designing Scalable 10G Backplane Interconnect Systems Utilizing Advanced Verification Methodologies" Proceedings *DesignCon* February 2006

ADVANCED PACKAGE PROTOTYPING USING NANO-PARTICLE SILVER PRINTED INTERCONNECTS

Sungchul Joo and Daniel F. Baldwin, Ph.D.
The George W. Woodruff School of Mechanical Engineering
Georgia Institute of Technology
Atlanta, GA, USA

ABSTRACT

To reduce manufacturing cost, lead time, process complexity, and enhance electrical and mechanical reliability performance, an embedded-active approach that targets rapid prototyping and low-volume production in micro-system packaging is being developed. The approach involves a rapid prototyping of micro-system packaging by a data-driven chip-first packaging process using direct printing of nano-particle metals. In the chip-first process, bare dice are first embedded into a copper or stainless steel carrier substrate, fixed by filling up the gap between the chips and the substrate with thermoplastic adhesives, and planarized to a common planar surface. On the coplanar substrate, polyimide or liquid crystal polymer (LCP) film is laminated to form a dielectric layer. Through the dielectric layer to the chip metal pads, micro vias are drilled by laser ablation. The vias are filled with nano-particle silver (NPS), which has high conductivity and good adhesion to copper, polyimide, and LCP. The NPS is deposited by screen printing and a three dimensional electrical circuit is formed. This packaging approach is data-driven, so requires no masks and reduces packaging turn-around time from months to days. It is also less limited by substrate composition and morphology, eliminates the need for special chip processing such as the need required for flip chip solder bumps, and permits using any chip technology and any chip supplier allowing mixed devices. The embedded-active process with nano-particle metals avoids the extreme processing conditions required for standard IC fabrication such as wet chemistry processing and vacuum sputtering. The nano-particle conductors typically measure around 5nm in diameter and can be sintered at plastic-compatible temperatures as low as 220C to form material nearly indistinguishable from the bulk metal. The embedded-active packaging shows an excellent reliability performance in terms of thermal shock, which is performed in the range of -45 and 125 degree Celsius. These results represent an important step to a system packaging characterized by high density, low cost, and data-driven fabrication for rapid package prototyping. This paper presents an electrical performance characterization of the nano-particle conductors, details of the rapid prototyping process sequence, an initial reliability characterization of the package architecture, and a failure mode analysis of the packages.

INTRODUCTION

The electronics packaging process becomes more complex, causing increased manufacturing cost and lead time and difficulties in improving mechanical and electrical reliability performance as more functions are needed in a limited size of an electronics package.

For the purpose of increasing the functional density of the electronics package, General Electric (GE) and Texas Instruments (TI) developed a chip-first approach, which is also called high density interconnect (HDI). [1] In this approach, ICs are placed in the cavities of a ceramic or plastic substrate and bonded to the substrate. Above the substrate, a dielectric layer is formed, micro-vias are created by the laser drilling technique, and Ti/Cu/Ti metallization is used to form a multilayer interconnect.

As a similar HDI technology, Intel developed a bumpless build-up layer packaging process that involves chips embedded in a bismaleimide triazine (BT) laminate or a copper heat spreader, which then has one or more build-up layers formed on top. [2] A standard microvia formation process makes the connections between the build-up layers and the chip pads. The embedding of the chips in the panel may be done with molding or dispensed encapsulation material.

In addition, the Technical University of Berlin and the Fraunhofer Institute IZM developed somewhat different HDI process. [3] This process consists of die-bonding to laser-cut windows of ceramic or silicon substrates by filling the gap between chips and substrate with an epoxy, planar embedding of multi chip module (MCM) by mechanical polishing of its backside, a standard thin film multi-layer process with a photosensitive polymers on the surface of the embedded substrate, and interconnects of bare dies or standard passive components to the copper routing of the MCM.

Although the traditional HDI approaches can meet the requirements of increasing functional density, they cannot avoid the effect of extreme processing conditions required for standard IC fabrication such as vacuum sputtering and wet chemistry processing, critical factors influencing process complexity, the manufacturing cost, and the lead time of the process.

An innovative chip-first approach using a nano-particle metal (NPM) as an interconnector, specifically targeting rapid package prototyping and low volume production in microelectronic packaging, is being developed not only to reduce process complexity and manufacturing cost and lea time but also to enhance electrical and mechanical reliability performance. [4] This paper addresses an electrical performance characterization of the nano-particle conductors, details of the rapid prototyping process sequence, an initial reliability characterization of the package architecture, and a failure mode analysis of the packages for its successful widespread implementation.

RAPID PACKAGE PROTOTYPING PROCESS

The assembly process starts with a bare substrate made of metal such as copper or stainless steel, where cavities for inserting chips are formed (Figure 1). Chips are upside-down bonded into the cavities by compression bonding of a thermoplastic adhesive film or liquid crystal polymer (LCP). Prior to the compression bonding, for a coplanar surface of the chips with the substrate and a temporary maintenance of the initial position of the chips aligned by the placement machine, a polyimide tape is attached to the bottom of the substrate. After the chips are bonded and the tape is separated, polyimide or LCP film is laminated on the active surface of the chips and the substrate to form a thin film structure used for a dielectric layer. Through the dielectric layer micro-vias are drilled to the chip pads by an excimer-laser ablation technology. Finally, nano-particle metal is deposited to complete an electrical circuit by screen printing.

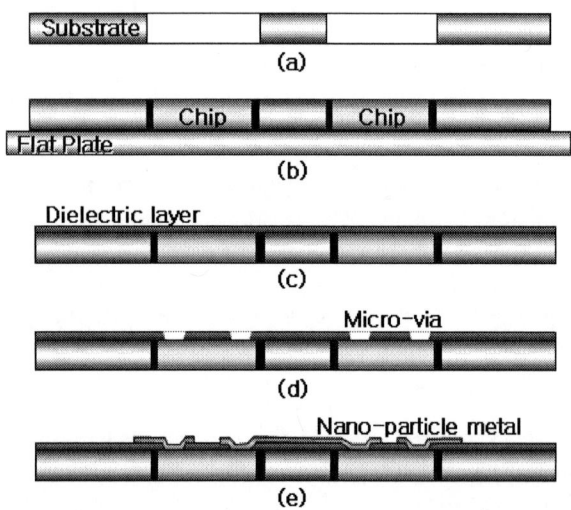

Figure 1. Schematic outline of a rapid package prototyping process: (a) Make cavities in a substrate (b) Attach die to the substrate (c) Laminate dielectric layer (d) Create micro-vias (e) Deposit nano-particle metal

Figure 2 shows the photographs of package prototyping by chip-first process. An array of 4 by 5 cavities with each side length of slightly larger than a die is milled in the stainless steel substrate. Dies placed in the cavities are bonded to the

substrate by filling the gap with LCP between chips and substrate. Polyimide film is laminated over the chips and the stainless steel substrate.

Primary defect occurred in the lamination step is void resulting in delamination in the subsequent process step (Figure 3). The void in polyimide lamination is typically caused by moisture that the polyimide film has absorbed. To remove the moisture in a polyimide film, pre-drying in forced air ovens at 100C for 24 hours are indispensable. Polyimide film dried in this way contains up to 0.07% residual moisture, which is enough to produce void free laminations. In addition, exposure to a 70% relative humidity environment will cause two-side coated polyimide film to regain sufficient moisture in 30 to 60 minutes to produce voids during lamination. Lamination pressure, temperature, and the time above melting point should also be large enough for the Teflon adhesive to bond to the chip and the polyimide dielectric layer to prevent delamination. The optimized lamination conditions for the rapid package prototyping are set as in Table 1.

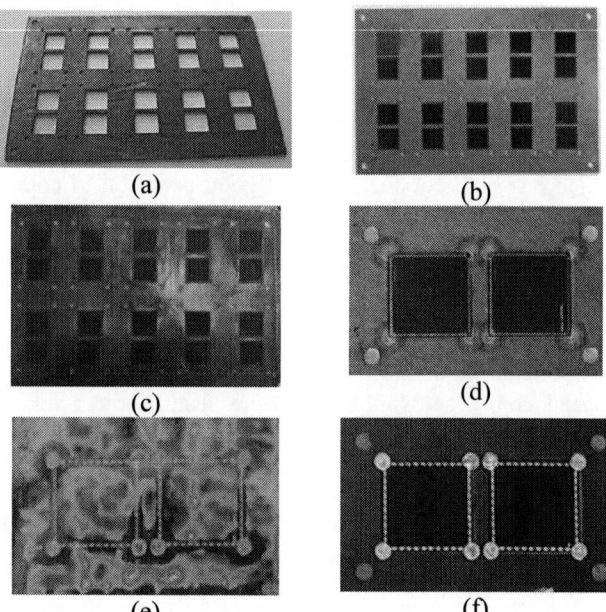

Figure 2. Photographs of package prototyping by chip-first approach: (a) Carrier substrate (b) Die bonding (c) Polyimide lamination (d) Laser ablation for two chips (e) Screen printing of NPS (f) Stripping polymer mask

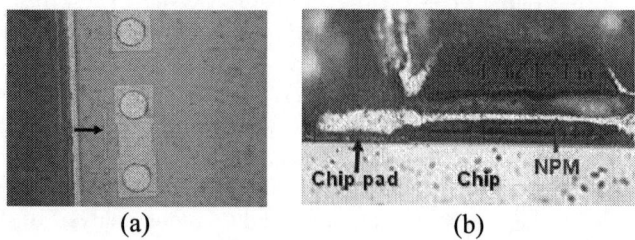

Figure 3. Potential defects in chip-first approach: (a) voids in lamination process and (b) delamination caused by voids (cross section)

Table 1. Lamination conditions

Pry drying	100C, 24 hrs		
Process start	Within 30 min after drying		
Process conditions	Pressure	Maximum Temperature (Tm)	Time at Tm
	240 psi	575F	20 min

Micro-via formation and circuit patterning are accomplished by using a computer controlled excimer laser. In the excimer-laser ablation technology, laser photons directly excite and break the chemical bonds of the solid, causing ejection of material. The excimer laser machine uses KrF that produces a wavelength of 248nm. The primary factors for laser ablation are fluence (energy density), repetition rate (pulse frequency), the number of pulse (bursts), transmission of light, and demagnification. These factors are set to optimize the laser ablation process using 600um mask under constant energy of 200mJ as in Table 2 and to get good quality vias as in Figure 4.

Table 2. Excimer laser ablation conditions

Factor	Level	Factor	Level
Repetition Rate	100Hz	Transmission of light	80%
The number of pulse	Via: 250 Bridge: 140	Demagnification	4.5

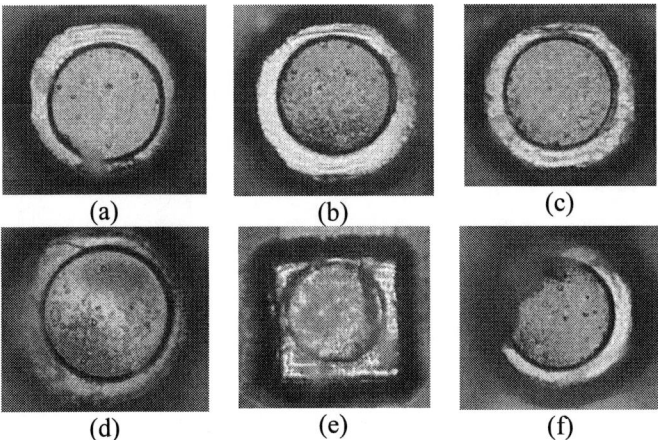

(a)   (b)   (c)

(d)   (e)   (f)

Figure 4. Good quality vias and bad vias in micro-via formation process: (a), (b), and (c) good quality via (d) copper pad etching (e) shallow etching in square via (f) Teflon hang-on

NANO-PARTICLE METAL
Typically, the particle size of nano-metal is in the range of 1 to 100 nm. Due to their small size, it has been found that nano-particle metals have much low sintering temperature than their bulk materials. [5] The size effect of nano-particle metal results from a considerably high surface area to volume ratio in nano-particle. Nano-particle silver (NPS), which measures 3 to 7 nm in diameter, is coated with a dispersion agent for separation of particles at room temperature. At its sintering point of 220C, heat activates supplementary material that removes the dispersion agent from the surface of the particles and initiates the connection among particles by welding and fusion mechanism. [6]

Adhesion property
In order to be suitable for use in electronics package prototyping as an interconnection, the NPS should have a strong adhesion characteristic specifically to dielectrics and the metal pads of ICs. Dielectric materials, polyimide, liquid crystal polymer (LCP), benzocyclobutene (BCB), and silicon nitride (Si3N4), and chip-pad metals, copper and aluminum, were chosen for this study because of their popular use and commercial availability. According to the ASTM D 3359-02, tape test to see if the adhesion of NPS to the selected materials is at a generally adequate level is performed by applying and removing pressure-sensitive tape over cuts made in the films [7].

Regarding the mechanism of adhesion of NPS to dielectrics or metals, the surface of these materials is roughened by dispersion agent in NPS that chemically erodes material surface during the sintering process of 220C for one hour, and NPS sinks within such a rough surface, resulting in mechanical interlocking effect. Therefore, surface pretreatment of specimens is very important for improving adhesion. To remove the oxide layer on copper surface, hydrochloride of 10% is mixed with 90% of de-ionized (DI) water; to eliminate organic contaminants Acetone-Methanol-Isopropyl-DI water is used and to pretreat the surface of aluminum and BCB, plasma cleaning is used.

By rating the adhesion according to the standard, we obtain the adhesion test results in Table 3. Polyimide, LCP, and copper show the strongest adhesion characteristics, which indicates that polyimide and LCP are suitable as dielectrics and copper as chip-pad metal for electronics packaging. NPS cannot remove the oxide layer on the surface of aluminum film, resulting in the weak adhesion result of aluminum to NPS.

Table 3. Tape test results

	Material	Adhesion Classification*
Dielectrics	Polyimide	5B
	LCP	5B
	BCB	4B
	Si3N4	0B
Metal	Cu	5B
	Al	0B

*Percentage of area removed; 5B=0%, 4B<5% 0B>65%

Resistivity
Different thickness levels of NPS interconnect produce significant effects on the resistance of the interconnect

because the surface of NPS sinters first and volatile solvent tends to remain inside of the material at higher thickness, causing much higher resistivity. To achieve low resistivity and an optimum thickness, resistances of the films with varied thickness are measured. Basically, the thickness of NPS shrinks approximately ten-fold from the initial thickness and the conductivity slowly increases with time during the sintering process.

Four-wire method is used for measuring resistance to remove lead resistance of a resistance tester. The formula R=ρ*L/A, applied on a rectangular test vehicle, shows the printed silver to have a resistivity of between 5 and 7 $\mu\Omega \cdot cm$ within 1.2 mil (30um) and much higher value at 5 mil (125um) of initial thickness as in Figure 5, which is comparable to 1.6 $\mu\Omega \cdot cm$ of bulk silver. The high resistivity probably results from the incomplete sintering of the particles due to the solvent remaining inside of the metal. [8] Therefore, the initial thickness should be less than about 30um to obtain a low resistivity interconnect, and redeposit of NPS over the first cured silver is recommended when a thicker interconnect is required.

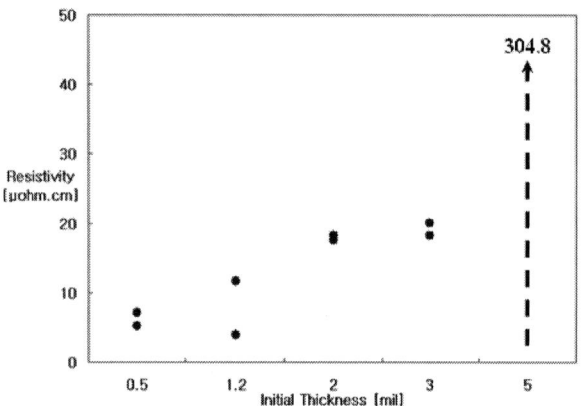

Figure 5. Resistivity change of nano-particle silver

Leakage current

The electrical conductivity of all materials including polymers ranges from 10^6 of graphite to 10^{-18} $\Omega^{-1}.cm^{-1}$ of quartz. The low but finite conductivity of many polymers has been frequently explained by ionic conduction [9]. In addition, one serious concern of NPS used in electronics packaging is the electro-migration of silver ions because such migration may cause a reduction in electrical spacing and ultimately a short circuit between interconnects when voltage is applied under high humidity and temperature conditions [10]. Therefore, leakage current for the combination of LCP and NPS should be measured with increasing potentials to determine its leakage-current level and a voltage that results in excessive current leakage.

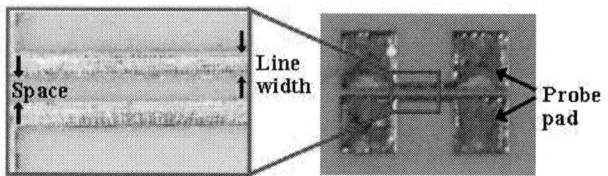

Figure 6. Test vehicle for leakage current test

The opposite electrode pattern with 10 mm in length and varied space and line width, as in Figure 6, is formed on the LCP substrate. Leakage current between the two electrodes attains only a few Pico-ampere levels before the applied potential reaches the breakdown voltage of the test pattern, which falls within a range of 400 to 500 V, as shown in Figure 7. It is known that the rate of silver migration is increased by an increase in the applied voltage and a decrease in the distance between electrodes of opposite polarity [10]. Thus, the breakdown voltage of the test pattern is predicted to decrease as the pitch decreases.

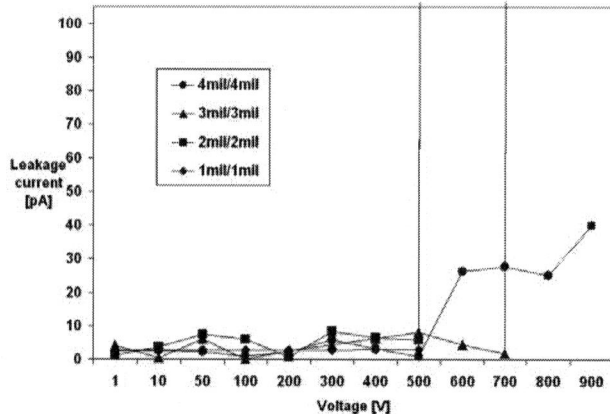

Figure 7. Leakage current of nano-particle silver with various line/space and final thickness

RELIABILITY TEST
Test vehicles

The test vehicle used in the present study is based on a 0.2 inch by 0.2 inch test die, which has 88 I/O pads giving 4 resistance nets on four sides composed of 44 daisy chain pairs on 8 mil (204 micron) pitch located on the peripheral of the die, assembled on a bare stainless steel substrate. Figure 8 illustrates the micro-via daisy chain structures utilized for reliability test. The die has a nitride passivation and a polyimide (PI) final passivation with 75 micron openings, and Cu/NiV/Al under-bump-metallization (UBM) structure of 102 microns diameter.

Experimental results

Fourteen resistance nets are visually inspected under microscope for process defects and initial resistance measurements are recorded. A failure criterion of 20 Ω for the initial resistance, reflected on a preliminary test for resistance, is used to identify defective parts. Test vehicles that meet this criterion are sent to reliability testing and are subjected to air-to-air thermal shock cycling ranging from -40C to 125C in a THERMOTRON thermal shock chamber ATS-320-DD. Each cycle lasts 20 min and the transfer time from air to air is less than 10s. [11] Resistance measurements are recorded after every a few hundreds cycles and 100% change in resistance is considered as a failure. The samples are cycled to 1115 cycles.

The average initial resistance of a 4.5 Ω with a standard deviation of 2.7 Ω is measured for the test vehicles. The rapid package prototyping test vehicles demonstrate excellent robustness during the reliability test. No failures are observed during 1115 thermal shock cycles for all the resistance nets. It should be noted that the resistance measured is for 22 vias in series, and any one via failure in the net would result in the failure of the entire net. The resistance variation for each cycle is shown in Figure 9 and the reliability test results for the fourteen resistance nets are summarized in Table 4.   Figure 10 shows cross-section of the daisy chain structure that survived 1115 thermal shock cycles.

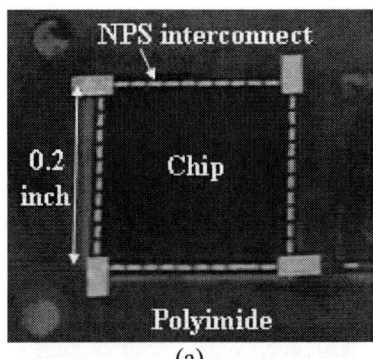
(a)

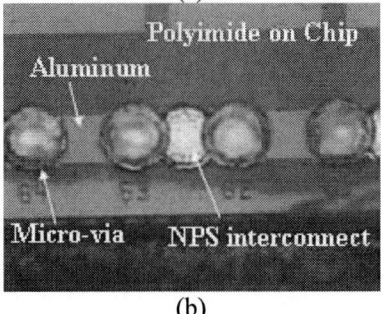
(b)

Figure 8. Reliability test vehicles: (a) four resistance nets and (b) daisy chain circuit of a resistance net

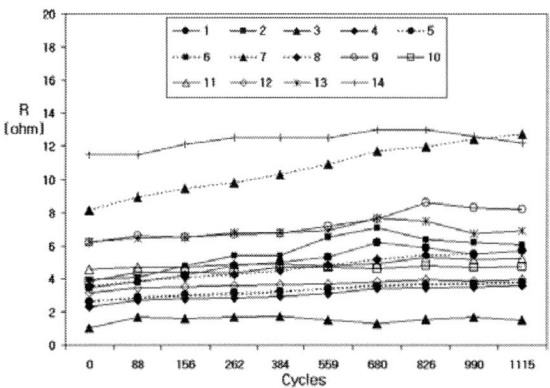

Figure 9. Resistance variation during thermal shock cycling

A potential failure mode might be delamination between a dielectric layer and a chip. For a comparison with the survived sample, a cross section of a failed sample in a previous test is shown in Figure 11. The sample failed at 560 cycles due to delamination between polyimide film and Teflon adhesive caused by insufficient pre-drying conditions of 20 min at 120C in lamination process step. As previously stated, pre-drying at 100C for 24 hours are recommended to remove the moisture in a polyimide film before laminating.

Table 4. Thermal shock test results

Net No.	Resistance		Resistance change [%]	Pass /Fail
	@ 0 cycle	@ 1115 cycles		
1	3.44	5.74	66.9	P
2	3.95	6.09	54.2	P
3	1.04	1.5	44.2	P
4	2.31	3.6	55.8	P
5	2.62	3.78	44.3	P
6	2.67	3.81	42.7	P
7	8.14	12.71	56.1	P
8	3.53	5.7	61.5	P
9	6.22	8.2	31.8	P
10	3.88	4.73	21.9	P
11	4.56	5.24	14.9	P
12	3.17	3.95	24.6	P
13	6.27	6.9	10.0	P
14	11.5	12.2	6.1	P

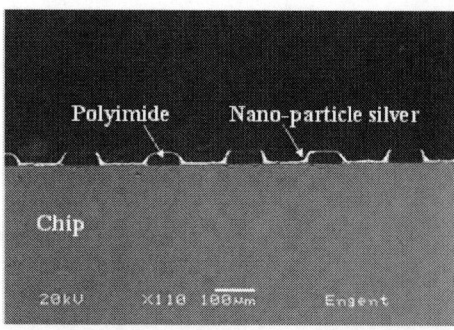

(a)

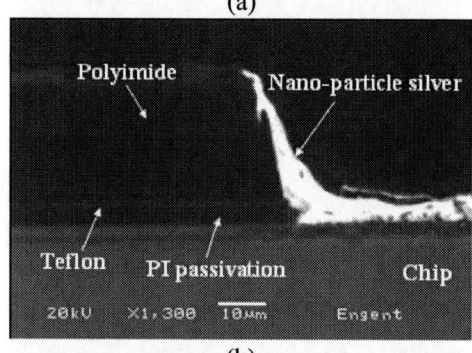

(b)

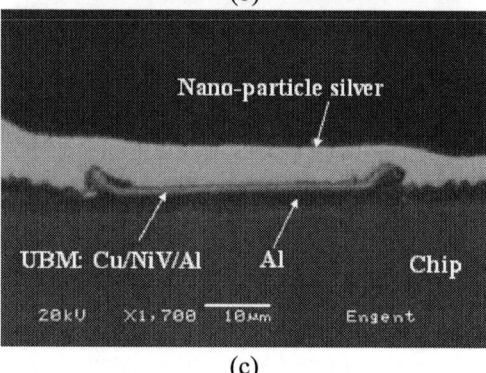

(c)

Figure 10. SEM images of a test vehicle survived 1115 cycles: (a) A part of a daisy chain resistance net (b) Magnified view of the interface between NPS, polyimide, and chip (c) Interface between NPS and UBM

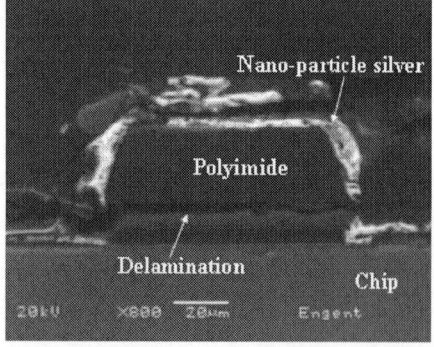

Figure 11. Delamination between Teflon and polyimide passivation layer after 559 cycles due to insufficient pre-drying condition

CONCLUSIONS

The innovative rapid package prototyping technology based on chip-first process using nano-particle silver, which is regarded as a suitable interconnect material in electronics packaging in terms of adhesion, resistivity, and current leakage characteristics, has been evaluated using air-to-air thermal shock test. The experiments show that the rapid package prototyping test vehicles are quite robust and reliable according to a current industry standard. All the fourteen test vehicles having 200um pitch have passed 1115 cycles without any failures. Insufficient pre-drying of polyimide, however, may be a potential cause of package failure in the continuing test after 1115 cycles. Improvement and control of lamination process is a key factor for improving the reliability of the rapid package prototyping technology, which will be pursued using temperature and humidity testing in the future.

ACKNOWLEDGEMENTS

The authors would like to thank Georgia Tech's Manufacturing Research Center, Center for Board Assembly Research, and the Packaging Research Center for supporting this work and laboratory space. The authors would also like to thank Engent AAT Incorporated for use of SEM, MSMA lab in Georgia Tech for use of excimer laser machine, Dupont for providing polyimide films, and Rogers Corporation for supplying LCP films.

REFERENCES

[1] Daum, W., Burdick, W. E., and Fillion, R. A., "Overlay high-density interconnect: a chips-first multichip module technology," Computer, Volume 26, Issue 4, April 1993, pp. 23-29

[2] Towle, S. N., Braunisch, H. C., et al, "Bumpless build-up layer packaging," Proceedings of ASME International Mechanical Engineering Congress and Exposition (IMECE), 2001, pp. EPP-24703

[3] Topper, M., Buschick, K., Wolf, J., et al, "Embedding technology-a chip-first approach using BCB," Advanced Packaging Materials. Proceedings, 3rd International Symposium on, 1997, pp. 11–14.

[4] Sungchul J. and Baldwin, D.F., "Demonstration for rapid prototyping of micro-systems packaging by data-driven chip-first process using nano-particles metal colloids," Proceedings of Electronic Components and Technology, 2005, pp. 1859 – 1863.

[5] Allen, G. L., Bayles, R. A., Gile, W. W., and Jesser, W. A., "Small Particle Melting of Pure Metals," Thin Solid Film, 1986, Vol. 144, pp. 297-308

[6] Moon, K., Dong, H., Pothukuchi, S., Li, Y., and Wong, C. P., "Nano Metal Particle for low temperature interconnect technology," Proceedings of Electronic Components and Technology, 2004, pp. 1983 – 1988.

[7] Standard test methods for measuring adhesion by tape test, ASTM D 3359-02

[8] Fuller, S. B., Wilhelm, E. J., and Jacobson, J. M., "Ink-Jet Printed Nanoparticle Microelectromechanical Systems," Journal of Microelectromechanical systems, Vol. 11, No. 1, Feb., 2002.

[9] Seanor, D.A., Electrical Properties of Polymers, Academic Press, 1982

[10] Boden, P. J., "Surface mount technology-a study of safety considerations: silver migration and adhesive flammability," IEEE Transactions on Components, Packaging, and Manufacturing Technology, Part B: Advanced Packaging, Volume 17, Issue 1, pp. 83 – 90, Feb., 1994.

[11] Thermal Shock, JESD22-A106B

DRIE WITH HIGH RATE AND UNIFORMITY FOR MEMS AND WLP

Leslie Lea
Surface Technology Systems plc
Newport, South Wales, United Kingdom
Leslie_Lea@stsystems.co.uk

ABSTRACT

The latest generation of plasma processing equipment designed for precision etching of MEMS devices such as accelerometers or pressure transducers is also ideal for device packaging applications.

Limitations of current packaging are driving the investigation of advanced packaging schemes. System-in Package (SiP) focuses on fabricating 3D stacks of chips in a single package. Systems might include processing capability, memory and sensors. Via holes must be etched through wafers and then metallized, in order to form inter-layer contacts. Via holes can be precisely fabricated using the Advanced Silicon Etch process (ASE[R]), a version of the Bosch time multiplexed etch process.

Packaging of optical devices such as CCD image sensors is becoming particularly important. For Chip Scale Packaging (CSP), plasma etching may be used effectively to etch scribe channels, and additionally through wafer vias.

As the manufacture of devices moves from 150mm wafers to 200mm wafers it becomes increasingly important to design plasma processing equipment that is capable of achieving very high etch uniformity over the larger wafer size and high etch rate to reduce the cost of ownership. STS has followed a program of extensive continuous development of plasma process tools and silicon etch processes for the last decade.

INTRODUCTION

Plasma etching is becoming a potent enabling technology for wafer-level device packaging, which is steadily moving from the development and pilot production stages to full scale production[1].

The application may be the etching of scribe channels for Chip Scale Packaging (CSP). The packaging of optical devices such as Complementary Metal Oxide Semiconductor (CMOS) and Charged Coupled Device (CCD) image sensors is becoming increasingly important. A practical application of such sensors is in the camera modules used in cell phones, figure 1.

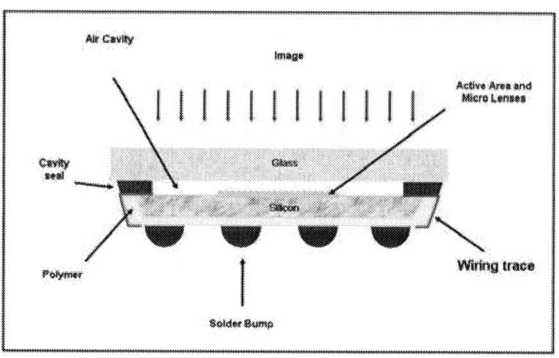

Figure 1. Schematic of a camera module as used in some cell phones
Courtesy of Tessera Inc.

Plasma etching may be used to etch scribe channels, figure 2, and additionally in some cases through wafer vias, for CSP, thus providing an effective enabling technology.

Figure 2. SEM of camera chip showing at left hand side the edge of the etched scribe channel
Courtesy of Tessera Inc.

Usually for this application the etch process will use a mixture of etchant and passivant precursor gases to achieve the required sloped or tapered characteristics.

Alternatively the requirement may be for the etching of vias for metalization for interconnects for System-in-Package (SiP). Novel device designs and the limitations of current packaging concepts are driving the IC industry to investigate advanced packaging schemes. One such technology focuses on fabricating 3D stacks of chips in a single package. Such systems might include processing capability, memory, analog to digital conversion and sensors. Via holes must be etched through wafers and then metallized, in order to form inter-layer contacts, figure 3.

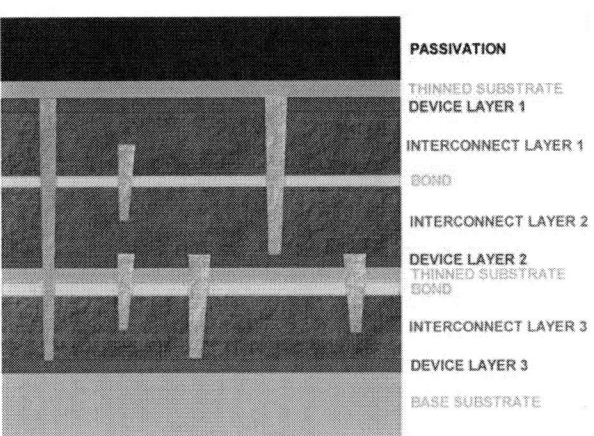

Figure 3. Schematic showing via holes etched through substrates then metallized to form inter-layer contacts

Such "via" holes can be precisely fabricated using the Advanced Silicon Etch process (ASE[R]), figure 4, a version of the Bosch time multiplexed etch process[2] that has been progressively developed over more than ten years by Surface Technology Systems to give significantly increased functionality.

Figure 4. SEM of via etched using ASE[R] DRIE process

As the manufacture of devices moves from 150mm wafers to 200mm wafers it becomes increasingly important to design plasma processing equipment that is capable of achieving very high etch uniformity over the larger wafer

size and high etch rate to reduce the cost of ownership. STS has followed a programme of extensive continuous development of plasma process tools and silicon etch processes with the fourth generation of silicon etch tool launched in 2005. Work is continuing to provide plasma etch capability with further enhanced etch uniformity and profile control whilst delivering an increased etch rate, figure 5.

Figure 5. An example of a 100 μm diameter via etched at a rate of 35 μm/min using the ASE[R] process on a 200 mm diameter wafer with an open area of c 10%

For high etch uniformity it is necessary to achieve the appropriate spatial profile of neutral radicals reaching the wafer to match silicon loading characteristics. The requirement on the spatial density profile of ions above the wafer is in general different from that of the radicals, in order to ensure precise control of the etch direction. STS has developed a technique which provides a level of controlled de-coupling of the spatial profiles of the two species, enabling very high etch rate uniformity and feature profile control over 200 mm wafers.

TIME MULTIPLEXED ETCH PROCESS
The time multiplexed etch process developed by Laermer and associates at Robert Bosch enables the highly anisotropic etching required for many MEMS applications. Surface Technology Systems were the first process tool manufacturer to take a license for this process and have further developed it into the Advanced Silicon Etch[3,4,5] (ASE[R]) process. This process works by the use of repeated cycles of passivation and etch. In each passivation step a precursor gas is utilised that when subjected to the plasma results in a polymer layer being deposited on all surfaces of the features being etched. In the first part of the following etch step, ions accelerated towards the wafer preferentially remove the polymer from the base of features. In the remainder of the etch step neutral fluorine radicals react with the exposed silicon at the base of the features, isotropically etching the material. The next passivation step then causes polymer to be deposited on all surfaces of the features including the silicon exposed during the previous

175

etch step. As the cycles of etch and passivation repeat, the features are etched ever deeper into the silicon. The etching of the silicon is overall highly anisotropic, but in each cycle the etch process is essentially isotropic. It is possible to achieve very high aspect ratios to features using this process, greater than 50:1 in some cases, with good average CD control and high selectivity of order 200:1 to the mask, however the sidewalls of the features exhibit a roughness described as "scalloping" due to the isotropic etch part of each cycle. If the length of the cycle time remains constant then the higher the etch rate the greater the depth of the scallops. For applications requiring low sidewall roughness, etch rate would normally need to be reduced, however by careful design of the process tool it is possible to reduce the length of the cycle time while maintaining etch rate near to constant.

ETCH TOOL

Plasma process equipment used to operate the ASE[R] process uses the Inductive Coupling of RF Power (ICP) from a power supply into the plasma using an antenna which is usually situated on the atmospheric side of a dielectric window that forms part of the plasma source region. ICP plasma processing equipment may be configured so that the plasma is formed in the same chamber in which the wafer is processed. Alternatively the plasma may be formed in a separate chamber (de-coupled plasma source) and allowed to diffuse into a second usually larger chamber in which the wafer is located.

The use of a de-coupled plasma source brings a number of benefits:

1) The reduced volume of the source allows more efficient breakdown of the precursor gas because of the higher power density that can be delivered from the RF power supply.
2) The geometry and volume of the source region can be tailored so that the ions and neutral radicals exiting it can diffuse down to the wafer with the desired profiles to obtain high etch uniformity.
3) Additional means may be included between the de-coupled plasma source region and the chamber in which the wafer is processed in order to alter the balance between numbers of ions and numbers of neutral radicals reaching the wafer to control selectivity to mask and/ or ion damage to feature profiles.

For etching of silicon using the ASE[R] process, STS manufactures both conventional ICP plasma processing tools for greater flexibility in less demanding applications and ICP plasma processing tools utilising de-coupled plasma sources for applications requiring precise control of particular aspects of the etching.

For a conventional de-coupled plasma processing tool, the plasma is formed in a smaller chamber by RF power coupled by an antenna through a dielectric window. The dielectric window is often a cylindrical tube located on the

same axis as a cylindrical chamber in which the wafer is processed, figure 6.

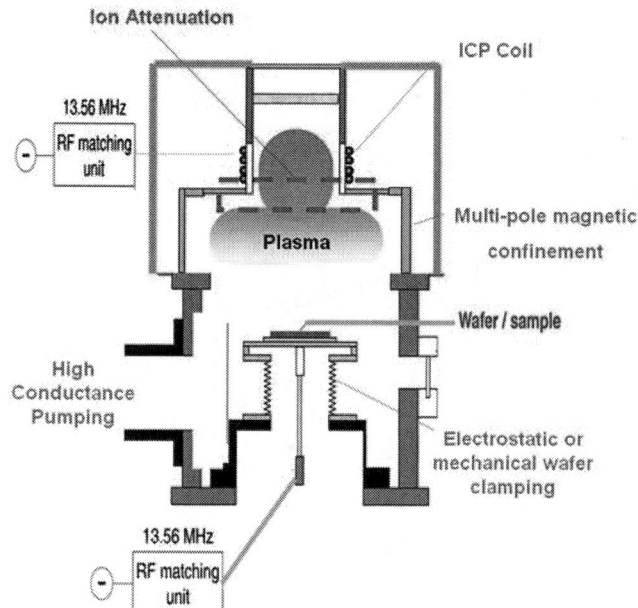

Figure 6. A conventional de-coupled source type plasma processing tool

Ions and radicals diffuse from the smaller chamber into the chamber in which the silicon wafer is processed. Because of the differences in deflection or loss probability between ions, electrons and neutral radicals when encountering electric or magnetic fields and material surfaces such as the chamber walls, the radial profiles of the charged species may differ from the radial profiles of the neutral radicals in the vicinity of the wafer.

For the conventional de-coupled plasma processing tool, the spatial profile of both ions and neutral radicals in the vicinity of the wafer will usually be center high, decreasing radially towards the walls of the chamber. The density profile variation of the ions will usually be more extreme than for the radicals, because on encountering the walls or an object within the chamber, ion and electron pairs have a high probability of recombining, while neutral radicals may survive a few interactions. For chemical etching of silicon with a reasonable exposed area (exposed silicon area on the wafer surface defined by the mask) to achieve a near constant chemical etch rate across the wafer requires more radicals to be present near the center where each feature is surrounded by other features than towards the edge where there are more features nearer the center. The conventional de-coupled plasma processing tool can often quite reasonably satisfy this criterion.

For the ASE[R] process there is a requirement to deposit a reasonably uniform passivation layer and then remove this layer from the base of features by ion bombardment before the chemical etching of the silicon can take place. Experiments show that while it is desirable to have reasonable uniformity to the thickness of the passivation layer across the wafer, it is not essential provided that the ion bombardment can remove the passivation from all areas in a similar time so that the overall etch rate determined by the chemical reaction of the neutral radicals with the silicon is similar across the wafer. As described above, the spatial profile of the ion density above the wafer for a conventional de-coupled plasma source is such as to be quite center high which will tend to cause the passivation layer to be sputtered away more quickly towards the wafer center, increasing the potential for a higher overall etch rate at the wafer center than towards the edge. Because for a conventional de-coupled plasma source the ion density profile above the wafer is center high, the Debye length and therefore the thickness of the sheath between the plasma and the wafer will be less near the center of the wafer than towards the edge. This non-uniformity of the sheath thickness across the wafer is likely to lead to a steering of ions as they are accelerated to the wafer. In the center of the wafer they will move perpendicular to the wafer surface, but towards the edge of the wafer they may move in such a way that they impact the wafer at an angle to the perpendicular. For a conventional de-coupled plasma source, the feature may be etched perpendicular to the wafer surface near the center of the wafer, but near the edge of the wafer the feature may tilt outwards (from wafer center) as it is etched deeper into the wafer.

An ideal plasma processing tool should have the capability of delivering neutral radicals to the wafer surface with a density profile to match the loading effect of the exposed area of silicon so that etch rate is very close to constant at all points on the wafer. The tool should also be capable of providing a near uniform ion density just above the wafer surface so that the sheath is of the same thickness across the wafer and ions are accelerated perpendicular to the wafer surface.

RESULTS

Work has recently been conducted in an experimental etch process tool at STS with the aim of achieving the dual requirements of near uniform ion density just above the wafer surface and neutral radicals delivered to the wafer to match the loading effect of the exposed silicon. Figure 7, shows etch rate of 50 um wide trench features as a function of position across the diameter of a 200 mm test wafer, while figure 8, shows the degree of tilting of the same features.

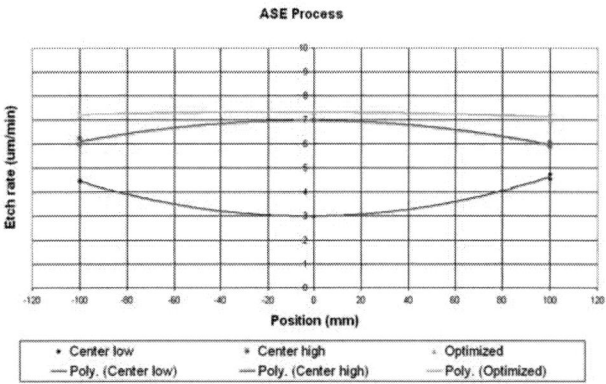

Figure 7. Experimentally measured etch rate as a function of position across a 200 mm wafer with the plasma source adjusted to give center etch rate low, center etch rate high and optimized etch rate across wafer

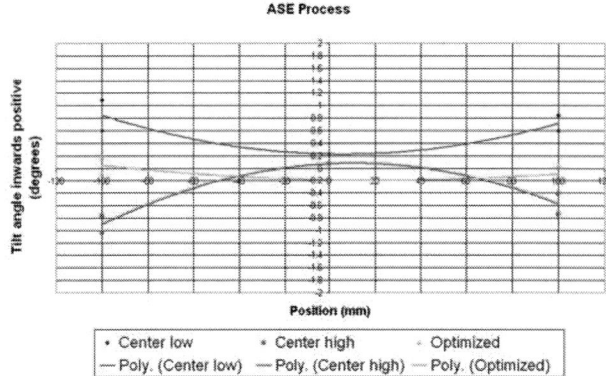

Figure 8. Experimentally measured feature tilt angle across a 200 mm wafer, positive indicates tilt inwards towards center when etching down into the wafer. Results are for center etch rate low, center etch rate high and optimized etch rate across wafer

In each figure there are 3 sets of data points with guidance lines fitted to each set of data. Measurements shown were made at the center of the test wafer and at the edges for the x and y co-ordinate directions (left edge and notch position shown as −100 mm, right edge and opposite notch position shown at +100 mm). The three sets of data points shown in figure 7, are for the process tool set up to give etch rate low at the center of the wafer, etch rate high at the center of the wafer and etch rate near constant across the wafer. The corresponding sets of data in figure 8, show feature tilting inwards at the edge of the wafer, outwards at the edge of the wafer and almost no tilting across the wafer (errors in tilt measurements up to c 0.3 degrees).. The three data sets were obtained by altering the radial distribution of power fed into the plasma generation region of the etch tool and show that it is possible to simultaneously achieve uniform etch rate across a wafer and a low degree of feature tilting towards the edge of the wafer.

CONCLUSION

Deep Reactive Ion Etching (DRIE), is an enabling technology for device packaging applications in addition to MEMS. By careful design of the plasma processing tool it is possible to obtain excellent etch uniformity across 200 mm diameter wafers at high etch rate, while reducing to a low level the tilting of features towards the edge of the wafer.

REFERENCES

[1] Chambers A. A., Silicon Micro-Machining as an Enabling Technology for Advanced Device Packaging, Semiconductor Manufacturing Magazine, November 2004

[2] Laermer F. and Schilp A. 1992, Method for anisotropically etching silicon. German patent DE4241045

[3] Bhardwaj J. K. and Ashraf H., 1995, Proc. SPIE, 2639, 224

[4] Bhardwaj J. K. and Ashraf H., 1997, Anisotropic dry silicon etching, Symp. Microstructures and Microfabricated Systems at the Annual Meeting of the Electrochemical Society (Montreal, Canada)

[5] McAuley S. A., Ashraf H., Atabo L., Chambers A., Hall S., Hopkins J. and Nicholls G., J. Phys. D. Appl. Phys. 34 (2001) 2769-2774

AEROSOL-JET PRINTING FOR 3-D INTERCONNECTS, FLEXIBLE SUBSTRATES AND EMBEDDED PASSIVES

Martin Hedges
Neotech Services MTP
Nuremberg, Germany
info@neotechservices.com

Mike Kardos, Bruce King, and Mike Renn
Optomec Inc.
Albuquerque, NM, USA
info@optomec.com

ABSTRACT

A new additive aerosol-jet printing technology called Maskless Mesoscale Materials Deposition ($M^3D^®$) is finding wide application in a number of electronic manufacturing applications. The M^3D technology, originally developed under DARPA contract, has been commercialized by Optomec. This paper will discuss how the M^3D process is being used to deposit a wide variety of materials onto a wide variety of substrates without conventional masks or thin-film equipment. The process is non-contact, allowing traces to be printed over steps or curved surfaces. Printed features can be as small as 10 microns.

The M^3D printer utilizes low viscosity inks in the range of 1-1000 cP. Typical materials that can be printed include nanoparticle metal suspensions, polymers and adhesives. Conductor traces can be printed using gold or silver nanoparticle inks. Conductors can also be formed by printing a seed layer, followed by electroless copper plating. Polymer thick film pastes can be printed to form embedded resistors. Polyimide and various epoxies can be printed for adhesives, overcoat dielectrics, etc.

Substrates include silicon, polyimide, glass, FR-4 and aluminum oxide. In principle, virtually any substrate can be used provided the ink is compatible with it.

This new technology has been applied to areas such as embedded passives, flexible substrates, mask and circuit repair and conformal electronics. It is expected that it will find further applications in 3D packaging.

INTRODUCTION

In recent years, a new class of manufacturing techniques has become available which offers manufacturers significant cost, time and quality benefits across a broad spectrum of industries. These new techniques are collectively known as additive manufacturing. During additive manufacturing, material is deposited layer by layer to build up structures or features. This is in contrast to traditional subtractive manufacturing methods where masking and etching processes are used to remove material to get to the final

form. Features of additive manufacturing processes include direct CAD-driven, "Art-to-Part" processing which eliminates expensive tooling, masks and vertical/horizontal integration which leads to fewer overall manufacturing steps. These features combine to offer diverse benefits:

- *Better Product Designs* - Greater design and manufacturing flexibility offers the potential for revolutionary new end-products with improved performance based on novel size, geometries (including 3D Interconnects), materials and material combinations.
- *Time Compression and Increased Manufacturing Agility* - CAD driven, tool-less processes speed up product development and manufacturing, while allowing greater flexibility in mass customization.
- *Lower Costs* – This benefit arises because tooling and mask costs are eliminated. Process costs in terms of operator input, supplier chain complexity and work flows are reduced. Raw material is used more efficiently, thus reducing waste levels. Life-cycle costs are reduced by lower design development costs, increasing product quality and the ability to repair components.

This paper will introduce an additive manufacturing technique designed for the electronics industry that offers significant potential in the manufacture of 3D Interconnects: Maskless Mesoscale Materials Deposition (M^3D).

M^3D - MASKLESS MESOSCALE MATERIALS DEPOSITION

M^3D was originally developed to fill a neglected middle ground in microelectronic fabrication. Current techniques create very small electronic features, for example by vapor deposition, and relatively large ones for example by screen-printing. No technology was capable of satisfactorily creating crucial mesoscale-sized (1-100μm) production of interconnects, components, and devices. As electronic devices continue to shrink, thick-film fabricators are approaching the physical limits of stencil printing. Thin-film technology can deposit mesoscale features but requires a highly skilled workforce and a major investment in new manufacturing capability for each new application. Thick-

and thin film techniques are 2D processes and are not ideal for manufacturing 3D conformal electronic features needed for 3D Interconnects.

HOW M³D WORKS

M³D uses aerodynamic focusing for the high-resolution deposition of colloidal suspensions and/or chemical precursor solutions. An aerosol stream of the deposition material is focused, deposited, and patterned onto a planar or 3D substrate. The basic system consists of three key components, Figure 1:

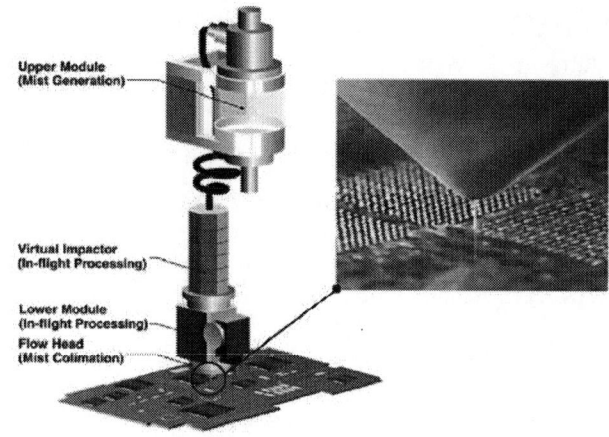

Figure 1. Schematic of the M³D process and photo of the deposition head.

- A module for atomizing liquid raw materials (Mist Generation).
- A second module for focusing the aerosol and depositing the droplets (In-Flight Processing).
- The final module is a laser for post treatment sintering of the deposits. This is used for sensitive substrates which cannot be conventionally sintered.

Mist Generation is accomplished using an ultrasonic transducer or pneumatic nebulizer. The aerosol stream is then focused using a flow guidance deposition head, which forms an annular, co-axial flow between the aerosol stream and a sheath gas stream. The co-axial flow exits the flow guidance head through a nozzle directed at the substrate. The M³D flow guidance head is capable of focusing an aerosol stream to as small as a tenth of the size of the nozzle orifice. The deposition process is CAD driven; the process directly writes the required pattern from a standard .dxf file. Patterning is accomplished by attaching the substrate to a computer-controlled platen, or by translating the flow guidance head while the substrate position remains fixed.

Thermal post processing of the deposited material is often needed to cure the material or increase properties such as electrical conductivity. Depending on the application, either conventional sintering or curing is used, or for low temperature substrate materials a continuous-wave Nd:YAG

or diode laser is used to locally heat the deposited material without affecting the surrounding substrate.

CAPABILITIES AND APPLICATIONS

M³D is already being applied to a range of 2D electronics applications (Table 1), and is now being investigated for 3D Interconnect applications.

Packaging and Assembly
High Density Interconnects
Flip-Chip / Direct Die Attach
Embedded / Integrated Passives
Flex Circuits
Meso-Dispensing
Electronic Components
Resistors, Capacitors and Inductors
Micro-Antennae
Micro-Batteries
Electronic Devices
Flat Panel Displays
Fuel Cells
Micro-Sensors
MEMS & RFID
Hybrid Manufacture
Smart Structures
BioTech
Bio-Sensors& Implantable Devices
Micro-Arrays

Table 1. M³D Application Areas

To further leverage the process for a wide range of 3D Interconnect applications the following key capabilities can be utilized:

- 3D Structuring Capability
- Low Temperature Processing
- High Quality Deposits
- Wide Range of Materials & Substrate Combinations
- Environmental Aspects
- Competitive Cost Basis

This section of the paper will investigate these capabilities and provide some general application examples that can be related to the design and manufacture of 3D interconnects.

3D Structuring Capability

M³D can precisely deposit materials on both planar and non-planar substrates. This is made possible by the relatively high (5mm+) stand-off point of the deposition head and long focal length of the material beam exiting the nozzle. There is no physical contact with the substrate by any portion of the tool (other than the deposition stream), and therefore conformal writing is easily achieved. This allows the process to build 3D conformal features on to shaped components, write into trenches (Figure 2), over steps and contours (Figure 3), or fill "vias" with high aspect ratios, for example 30mm diameter, 750mm deep.

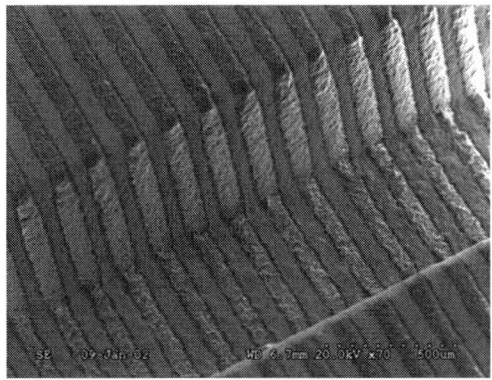

Figure 2. 60μm Ag lines written over a 500μm trench.

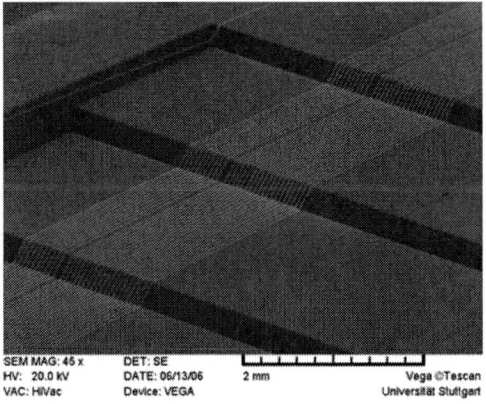

Figure 3. 20 μm Ag lines written over stepped injection molded LCP
(Courtesy HSG-IMAT).

Figure 4 shows an example of conformal packaging in a Smart Card application. This involves 3D direct writing on several different kinds of materials; the interconnect is made from the Cu pad, over the Kapton layer and epoxy adhesive and on to the Si chip. In this case the height difference is approximately 150μm between the Cu pad and the chip. This allows the replacement of the traditional wire bond and reduction of overall part thickness. An additional benefit is improved mechanical reliability as the relatively delicate wire bonds are eliminated. After the Ag interconnects were written the device is processed in an oven at 200°C to sinter the interconnects.

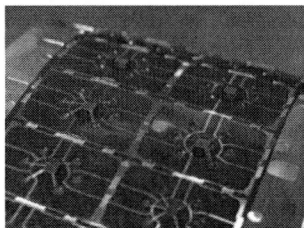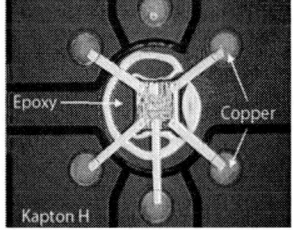

Figure 4. 150μm wide silver interconnect over an epoxy bump and Kapton.

For 3D surfaces with larger surface profiles the system makes use of 3 or 4 Axis deposition. This allows writing over steps of up to 50mm in the current M³D 300 system. Figure 5 shows an example of where this capability has been applied.

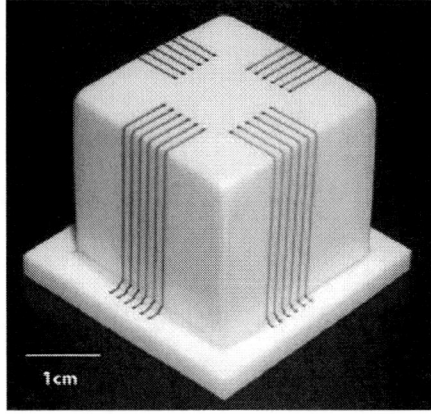

Figure 5. 3D Silver Interconnects (150μm line width) written over an alumina cube.

This low volume application requires conformal direct writing Ag interconnects on an alumina substrate over a 25mm height range. In this case the deposition head first tilted at a 45° angle and is moved in the Z-direction to be able to write on the vertical walls and accommodate the required height change. Conventional oven sintering is applied and the IC package is then flip-chip mounted onto the cube shaped alumina substrate.

Low Temperature Processing
Once the material has been deposited, conventional approaches for many commercial metal inks require high-temperature treatment often up to 250°C or higher. For non-sensitive polymer substrate materials such as LCP, PA6/6T, re-flow or cure ovens can be used to sinter the deposited material. However, certain substrates tend to have limited temperature capability, for example polycarbonate and polyester, have a temperature limitation of around 100°C. This sensitivity requires a manufacturing process that can deposit and process the material at low temperatures. M³D can locally process the deposition on substrates, using the integrated laser module that sinters the deposit while leaving the substrate unharmed. The end result is a high-quality thin film with excellent edge definition and near-bulk resistivity, typically 2-3x bulk but dependent on ink type. An example of low temperature processing is shown in Figure 6. This is a low cost polymer display application where temperature sensitive PMMA is used as the substrate to reduce costs. M³D was used to write the Ag gates and interconnects which where then laser sintered without damaging the PMMA.

Figure 6. Low Cost Polymer Display. Laser processed Ag gates and interconnects on PMMA, (Resistivity ~8 μOhm-cm).

Additionally, a new low temperature processing methodology had been developed for use with M³D. In this case, it is possible to carry out sintering at very low temperatures (<50°C). This will allow the development of new applications with ultra sensitive substrates.

High Quality Deposits

Deposit quality is dependent on the ink type used, the ink-substrate combination and other factors such as substrate roughness. M³D does not change the chemical or physical properties of the materials deposited or the substrates. In general terms, M³D can deposit with:

- Feature sizes down to 10 microns with +/- 10% edge roughness and pitch down to 20 microns
- Good conductivities
- Thickness as low as 100nm or as high as 5 microns (single layer deposit)
- Low surface roughness
- Good adhesion

M³D reliably produces ultra fine feature circuitry well beyond the capabilities of thick-film and ink-jet processes. Most materials can be written with a resolution of down to 20μm. For Ag, electronic features as small as 10μm with a 20μm pitch can be written.

This capability offers a solution for the production of smaller, high performance components critical to size-sensitive applications like those in the wireless and hand-held device markets where component density is increasing dramatically. The ability of M³D technology to create fine features with complex geometries in 3D from a wide range of materials makes it suitable for the production of both passive and active components, including resistors, inductors, capacitors, filters, micro-antennae and micro-batteries. The precise edge definition and repeatability of the process are particularly relevant to high frequency requirements. In comparison to screen printing, embedded resistors can be made smaller and more accurately with M³D such that no laser-trimming is needed to tune the resistor to the right value.

Gold and Silver inks generally display conductivities approaching bulk properties with conventional sintering and 2-3x bulk with laser sintering. Low viscosity inks can produce mirror-like surfaces while thick film inks have micron scale roughness. Due to the very fine droplet size of the aerosol (typically 1-5μm), even surface profiles in the deposit are common with none of the "coffee staining". Deposit adhesion is highly dependent on ink-substrate combination. For example, gold inks adhere to a wide range of substrates, including glass, ceramics and various polymers. Silver is more sensitive, but also has good adhesion to a wide range of substrates. Typically M³D deposits satisfy the standard tape test.

Wide Range of Materials & Substrate Combinations

M³D can deposit a wide variety of materials, including metals, conductors, insulators, ferrites, polymers, adhesives and biological materials. Deposits can be made on virtually any surface material – polymers, silicon, glass, metals and ceramics. This flexibility opens the way for many different 3D Interconnect applications using a single process. M³D uses a wide range of commercially available inks from many different sources.

Many devices that are manufactured for electronics products require multi-layer manufacturing techniques. The ability of the M³D system to deposit conductive, insulating, and adhesive materials layer-by-layer within a single system makes it an attractive solution for the production of embedded passives. A simple 2D example of this multi-layer capability is the deposition of a dielectric and then a Ag layer onto a Cu circuit board pad to create a basic capacitor, Figure. 7. Other examples of multi-layer applications include sub-micron layers for fuel cell applications, high-density interconnect backplanes (organic and metal) for flat panel displays, and micro-sensors for avionics. Other successes with multilayer deposits have been in the biosciences area, such as the generation of bio-sensor structures.

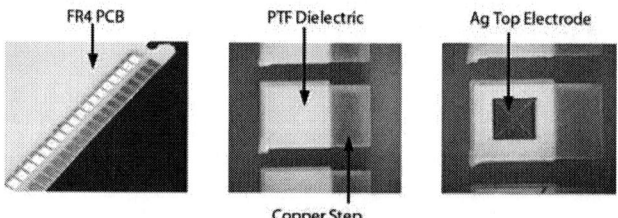

Figure 7. Multilayer deposition, Polyimide dielectric and Ag deposited onto Cu pads to make a simple capacitor. Capacitance: 19 pF/mm2 @ 15mm Dielectric thickness.

Environmental Aspects

M³D works without the need for masks or resists, which results in minimal waste and less environmental impact. As the process writes very finely and precisely, this reduces the amount of material required and waste generated for a given application. If direct metal deposits are used then plating bath is required reducing the quantities of waste material.

Competitive Cost Basis

One of the main drivers for cost reduction using M^3D is the elimination of physical tooling. M^3D software creates the deposition tool paths direct from standard .DXF CAD data. This digital tooling approach also offers manufacturing agility by allowing designers to quickly and cost effectively test new design alternatives and prototypes. This also offers manufacturing agility which allows designers to quickly and cost-effectively test new prototypes and products. It eliminates the delays and costs associated with tooling sets and other upfront capital required by conventional electronics manufacturing techniques. This feature also makes it much easier to carry out cost effective Rapid Product Development and to validate design changes without the need for "re-tooling". The result is reduced cost and faster time-to-market for new products.

The M^3D process can reduce the overall number of processing steps, which in turn can help to reduce both capital and operating costs. Since the system can process a wide range of materials and substrates, greater utilization of the capital equipment can be obtained. Process costs in terms of operator input, supplier chain complexity and work flows are reduced.

Material efficiency can often play a key role in reducing the cost of manufacturing operations. The tiny droplets dispensed by M^3D allow for very thin coatings which also allow for good interaction between differently applied layers. These same femto-liter sized droplets allow for very careful control of dosages dispensed. Since many electronics materials are expensive, their technology is a key enabler for reducing the cost of each device by reducing materials use and waste.

Another alternative for reducing processing steps and cost, compared to traditional mask-etch techniques, is the catalyst-layer approach for producing interconnects or other features. M^3D can directly deposit an activator/catalyst solution in the exact pattern required. The process is then completed by curing at 80°C, and then followed by a standard electroless Cu plating step (Figure 8).

Figure 8. Demonstrator test pattern (Cu plated on PI) created by the catalyst-layer approach.

In this sample, the catalyst test pattern has been printed onto a polyimide film and conventional electroless plating for

two hours has been used to plate approximately 10mm thick Cu onto the pattern. The traces in Figure 8 are 50 mm long and range in width from 10mm down to 500mm. Gap spacing ranges from 1.8 mm down to 300mm. All traces are highly adherent to the substrate and pass the standard tape test. Deposit conductivity is near bulk and similar to standard electroless Cu deposits. The process has also been demonstrated on polymer matrix composites and PET. This technique can be used to reduce cost, especially in patterns requiring combined fine and large area deposits.

PROCESS SCALABILITY

The current M^3D 300 system is aimed at low volume manufacturing and rapid prototype product development. The system is equipped with a single deposition head and a single deposition nozzle. It can write at speeds up to 200m/s with a high level (5mm) of dynamic accuracy. As higher volume applications are developed, there is a need to scale up the speed of M^3D manufacturing. M^3D systems are involved with ongoing projects in scaling atomizer throughput, development of multiple nozzle deposition heads, and closed loop control of the deposition process. These developments are being driven by high volume (millions of parts p.a.) production applications which will be released in 2007 as a replacement to screen printing. These applications are 2D, but the developments will provide the foundation for extension into high volume 3D applications as they evolve.

CONCLUSIONS

This paper has introduced the novel Maskless Mesoscale Materials Deposition process and outlined its features, benefits, and some select application areas. This CAD driven, direct write process is currently being used in a wide range of 2D electronics applications. However, the process also has the capability to write conformally, providing an enabling technology for 3D production. Designers of 3D interconnects can now harness the unique features of M^3D to create designs which offer a wide range of time, cost and quality benefits. Since many MID applications are evolving, M^3D can also be a powerful, flexible product development tool, as well as a viable production solution.

STUDY ON ADHESION OF DICING DIE ATTACH TWO-IN-ONE FILM FOR 3-D STACK PACKAGING

Shijian Luo, Ph.D. and Tom Jiang, Ph.D.
Micron Technology Inc.
Boise, ID, USA
sluo@micron.com and tjiang@micron.com

ABSTRACT

Recently, dicing die attach two-in-one film (DDF or DDAF) has been developed for 3-D stack packaging application. When DDAF is used, a wafer is laminated on DDAF film, then diced, UV cured (if the dicing tape is a UV tape), and die-picked and attached. During die picking, easy release and clean separation between the die bonding film and dicing film is critical. In this paper, the adhesion between the die attach film layer and UV dicing film layer of one DDAF material was investigated. UV intensity and UV dosage affected the adhesion. When the tape was cured under high UV intensity light, there were some hot spots on the two-in-one film, and die pick became more difficult for these locations than the other areas. The hard to pick areas were randomly located on the wafer. For a low intensity UV curing system, the UV dose (curing time) had a significant affect on adhesion. There was an optimal UV dose for easy release of the die bonding adhesive layer and UV dicing tape layer. Wafer lamination temperature and time affected the adhesion as well. When wafers were laminated onto the DDAF at a higher temperature (65°C), and left on the chuck at this temperature for a long time, the separation between die attach adhesive layer and UV dicing film layer became more difficult. The die picking force increased continuously with the tape age at room temperature before lamination. This indicates that inherent adhesion between die attach adhesive layer and dicing film, which can not be diminished by UV curing, increases continuously with the tape age at room temperature before UV curing. The increased adhesion between the die bonding film layer and UV dicing tape vs. aging time before UV curing is mainly due to the interaction (inter-diffusion and mixing) of the UV adhesive layer on UV dicing tape with the die bonding adhesive layer. The adhesion between the two layers did not increase with aging time after UV curing, as the adhesive layer on the UV dicing tape had been cured, thus limiting the inter-diffusion of the two layers.

Key words: adhesion, die attach film, dicing film.

INTRODUCTION

In early days of IC packaging, liquid die attach adhesive was applied onto individual die sites on substrates or lead frames immediately prior to the die attach process. Die attach tape (or B-staged adhesive) was introduced later. Die attach tape could be applied onto the lead frame or substrate either at the substrate (lead frame) supplier site or in the semiconductor package house. This could be done some time before the die attach process.

With the emerging of thin die, thin packages, and 3D stack die packages, die bonding material has been developed to be applied on the backside of a wafer. Then the wafer is mounted onto wafer dicing film, diced, die-picked, and die-attached on a lead frame or substrate. The wafer level pre-applied adhesive has several advantages over the traditional liquid die attach paste. It is a wafer-level batch process in which die attach material is applied onto all dice on the wafer in a single process. It has much better bondline thickness and uniformity control. It has less bleeding onto bonding pads on the die and on the substrate, less voiding in the bond line.

Very recently, dicing die-bonding two-in-one film has been developed to save processing steps at IC packaging and further improve the process efficiency for semiconductor packaging industry. When this kind of two-in-one film is used, a wafer is laminated on the two-in-one film, then diced, and die-picked. As the die is becoming thinner and thinner for thin package and 3-D die stack package applications, die picking is becoming more difficult even on regular wafer dicing film [1]. During die picking for this kind of two-in-one film, easy and clean separation between die bonding film and dicing film is critical. Difficulty at die picking can result in the breaking of thin dice. In this paper, the adhesion between die attach film layer and UV dicing film layer of one DDAF material was investigated.

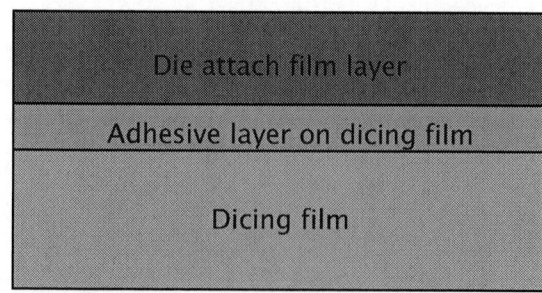

Figure 1. Construction of DDAF film

EXPERIMENT

Wafers were thinned first to the desired thickness (100μm), then they were laminated onto a 2-in-1 DDAF film. The construction of this DDAF film is shown in Figure 1. The wafers were diced into pre-defined size (~ 16 X 9mm).

During dicing, the wafer saw cut through the wafer, die attach film layer, and cut 10μm into the wafer dicing film. After dicing, UV curing of the dicing tape was done. Two different UV curing systems were used for UV curing: one with a low UV intensity and one with a high UV intensity. After UV curing, the die pick force test was performed. For the die pick force test, standard buttons (5.5mm diameter) were glued to the die surface with Loctite 409 Superbonder Gel, and the adhesive was cured at room temperature for 1 hour. The die picking force was tested on a Texture Tech system. In addition, die were also picked with the die attach system to check for cracked dice and the number of no-pick dice.

RESULTS AND DISCUSSION
1. Effect of UV Curing
The effect of UV curing on the adhesion between die attach film and dicing film was investigated. Two different UV curing systems were used: one with high intensity and one with low intensity. Three different UV doses (120mJ, 180mJ, and 360mJ) were used. Listed in Table 1 and Table 2 are the die pick force test results (die size: 16 X 9 mm, 100μm thick). The DDAF material used in this experiment was fresh. Wafers were laminated onto the DDAF within 1 day after the DDAF material was taken out from the cold storage condition (5°C). The die pick force from one regular dicing film was listed in Table 1 for comparison as well.

Table 1. Die pick force (Gram) on fresh DDAF after UV curing on a high intensity UV curing system.

UV Dose (mJ)	120	180	360	360 (regular dicing film)
Average	1427.3	1387.2	1302.8	848.1
Maximum	1889.5	1864.2	1660.1	1139.0
Minimum	914.5	903.0	838.3	336.4
Std. Dev	239.0	253.5	195.2	210.9

Table 2. Die pick force (gram) on fresh DDAF after UV curing on a low intensity system

UV Dose	120 mJ	180 mJ	360 mJ
Average:	1344.3	1174.9	1184.9
Maximum:	1480.1	1421.7	1574.1
Minimum:	1183.6	789.7	520.0
Std. Dev:	75.2	138.6	241.6

Comparing the DDAF film with a regular UV dicing film, die pick force was much lower from a regular dicing film than from the DDAF film. This indicates that die pick from the DDAF film would be more difficult than die pick from a regular UV dicing film. The separation interface is very different for the two cases, and thus one should manage this when using DDAF material. As the non-traditional pick method is proposed for thinner and thinner die, the non-traditional pick method becomes even more necessary when a thin die is picked up from DDAF material.

From Table 1 and Table 2, it can be seen that the average die pick force decreased with the increase of UV dose for both the high intensity UV curing system and low intensity UV curing system. At the same dosage, the low intensity UV curing system provided a lower average die pick force than the high intensity UV curing system. The standard deviation of the die pick force of the low intensity system increased with UV dose.

Listed in Table 3 and Table 4 are the observations at die picking with die attach equipment. The number of cracked dice and number of no-pick dice were recorded. For regular dicing film, there was no problem to pick the 100μm thick dice with no crack die or no-pick. However, when picking this same size and same thick die on DDAF film, the results were quite different. For the DDAF film cured with the high intensity UV curing system (Table 3), some dice could not be picked up with the standard die picking parameters. Some dice cracked due to the difficulty at die picking. For DDAF film cured with the low intensity UV curing system (Table 4), none of the dice could be picked up for the tape cured with 120mJ UV dose. When the UV dose increased, the die picking became easier. When the UV dose of 360mJ was used on low intensity UV curing system, all the dice could be picked from the tape, although with some dice cracked.

Table 3. Number of no-pick and cracked dice from 2 wafers mounted on DDAF film cured with a high intensity UV curing system

UV Dose (mJ)	120	180	360	360 (regular dicing film)
Number of cracked dice	21	26	30	0
Number of no-pick dice	16	23	7	0

Table 4. Number of no-pick and crack die from 2 wafers mounted on DDAF film cured with a low intensity UV curing system

UV Dose (mJ)	120	180	360	360 (regular dicing film)
Number of cracked dice	n/a	18	19	0
Number of no-picked dice	All	47	0	0

Although both high intensity UV curing and short time/low intensity UV curing had no-picks, they had different mechanisms. For the low UV dose on the low intensity UV curing system, the no-pick dice were due to insufficient UV curing of the adhesive layer of dicing film. During UV curing, the adhesive layer on the dicing film was cured, and the tackiness of the dicing film was significantly reduced.

Thus, the adhesion between the DDAF layer and dicing film was significantly reduced, making die pick easier. If the UV curing of the adhesive layer on the dicing film was not sufficient, the separation between DDAF and dicing film would not be easy. When the UV dose increased to 360mJ, the adhesive layer on the dicing film was cured completely, the die pick force reached the lowest value, and all of the dice can be successfully picked by die attach system.

While for high intensity UV curing, particularly with a dose of 360mJ, the no-pick was not due to insufficient cure of the dicing film, but due to some strong bonding between the DDAF and dicing film (chemical bond, curing of DDAF film) either caused by high UV intensity or hot spots. In this particular DDAF material, there was acrylic polymer in the die attach film layer. Some reaction of the residual carbon-carbon double bond with the dicing film through free radical mechanism is possible under high UV intensity or temperature. A thermocouple was used to measure the maximum temperature during UV curing on these two UV curing systems. On the low intensity system, the temperature increase due to UV curing was not significant, and the maximum temperature measured was less than 30°C. While on the high intensity system, the maximum temperature measured was as high as 60°C.

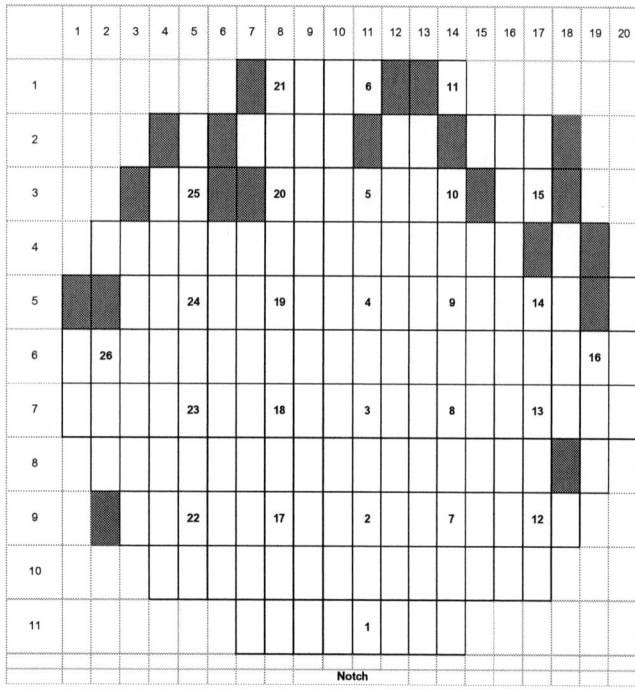

Figure 2. Location of no-pick die (marked red) for one wafer cured on a low intensity UV curing system with 180mJ UV dose.

The no-pick die sites were mapped on wafer. For low intensity UV curing with UV dose of 180mJ, the no-pick dice were mainly located at periphery and the end opposite to notch (Figure 2). The reason was that the wafer was pushed into the UV chamber with the notch end first while the UV light was on. Thus, the end opposite of the notch

was exposed for a shorter time. The actual UV intensity the tape received was expected to be lower at the periphery and the end opposite to notch than at the center, due to the geometry effect. For the high intensity UV curing system, the no-pick sites were randomly located on wafer. For this UV curing system, the wafer traveled underneath an open slot, through which the UV light was exposed, and the UV dose was controlled by the wafer traveling speed. From the results above, it can be seen that low intensity UV curing is better than the high intensity UV curing system for this DDAF material, and high UV dose (360mJ) is better than low UV doses (120mJ and 180mJ)

The die pick force test was measured for the DDAF tape when it was cured on the low intensity UV curing system with even higher UV dose (longer UV exposure time) to further optimize the UV curing condition. The results are listed in Table 5. When the UV dose increased from 360mJ further to 720mJ and 1080mJ, the die pick force increased further. In addition, the number of no-pick dice increased also (Table 6). When UV dose of 1080mJ was used, 80% of the dice on the wafer could not be picked up by the picking head on the die attach machine. From these results, the DDAF material should be cured on low intensity UV curing system with UV dose of around 360mJ. By comparing Tables 2 and 4 and Tables 5and 6, it could also be noticed that when die pick force and number of no-pick die increased dramatically for 11 days old DDAF material. The effect of DDAF aging will be discussed later.

Table 5. Die pick force for DDAF tape cured on the low intensity UV curing system (100μm thick die, DDAF was 11 days old before lamination, lamination temperature: 65°C)

UV Cure dose	360mJ	720mJ	1080mJ
Average:	1849.7	2095.7	2220.5
Maximum:	2232.2	2549.9	2555.8
Minimum:	1394.2	1095.0	1768.7
Std. Dev:	233.4	317.7	198.8

Table 6. Die pick observation for DDAF tape cured on the low intensity UV curing system (100μm thick die, DDAF was 11 days old before lamination, lamination temperature: 65°C)

UV Dose	360mJ	720mJ	1080mJ
Number of cracked dice	23	39	11
Number of no-pick dice	1	8	80%

2. Effect of Wafer Lamination Condition

During wafer lamination, a raised temperature was needed to get rid of voiding between the wafer and the DDAF film. The chuck temperature could affect the die picking at die attach as well. Wafer lamination was performed with standard force and speed but with different wafer chuck temperatures, and the wafers were removed either from the chuck immediately or removed from chuck 10 minutes later. Die pick force for different wafer lamination temperatures and duration time are listed in Tables 7 and 8. Number of no-pick dice and number of broken dice for the wafer laminated at these conditions are listed in Table 9.

Table 7. Die pick force (gram) for wafers laminated at 65°C for different duration time

Duration time	< 1 min	< 1 min	10 min.	10 min.
Average	1734.4	1736.0	1854.2	2039.2
Maximum	2095.6	2316.7	2276.4	2630.9
Minimum	1286.7	1260.7	1425.9	1172.8
Std. Dev:	230.9	244.3	231.8	318.6

Table 8. Die pick force (gram) for wafer laminated at 50°C for different duration time on chuck

Duration time	< 1 min	< 1 min	10 min.	10 min.
Average	1946.9	1882.1	1946.8	2002.2
Maximum	2412.4	2445.2	2240.5	2355.6
Minimum:	1316.7	1071.9	1494.9	1449.6
Std. Dev:	289	353.08	176.19	183.23

Table 9. Die picking performance for wafer laminated at different conditions

Lam Temp (°C)	Lam Time (chuck)	No. Broken Die	No. No-Picks
50	10 min	6	48
50	< 1 min	0	52
65	10 min	0	172 (whole wafer)
65	< 1 min	0	15

The longer duration lamination (10 min) at 50°C did not affect the die pick. The longer duration at 65°C definitely negatively affected the die picking, resulted in higher picking force and increased number of no-picks. When the wafer was laminated at 65°C and left on the chuck at that temperature for 10 min, the dice on the whole wafer could be successfully picked with standard die pick parameters. This was because that the strong bonding was established between DDAF layer and dicing film at higher temperature and longer time, and this strong bond could not be significantly diminished by UV curing of the dicing film. Thus, for easiest die pick purpose, the lower wafer lamination temperature was preferred. However, to get rid of voiding after lamination, higher temperature was preferred. An optimal temperature existed for balanced results (less voiding and easy die pick).

3. Effect of Aging Time

DDAF tape was pulled out and aged at room temperature for different time, and then laminated onto wafer at 55°C with standard process condition. The wafers were then diced, UV cured next day, and the die pick force was tested. The results are shown in Table 10.

Table 10. Die pick force (gram) versus tape age at room temperature

Days before lamination	0	7	10	24	31
Average:	1106.0	1512.8	1757.8	2020.7	2259.3
Maximum:	1358.3	2265.6	2196.9	2587.2	2831.1
Minimum:	667.4	859.8	978.1	1541.3	1723.7
Std. Dev:	128.6	309.5	352.7	267.9	256.0

Evident from Table 10, the die pick force increased dramatically with the tape aged at room temperature before wafer lamination. After 31 days, the die pick force doubled. This dramatic increasing in die pick force correlated to the increased die pick difficulty at die-attach process.

The wafers from the same experiment were die-picked with die attach equipment using the standard needle and dome configuration. For tape of 0 day old pre-lamination, there was no problem in picking with normal die-picking parameters. For tape of 7-10 days old pre-lamination, some dice could not be picked at the first try, and were picked after multiple trials using normal pick parameters. With slightly increased needle height and delay, all of the dice on these two groups were able to be picked successfully. However, a loud snap sound was heard as the die released from the dicing tape. For tape of 24 days old before lamination, needle height and delay had to be greatly increased beyond the normally acceptable range in order to pick up the die. All of the dice were picked, but some were cracked. No die fragments were left behind on the wafer despite the cracks. For tape of 31 days old before lamination, even with the excessive settings used for the 24 day samples, not all dice could be picked and some were severely shattered on the wafer.

The results above indicated that DDAF tape has to be processed (through dicing and UV curing) within a certain period of time after it is pulled out from its cold storage conditions.

The die pick force was tested versus time on tape post UV curing for up to 90 days (Table 11). The pick force did not increase with the time post UV curing. This means that DDAF material has a long work life post UV curing. The die picking would not become more difficult versus time post UV curing. This conclusion was further confirmed on die attach machine.

Table 11. Die pick force (gram) versus time post UV curing

UV Cure dose	360mJ	720mJ	1080mJ
T = 0 day			
Average	1849.7	2095.7	2220.5
Maximum	2232.1	2549.9	2555.8
Minimum	1394.2	1095.0	1768.7
Std. Dev	233.4	317.7	198.8
T = 30 days			
Average	1345.9	1456.8	1446.9
Maximum	2141.2	2472.2	2553.8
Minimum:	785.0	65.8	293.4
Std. Dev:	421.6	629.3	712.2
T = 90 days			
Average:	1736.8	1612.6	1242.7
Maximum:	2313.2	2063.4	1766.5
Minimum:	1280.6	1154.1	970.1
Std. Dev:	252.7	250.2	208.3

SUMMARY

UV intensity and UV dosage affected the adhesion. When the tape was cured under high UV intensity light, there were some hot spots on the two-in-one film, and die pick became more difficult for these locations than the other areas. For low intensity UV curing system, the UV dose (curing time) had a significant affect on adhesion. There was an optimal UV dose for easy release of die bonding adhesive layer and UV dicing tape layer. Wafer lamination temperature and time affected the adhesion as well. When wafers were laminated onto the DDAF at a higher temperature (65°C), and left on the chuck at this temperature for a long time, the separation between die attach adhesive layer and UV dicing film layer became more difficult. The die picking force increased continuously with the tape age before lamination at room temperature. This indicates that inherent adhesion between die attach adhesive layer and dicing film, which can not be diminished by UV curing, increases continuously with the tape age at room temperature before UV curing. The increased adhesion between die bonding film layer and UV dicing tape vs. aging time before UV curing is mainly due to interaction (inter-diffusion and mixing) of UV adhesive layer on UV dicing tape with the die bonding adhesive layer. The adhesion between the two layers did not increase with aging time after UV curing, as the adhesive layer on the UV dicing tape had been cured, thus limiting the inter-diffusion of the two layers.

ACKNOWLEDGEMENTS

The authors would like to thank following coworkers their contribution to the work presented this paper: Mr. Randy Park, Mr. Mike Ball, Mr. Nathan Draney, Mr. Paul Clawson, Mr. Shawn Hummel, Mr. Ron Pierce, Mr. Larry Attebery, Mr. Todd Schreib, and Mr. Eric Mueller.

REFERENCE

[1] Andreas Marte, Wolfgang Herbst, and Joachim Trinks, "New Needless Picking Technology for Thin Dice", Proceedings of 38th IMAPS, September, 2005, Philadelphia, PA.

PLACING WAFER LEVEL DEVICES IN A HIGH SPEED WORKFLOW

Gheorghe Pascariu
Hover-Davis Inc.
Rochester, NY, USA
gpascariu@hoverdavis.com

The electronics industry has long anticipated the arrival of advanced chip packaging that leverages the design advantages of bare die. With suppliers and manufacturers now facing the greater challenge of placing more die at higher speeds, the need to meet this demand has been overwhelming.

The concept of eliminating intermediate packaging costs while feeding flip chip devices at high speeds from wafer format, plus combining the **SMT and Semiconductor** placement process is the future for meeting tomorrow's demands.

Wafer level packages using flip chip technology can be solely comprised of flip chip devices or are often combined with surface mount devices. In either case, today's assembly schemes generally utilize a dedicated flip chip assembly line, with placement machines that tend to be relatively slow (2,000 – 4,000 real CPH) and very expensive. In addition, when the packages also include surface mount devices, a second assembly line dedicated to them is required. These assembly schemes are costly, complex, and require excessive floor space.

The advancement in SMT placement machine accuracies now enables Semiconductor "Advanced Packages" to be integrated into the SMT workflow. Wafer level packages combined with SMT placement machines create a **High Speed / Low Cost** solution, ideally suited for High Volume Manufacturing. This innovative approach is being implemented by leading Semiconductor OEMs and Subcontractors, and this cutting edge technology is giving them a much needed advantage in the extremely competitive world of Advanced Packaging. This technology is now being used for the production of high volume, System in Package, Chip On Board, Consumer Electronics, Next Generation Mobile Phones, Multi Chip Modules, Advanced Memory Devices, MEMS, High Speed Telecommunications and so on.

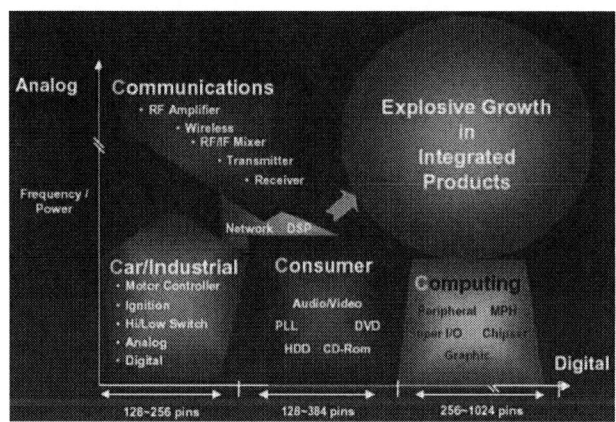

(Ref: W. Chen, ASE)

The Hover Davis Direct Die Feeder (DDF ULTRA) is an innovative modular design feeder that meets many die placement needs within the SMT process. It can be mounted on a variety of Pick & Place systems by means of a bayonet that provides a universal mounting interface. The bayonet combined with a simple communication protocol ensures a seamless integration that enables high speed delivery of components directly form a diced wafer.

DDF ULTRA is a **Universal** Die Feeder that can be easily moved between platforms enabling a highly flexible production flow that meets the most stringent demands of today's cutting edge technology and short product cycles.

The fastest growing market segment is SIP (System In Package), it is driven by the explosion of innovation in the Mobile telecommunication sector, especially hand phones.

Demand for high speed, complex features, form factor and connectivity is driving the integration of multiple devices in to the next generation communication devices enabling them to handle a lot more then just phone calls.

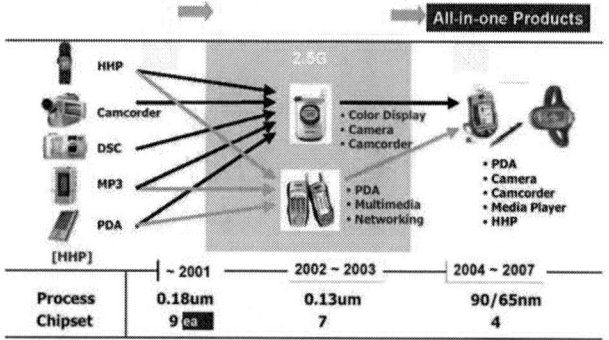

(Ref: Samsung Electronics)

Traditional assembly processes, in the past, had 2 types of lines.

1. Semiconductor "high Accuracy" where the silicone chips were packaged, either in batch mode for the wire bonded type or in line for the flip chip.

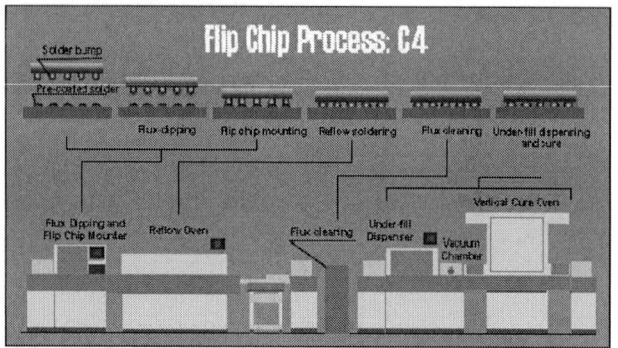

2. SMT where the passive components were added.

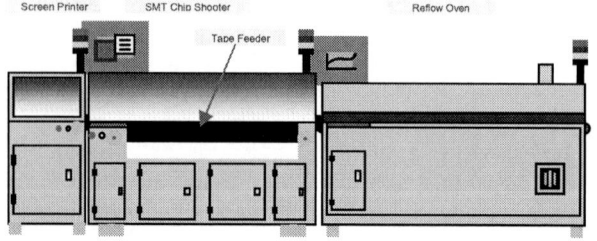

This type of setup is very expensive and not very cost effective.

By switching to the new technology the cost per placement is reduced by a factor greater then 10. This is achieved by combining the 2 processes in ONE line:

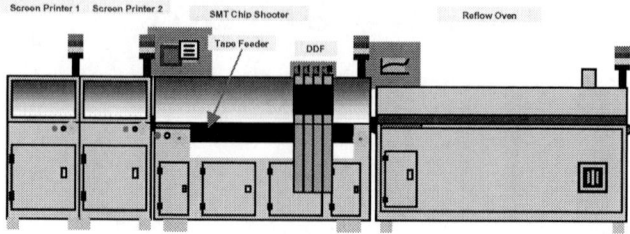

Below is a typical example of a SIP package.

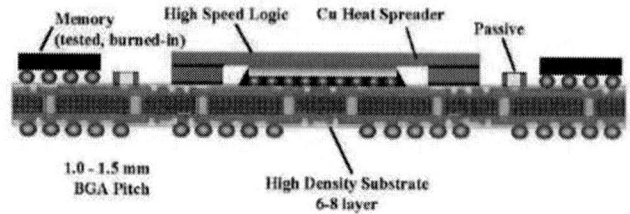

(Source: AMKOR, ITRS)

The old way of assembling this package, using conventional Semiconductor Equipment, requires multiple passes (due to multiple components, including several Si & GaAs dies) through the reflow oven leading warped substrates, process control is very difficult and yields are quite low.

By using a SMT high speed line with direct die feeding capability enables one pass process: the substrates are loaded into the Screen Printer 1 for flux application then transferred to the Screen Printer 2 for solder paste application, now the substrates are introduced into the SMT High Speed Placement machine where the fiducial marks on the substrate are versioned and the passive components followed by the Si & GaAs dies are processed (placed) directly from wafer, once all the modules are processed the finished substrates are indexed into the reflow oven.

This is an inline process that allows the manufacturers to take back control over the process and increase their yields while reducing costs via Super High Speed handling provided by the innovating combination of High Sped SMT System and DDF.

The need for the DDF is being driven by the market demand to process the packages faster and cheaper while improving the form factor and the performance. In order to improve the form factor and the performance the manufacturers must switch to Flip Chip packages, this being especially true in consumer electronics.

The challenge in Flip Chip applications was always the need for higher accuracy, while this was true in the past; it is no longer the case. Advances in packaging materials, more accurate substrates, better bumping technology and more accurate dicing saws with thinner dicing blades, reduced the need for high accuracy.

Flip Chip as well as Direct Die components are processed on the DDF ULTRA without the need for conversion or special tooling, as the system is equipped to handle both chip types. The chip type and process are chosen through a User Friendly Touch Screen Interface.

Wafers up to 200mm can be handled and component between 0.5mm up to 5mm square can be processed with normal setup, bigger components handling is optional. Wafers are loaded and processed in a vertical position providing much needed floor space savings.

The DDF ULTRA has a very compact form factor:
Height 1162 mm
Length 1125 mm
Width 120 mm
Weight 39 kg

Wafers are mounted on a metal ring (Wafer Frame) by means of an adhesive tape, enabling the wafer to be diced into individual components (dies). In order to attach the components to the pc board substrate they have to be removed from the wafer. In the past this required a separate process step and an entire industry was dedicated to it known as "Back End Semiconductor". Specialized Equipment such as Die Bonders and Flip Chip Bonders were used to prepackage the components so they can be placed in Tape and Reel for use in SMT Pick & Place equipment.

The arrival of Flip Chip Technology eliminated the need for the prepackaging of the individual components (except for special applications), thus creating the need for a Universal Direct Die Feeder that gives the SMT Pick & Place systems access to Semiconductor feeding technology for feeding components supplied in wafer format. By eliminating the Back End Semiconductor process step, tremendous savings are attainable.

The boards are loaded onto a feeder at the beginning of the line, this feeds them one-by-one as needed, the boards go first into the Screen Printer(s) where flux and adhesives (solder paste or epoxy) are applied, from here they enter the SMT Pick & Place system (High Speed chip attaching system with speeds in excess of 100,000 CPH). Once in the Pick & Place, the board is locked in place and the vision system locates the reference marks that enables accurate placement of the components. Once the vision processing is finished the Pick Heads move to the dedicated feeders to pickup components, locate them on the pick nozzles and accurately place them onto the board. After all the components are attached the board progresses into a Reflow Oven were the boards are cured.

With the DDF ULTRA fully integrated into the SMT pick & place line the entire board can be processed in a single pass increasing Yield and Productivity while Maximizing Throughput.

Once the wafer is loaded, the DDF ULTRA Vision System locates the components on the wafer making use of pattern matching algorithms. Ink Dot detection or Electronic Wafer Mapping are used to define the known Good components to be picked from the wafer by the DDF ULTRA and transferred to the Pick & Place machine for placement.

The Pick process is a very important element of the entire cycle. The Ejection Needle is synchronized with a Low Force Pick Tip, enabling the handling of very delicate components. This Advanced Process Control allows the DDF ULTRA to handle a variety of components from Si die and Solder bump flip chip die to delicate GaAs MEMS devices.

Flip Chip Mode
The components are presented on the wafer with the bumps up, the DDF ULTRA must pick the component and inverted the chip so that the Pick & Place can accurately place the Flip Chip with the bumps down on the board.

By using a synchronized combination of vacuum on the pick tip, which grips the bumped side (front) of the component and an ejection needle on the backside to release the tape from the back, the synchronized movement will gently remove the component from the wafer. With the component safely on the Pick Tip, the Pick Head will rotate 90° up and place the component on a Flipper.

The Flipper holds the components face down, with the bumps facing down, and transfers them to a Shuttle (delivery mechanism which presents components to the Pick & Place) which in turn will transfer the components to the presentation point, ready to be processed by the Pick & Place system.

The Flipper is designed to hold up to 6 components, the number is programmable. This enables parallel processing because DDF ULTRA continues picking components from the wafer while the pick & place nozzles remove the processed components from the shuttle.

Direct Die Mode
These components are presented on the wafer with the active surface (bond pads) up. With this type of chips the DDF ULTRA must pick the component and presented it to the Pick & Place with the active surface facing upwards, enabling the Pick & Place to accurately place it on to the board

The synchronized movement of the pick tip and eject system, using a combination of vacuum on the pick tip, attaching to the active surface (front) of the component and the ejection needle on the back, combine to gently release the component from the tape. With the component safely on the Pick Tip, the Pick Head will rotate 90° down and place the component on the Shuttle.

The Shuttle holds the component with the bonding pads facing up, and transfers it to the presentation point, ready to be processed by the Pick & Place system. After one component is placed onto the shuttle the pick head will pick the next component and index to place it on the next location onto the shuttle.

The Shuttle is designed to hold up to 6 components, the number is programmable. This enables pick & place systems with multiple nozzles on one head to maximize their productivity and achieve high CPH.

The components on the shuttle are replenished while the Pick & Place head accurately places the components onto the board.

The DDF ULTRA is taking the industry by storm as it changes the way next generation electronics are assembled. Throughput is going up, Costs are going down, Productivity and Yields are rising Increasing Profits, giving the savvy ones the competitive edge.

DDF ULTRA provides unrivaled performance at a fraction of the cost of traditional flip chip bonders while using the existing High Speed SMT equipment.

The future is here, welcome aboard.

METHODOLOGY FOR STACKING OF POWER SEMICONDUCTORS FOR THE HARSH AUTOMOTIVE ENVIRONMENT

Todd P. Oman
Delphi Corporation
Kokomo, IN

ABSTRACT

Power semiconductors require electrical interconnects to both sides of the die and a highly conductive thermal path. These requirements are typically satisfied by employing wire-bonds whether using direct chip attach to the substrate or a package solution. The wire-bond process is an expensive, serial process which precludes heat sinking from the top-side of the die. Additionally, power semiconductors typically have active circuitry under the metallization of the bond locations making them vulnerable to mechanical damage during the bonding process. A method to achieve the top-side interconnects without the shortcomings of the wire-bond process has been developed that also facilitates vertically stacking of the die. This method of interconnection is compatible with high power devices due to its low thermal resistance. The interconnect hardware, SPDC (Stacked Power Die Contact), is also compatible with existing surface mount process flows and equipment, thus simplifying implementation. The SPDC incorporates traces and vias through a dielectric material that provides the required isolation between the gate and source of a die and the drain of the adjacent die. Reliability testing has demonstrated that this method exceeds the automotive goal of a minimum life of 1000 thermal cycles for temperature extremes of -40 to +150°C.

Key words: power, silicon, stacked, vertical integration.

BACKGROUND

To achieve vertical integration for power devices, several shortcomings in existing packaging technologies must be overcome. From an electrical viewpoint, the interconnects for power semiconductor devices such as PFETs or IGBTs require contact to the front and back of the die. This requirement is typically satisfied with wire-bonds, but these bonds interfere with dual-sided heat sinking which is highly desirable for increasing the power dissipation of the die. Wire-bonded die may be in the from of DCA (Direct Chip Attach) where the die is bonded directly to the substrate or in an encapsulated package as shown in figure 1. Wire-bonds also have parasitic resistance and inductance reducing the overall system performance. Additionally, the small contact area of a wire-bond results in undesirable high current densities. Achieving high wire-bond yields is a challenge for power semiconductor devices since the bond sites are over active circuitry that can be damaged by excessive bond pressure. One alternative to wire-bonds is attaching a metal contact to the top of the die to make the

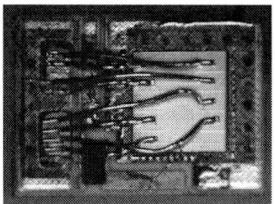
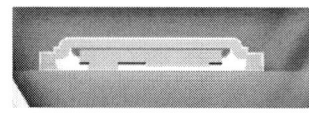

DCA with metal contact

DCA with wire-bond

DPAK   QFN MCM   SO - 8 Leadless

Figure 1: Typical Power Packaging Solutions
top: Direct Chip Attach
bottom: Encapsulated Packages

electrical connection to the substrate. This structure has two primary issues: 1) excessive solder joint fatigue from the large CTE (Coefficient of Thermal Expansion) mismatch ratio between the metal contact and silicon is approximately 17:3 when using copper and 2) no electrical isolation on top of the metal contact to facilitate top-side heat sinking. None of these structures support vertical integration or stacking of power die electrically or thermally.

A new packaging technology was developed that addresses these issues as shown in figure 2. The design goals were to develop a structure with complex electrical interconnect capability that facilitate the stacking of power die. The structure provides electrical isolation from the heat sink and is constructed with materials and a topology that result in low thermal resistance. The requirement to be compatible with existing process flows was met. Automotive reliability requirements require thermal excursions from –40 to 150°C which mandated that the design minimize stress on solder joints.

Referring to figure 2 which shows two stacked power die contacts in a parallel circuit configuration, each contact is constructed of a base ceramic material that serves as both electrical isolation and a mechanical restraint to reduce the CTE (Coefficient of Thermal Expansion) of the copper conductor that is brazed to the ceramic. Depending on the thickness of the copper which typically ranges between 0.2 and 0.5 mm, the effective CTE of the copper is reduced

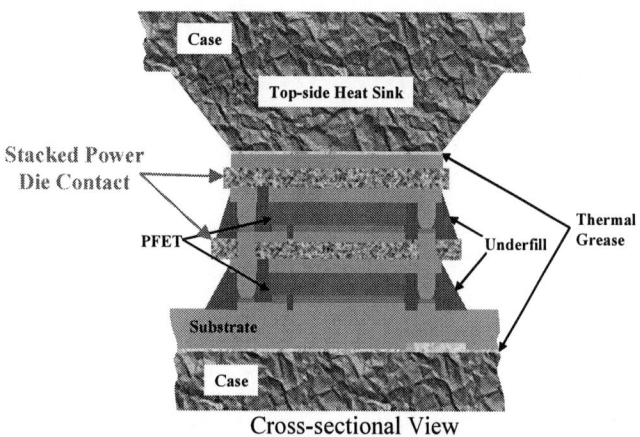

Cross-sectional View

Figure 2: Silicon Two Stacked Power FETs Connected in a Parallel Circuit with Dual-sided Heat Sinking

from 17 to approximately 8 ppm/°C. Copper vias have been employed to create the interconnect required through the ceramic to facilitate die to die interconnects. These vias are evident in the X-ray image shown in figure 3. To complete the electrical connection from the bottom contact to the substrate and from the bottom contact to the top contact, copper balls were brazed to the copper conductors. These balls precisely control the height of the contact since the small contact area allows the ball to directly contact the surface instead of "floating" in the solder joint. This tight control of the height is critical for controlling the solder joint thickness above and below the die since solder joint reliability is proportional to solder joint thickness [1]. The diameter of the balls is specified by summing the thickness of the die and the two desired solder joint thickness above and below the die. To reduce solder fatigue, a capillary underfill was dispensed to encapsulate the die, the copper balls, and their associated solder joints (see figure 2).

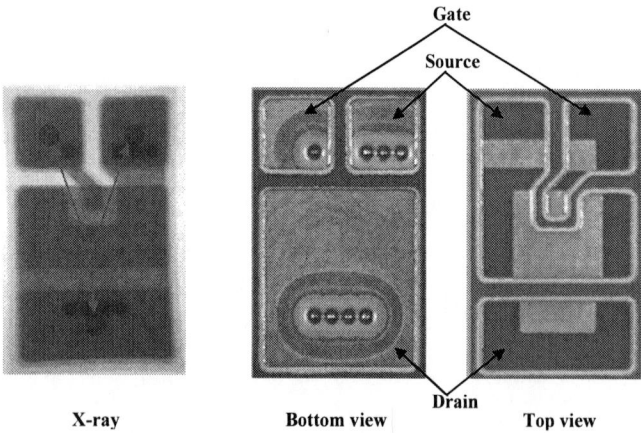

Figure 3: Stacked Power Die Contact for 3.8 x 3.8 mm IGBT

The product and process benefits for implementing the SPDC are:
Product:
- Electrically isolated for top-side heat sinking

- Large pedestal bond surface for distributed current and heat
- Substrate basis facilitates complex routing of conductors
- Increased reliability due to improved thermal performance
- Lower RDSon (short interconnects) and inductance (opposing current flow) for improved electrical and thermal performance
- Built-in EMI shielding with large copper plane
- Reduced foot print with near CSP and vertical integration
- Reduced stress on die solder joints from thermal expansion

Process:
- Uses standard assembly process
- Deliverable in standard die, package, and component methods (i.e. tape/reel, waffle pack)
- Self aligns
- Reduced yield loss for single Vs. multiple bond process
- Controlled die solder joint thickness with tightly controlled lead heights
- Reduced process cycle time by eliminating serial process of wire bonding

DESIGN & MATERIALS

Selecting a substrate material that provided the desired mechanical and electrical properties was a key decision at the beginning of the design phase. Thick copper conductors were highly desirable for heat spreading (~400 W/m°K) and current carrying capability (typically 0.06 mOhms/sq for a 0.25 mm thickness); however, the CTE mismatch between copper and silicon creates a very harsh environment during thermal cycling that can generate solder fatigue and cracking. Cracking of the solder joint degrades the thermal and electrical performance by reducing the conductive area. Based on this, the substrate material's mechanical property requirements were 1) high fracture toughness to avoid cracking during thermal cycle testing, 2) low CTE to reduce the effective CTE of the copper conductor from 17 to below 10 ppm/°C, and 3) high thermal conductivity to minimize die temperature. Silicon nitride (Si3N4) and aluminum nitride (AlN) meet the second and third criteria with AlN having the best thermal conductivity of 160 vs. 70 W/m°K. Extensive characterization of these two inorganic or ceramic materials demonstrated inadequate fracture toughness for aluminum nitride when using copper conductor thicknesses greater then 0.2 mm thick (figure 4). Another advantage of silicon nitride's fracture toughness (6.2 vs. 1.8 Mpa/m) is the ability to use thinner substrates which helps compensate for it's greater thermal conductivity. The typical minimum thickness for silicon nitride is 0.25 mm compared to 0.64 mm for aluminum nitride. The thermal resistance (Rth) contribution for the ceramic portion of the contact can be estimated using $R_{th}=L/A \cdot k$ where L is the substrate thickness, A is the substrate area, and k is the substrate thermal conductivity [2].

194

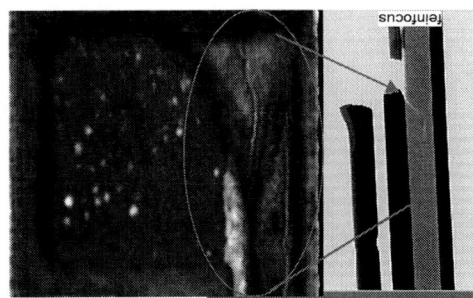

**Figure 4: Cracked AlN substrate
top: CSAM, bottom: lateral X-ray**

$R_{th\,(AlN)} = 0.00064/(0.008*0.0011*160) = 0.45\ °C/W$
$R_{th\,(SiN)} = 0.00025/(0.008*0.0011*70) = 0.41\ °C/W$

Prior experiments with similar silicon nitride structures had been performed with copper conductor thicknesses ranging from 0.2 to 0.5 mm. The thinner copper lowered the overall CTE of the contact to 8.45 ppm/°C which reduced the stress on the die solder joint, whereas the thicker copper raised the CTE of the contact to 11.9 ppm/°C which more closely matched the FR4 substrate being used reducing the stress on the copper ball solder joints. Although variations in stress were observed, all thicknesses exceeded the reliability requirements for a contact sized for a large PFET of dimensions 7x7 mm; therefore, a standard copper thickness of 0.3 mm (9.9 ppm/°C) was used to evaluate this concept.

As aforementioned, copper balls were employed to make the connections between the substrate and the contacts. Besides the advantage of tightly controlling the solder joint

thicknesses, this interconnect cannot be easily damaged like formed, compliant leads. The ball is also easily encapsulated and "gripped" by the underfill material to hold the structure together and minimize solder joint fatigue.

To facilitate vertical integration, the last feature added to the contact was copper vias. Initially, the vendor thought that the minimum manufacturable size would be 2 mm; however, this significantly increased the size of the SPDC. With additional work, the size of the via was reduced to 1 mm.

Assembly Process
The first step in the process is the printing of solder paste on the substrate which in this case was 0.031" thick FR4 (figure 5). The solder used for this build was eutectic SnPb and the board finish was OSP. Solder paste, board finish, and solder reflow parameters can significantly impact the quantity and size of solder voids generated in the large solder joints above and below the die. The acceptable levels of voiding are dependent upon performance goals. Next, the power die is placed on the substrate. A solder preform is placed on the die after applying liquid flux. The liquid flux was chosen over a fused flux to aid in keeping the assembly aligned until the solder is reflowed. A fluxed SPDC is placed on the solder preform. The SPDC also had manually placed solder preforms where the next contact's copper balls make contact. This solder placement could be automated by printing solder paste on top of the contact prior to placement. The double stack shown is completed by liquid fluxing and placing a preform, a power die, a preform, and the final contact. Since this is the last contact for

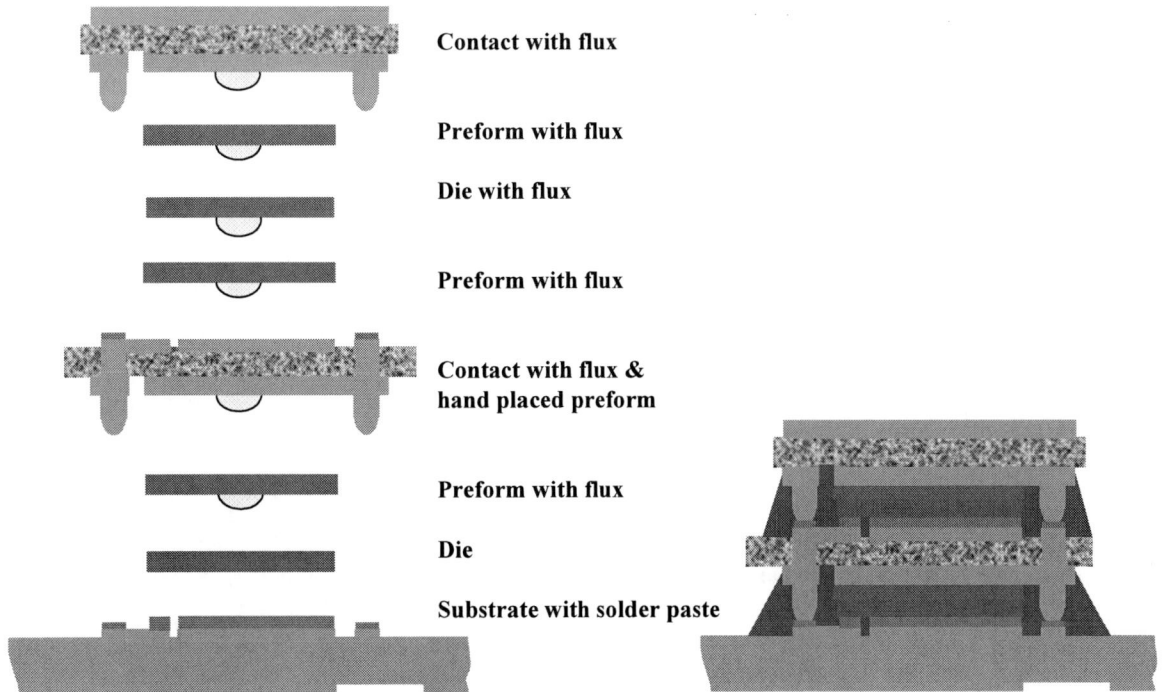

Figure 5: left: Semi-automated Assembly process, right: final underfilled assembly

a double stack, no vias are required in this contact. Once the desired number of power die have been stacked, the assembly is processed through a solder reflow oven. Prior to underfill, the assemblies are cleaned to remove any flux residue that would interfere with adhesion. A 1 mm exposed ledge is required for each SPDC to facilitate an area to deposit the capillary underfill. This is shown in figure 6 for the SPDC that was designed for a 7x7 mm PFET. The

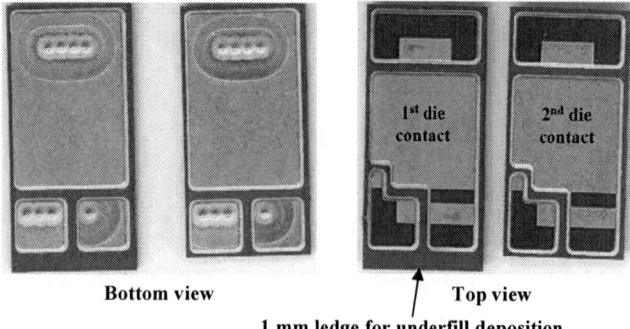

Figure 6: SPDC for a 7x7 mm PFET

assembled test board with three double stacks of the 7x7 mm PFETs is shown in Appendix A. The overall assembly yield achieved was 87% which was better then anticipated based on the engineering pick&place equipment's poor placement capability and the manual placement of solder on the SPDC instead of the desired automated printing of solder paste.

Characterization

The copper conductor thickness and via size are more then adequate from an electrical performance aspect. The primary concern was understanding the impact on thermal performance for this structure. A 3.8x3.8 mm thermal test die was used to characterize the thermal performance of the small SPDC shown in figure 3. A two layer board with minimal copper 1 ounce copper was used for a worst case scenario. A heat sink was applied to the top contact with a force of 25 PSI and a thermal grease was used to eliminate air gaps. Thermal resistance values were obtained for a bare die where the silicon was directly applied to the heat sink, a single die, a double die stack, and a triple die stack as plotted in figure 7. For this size die and contact, the thermal resistance increases 0.6 °C /W for each stack added. As shown in figure 5, each stack consists of a layer of solder, silicon die, solder, and SPDC.

Testing

The test boards containing thirteen samples were subjected to 3007 thermal cycles of -40 to 150°C with a cycle time of 80 minutes and dwell times of 15 minutes. The devices were continuously monitored electrically to identify the exact cycle of failure. A total of four failures were observed at 1151, 1522, 1888, and 2236 cycles; therefore, the goal of exceeding 1000 cycles failure free was met. Instead of using the JEDEC standard of 30 Ohms for the failure criteria, a more stringent limit of 1 Ohm was used. A

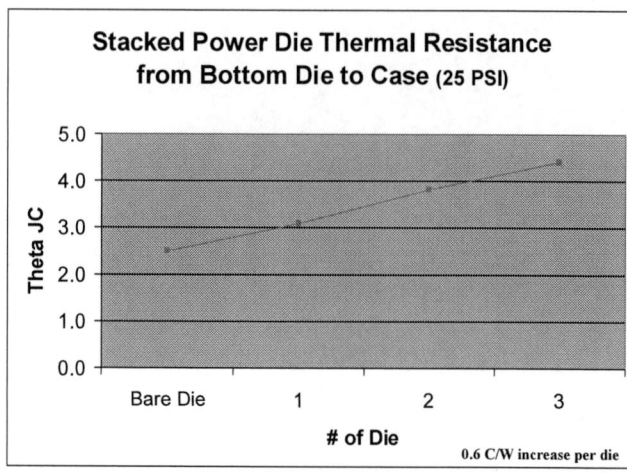

Figure 7: Thermal Performance for 3.8x3.8 mm Die

Weibull plot of this data is shown in Appendix B. From a Weibull viewpoint, the goal for beta is 3 and the goal for eta is 2500 cycles. Although the sample size was not optimal for estimating these values where a sample of at least thirty is desired, the beta predicted from this test was 2.7 which is slightly less then desired and eta was 3294 which is acceptable. In summary, this technology shows acceptable reliability, especially since this test case used the largest die size that is economical for automotive systems.

To evaluate the integrity of the large solder joint above the die where the shear strain is the greatest, acoustic microscopy images were obtained as shown in figure 8. This joint exhibits no cracks after being exposed to the 3007 thermal cycles which is a testament to the CTE compensation of the silicon nitride and the performance of the underfill.

Failure analysis identified the root cause of the failures as delamination of the top SPDC contact from the underfill resulting in an open circuit at the copper ball solder joint which is obvious from the cross section shown in figure 9. The large arrow is pointing at the open solder joint and the delamination of the right side of the top SPDC is evident by the dark area between the copper and the underfill. The bottom SPDC is mounted to FR4 which has a much greater CTE, but this SPDC is constrained by underfill both below and above. The top SPDC does not have a constraint

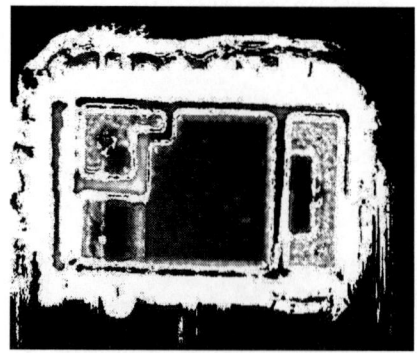

Figure 8: Acoustic Microscopy Image of Die Solder Joint

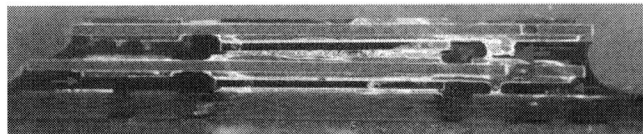

Figure 9: Cross Sectional View of a Double Stack after 3007 Thermal Cycles

above. For this particular test, the top SPDC was not optimized with a symmetrical metal pattern which would have minimized bowing, instead it was designed to facilitate the fabrication of a triple stack for the thermal characterization that was performed. Minor cracking was also observed at the high stress points generated by the copper trace edges (small arrows) as expected for 3000 thermal cycles. SEM analysis of the cross sections of the new copper vias were taken to investigate the integrity of this structure (figure 10). No cracking or fatigue of this structure was observed in the silicon nitride, copper, or brazing material.

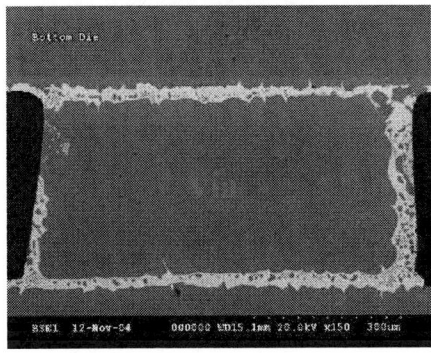

Figure 10: Cross Sectional View of a Copper Via

CONCLUSIONS

This technology provides a vehicle to exceed CSP packaging densities for power devices such as PFETs and IGBTs that meets automotive reliability requirements and exceeds those of the commercial sector. Based on the design of this structure and the characterization that was performed, the following can be concluded:

- Excellent thermal performance was attained with only a 25% degradation as compared to a bare die
- Ampacity can be selected by increasing the quantity of copper balls that is facilitated by this structure
- Lower RDSon (short interconnects) and inductance (opposing current flow) for improved electrical and thermal performance
- Built-in EMI shielding with large copper planes

Based on the design of this structure and the assembly process experiences for two die sizes, the following features were confirmed:

- Uses standard assembly process
- Deliverable in standard die methods (tape/reel, waffle pack)
- Self aligns
- Controlled die solder joint thickness with tightly controlled lead heights

Conclusions based on the reliability test results:

- Uses standard assembly process
- Exceeded target of a Weibull Eta greater then 2500 cycles
- Solder joint cracking was minimal
- Copper vias showed no degradation
- Increased device reliability due to improved thermal performance

REFERENCES

[1] Hyun Soo Park, "How Solder Joint Height Affects the Fatigue Life of the Solder Joint", EEE Links Vol. 1, No. 2, April 1995, pp 4-10.

[2] Young & Freedman, "University Physics," Addison Wesley, ISBN 0-321-20469-7.

Appendix A: Test Board for Thermal Characterization & Thermal Cycling Reliability Test

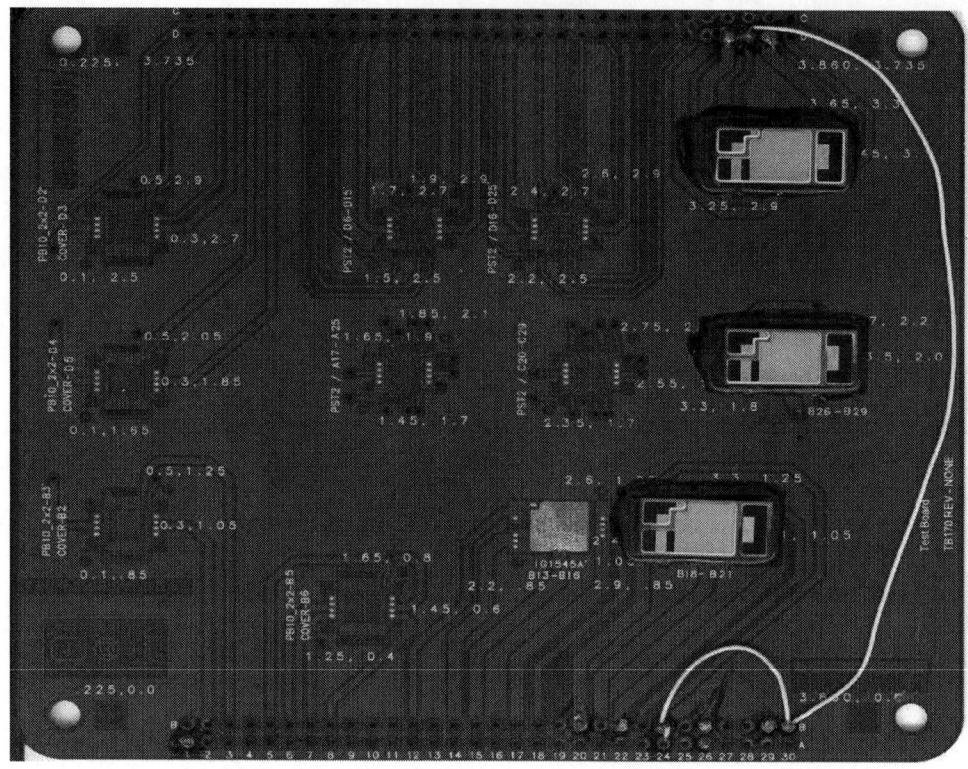

Appendix B: Weibull Plot for –40 to 150°C Thermal Cycle Test

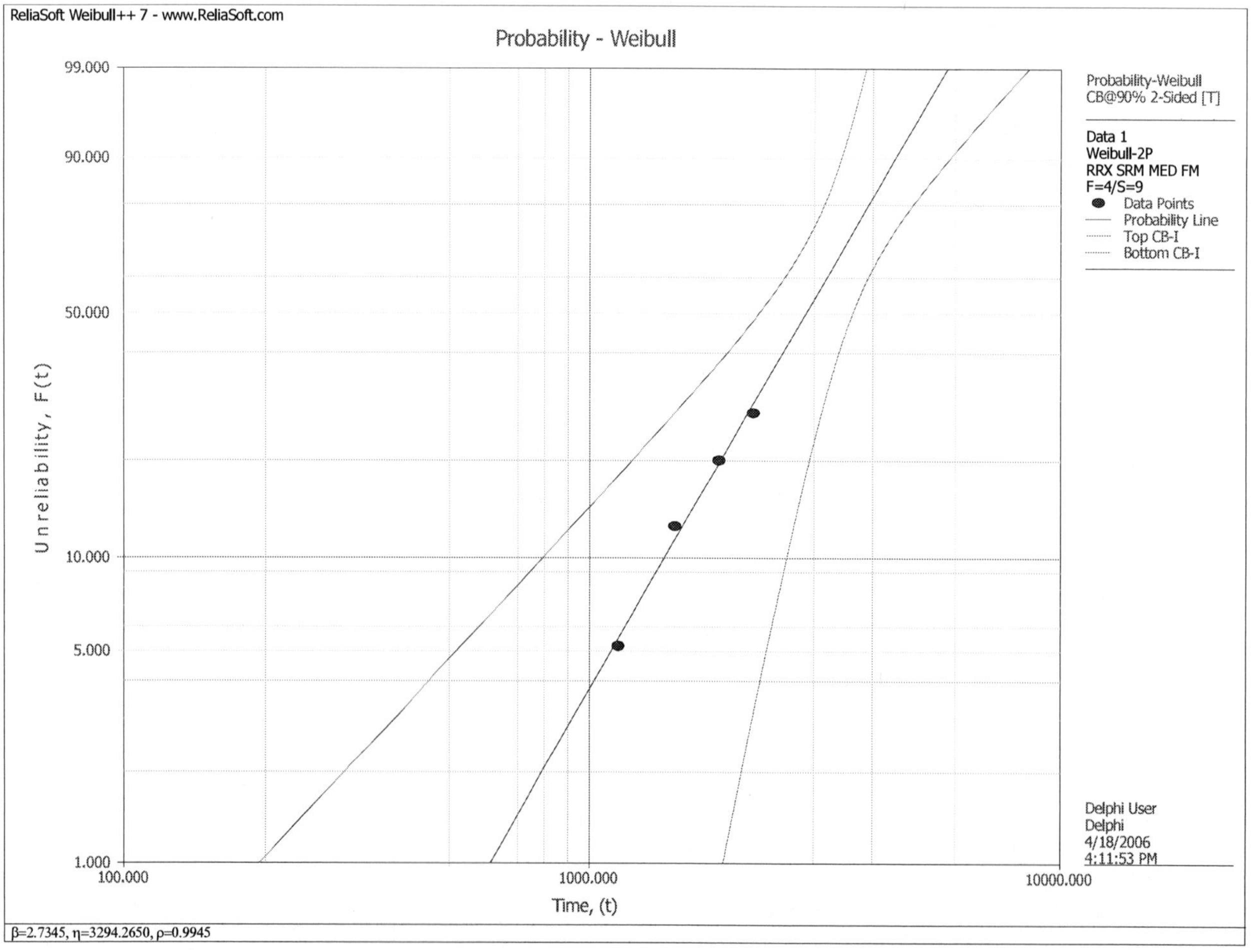

ReliaSoft Weibull++ 7 - www.ReliaSoft.com

Probability - Weibull

Probability-Weibull
CB@90% 2-Sided [T]

Data 1
Weibull-2P
RRX SRM MED FM
F=4/S=9
● Data Points
— Probability Line
···· Top CB-I
···· Bottom CB-I

Delphi User
Delphi
4/18/2006
4:11:53 PM

β=2.7345, η=3294.2650, ρ=0.9945

WHITE RING DEFECT FORMATION IN LEAD-FREE WAFER LEVEL PACKAGING

Kimberly D. Pollard, Ph.D.*, Raymond Chan, Ph.D., and Diane Scheele
Dynaloy, LLC
Indianapolis, IN, USA
kimpollard@dynaloy.com

ABSTRACT

FlipChip (FC) and Wafer Level Packaging (WLP) are two of the fastest growing segments in the integrated circuit (IC) packaging industry. The practice of incorporating FC or bump technology in devices is experiencing significant growth, due to the improvement it provides in power and ground distribution, with the attendant reduction in simultaneous switching noise, (SSN). The demand for WLPs is increasing due to their small form factor and the increased demand for portable products.

Both FC and WLP technologies impose significant new demands related to the materials used for permanent dielectric layers and photoresists for fabrication processes. Cleaning chemistry must be designed to remove highly cross-linked films such as photoresists and fluxes; but keep the solder bump, exposed metals, under-bump metallization (UBM), and dielectrics undamaged. Device reliability and performance rely heavily on the proper working of all these components. *In situ* or discrete analytical procedures are required for testing the reliability of the materials used.

Typical quality assurance and control procedures for fabrication monitor the appearance (surface condition, shape) of the solder bumps. Wafer or device inspections are commonly performed by optical microscopy, and indicate if the device is free of photoresist after cleaning, and whether or not metal remains after etching.

The focus of this paper is defect formation during lead-free wafer level packaging. The defect type that the talk will concentrate on is referred to as the 'white ring' defect. It is found at the base of electroplated bumps and first becomes visible to the line operator in the bumping process after the thick photoresist is removed.

Key words: wafer bumping, defects, photoresist removal

BACKGROUND

A typical packaging process to produce solder connections could include the following steps: (1) application of a passivation layer, (2) deposition of UBM materials for adhesion promotion and to deter migration, (3) deposition of an electroplating seed layer, (4) application of a laminated photoresist which is exposed and developed to generate the bump placement mold, (5) solder electroplating, followed by (6) photoresist removal and (7) inspection. Although in many fabrication facilities, many of these processes are performed by separate and distinct process groups, defect formation can occur many steps prior to detection. As an example, in the overall process defined above, the wafer surface is not visible for inspection between the steps of photoresist application and photoresist removal.

Historically, common defects include "photoresist lifting", ("resist lifting", or "photo lifting") in the PCB board industry and "photoresist delamination" in the semiconductor industry.[i] This defect has resulted in foot plating and in more extreme cases, metal films being deposited in areas which would otherwise have undergone seed layer removal *via* an etch process.

In the case of plating lead free solders, (for example, SnAg,) the metal that is deposited outside of the foot plating area is white in color and has been referred to as the 'white ring'. This ring is visible to the line operator only after the thick photoresist is removed.

Other causes of the white ring have been examined including redeposition of photoresist, and local mixing of electroplating solution and photoresist removal products. The results will be discussed briefly, but our results indicate that they are not the cause of the white ring.

The delamination and underplating results reported here have been shown to occur during electroplating of eutectic solders when the photoresist processing has not been optimized. In this instance, the excess metal deposit is blue in color and referred to in the industry as "blue film".

EXPERIMENTAL PROCEDURES

Testing was carried out on test samples that were composed of a Ti/W adhesion layer, a Cu seed layer, patterned thick photoresist [thickness = 80 μm, bump diameter ~ 100 μm, space ~ 300 μm], a Ni stud layer, and a lead free [$Sn_{93}Ag_7$] electroplated solder bump. Samples were cleaned with Dynastrip™7000 at 97 °C for 40 minutes. Samples were examined using a Hitachi 3000N Scanning Electron Microscope (SEM), a PV7743 energy dispersive x-ray analysis (EDX) detector from Ametek, and a Physical Electronics model PHI 660 scanning Auger microprobe with a LaB_6 cathode scanning electron gun operated at 10 keV. The base pressure was 3.4×10^{-10} Torr. The parameters used in the depth profile analysis were 3 keV, with 70

$\mu A/cm^2$ Ar+ plasma. The etch rate was calibrated using SiO_2/Si and found to be 37 nm/min.

RESULTS OF TESTING

Test samples were cleaned and then examined for resist removal. Figure 1 shows optical images of the bumps, including the area described as the 'white ring'.

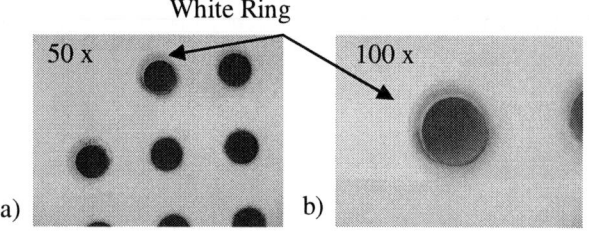

Figure 1. Optical microscope images of the white ring.

The white ring is characterized by a metal layer which extends out beyond the typical distance of the foot plate. In the images in Figure 1, the ring appears to have a white color.

Figure 2 shows SEM images of the ring.

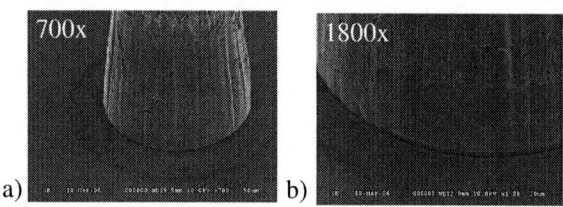

Figure 2. SEM images of a typical bump showing the formation of the metal ring.

The images show that the metal film propagates outward from the bump, over the copper field metal.

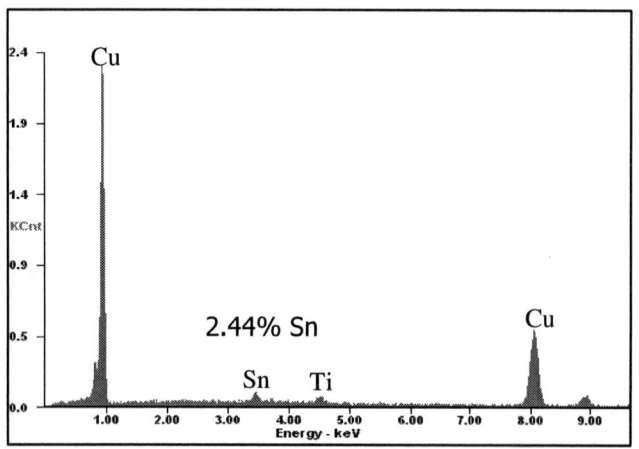

Figure 3. EDX spectrum of the white ring at 15 kV.

The EDX spectrum was obtained for a sample on the white ring and is shown in Figure 3. The copper and titanium are due to the field metal while the tin is due to the white ring. If the composition of the ring is tin, the only sources of tin are from the electroplated bump, of which it is the major

component, or from uptake of tin ions into the photoresist during electroplating.

One drawback to this technique is that the sampling size of the beam in EDX is large enough (up to 3 μm) that analyses at areas close to the bump may have contributions from the solder bump. As a result, further data were collected to ensure that they were representative.

To further confirm the results, we obtained cross-sectional SEM images of samples prior to photoresist removal, (Figure 4). Samples were visualized after sputter deposition of approximately 10 Å of Pd metal.

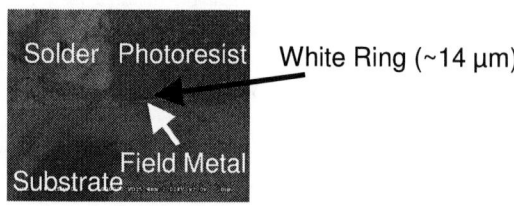

Figure 4. SEM micrographs of the electroplated metal underneath the photoresist (PR). (a) 2000x magnification (b) 6000x magnification.

In Figure 4, it may be seen that there is metal deposition underneath the photoresist. In Figure 4 (a), the solder bump, substrate, field metal and photoresist can be seen. The 'white ring' is the lighter colored film (due to the heavier element, Sn) on top of the seed layer (Cu) and under the resist. In Figure 4 (b), the higher magnification image shows the 'white ring' film on top of the field metal. These results are consistent with delamination of photoresist, allowing further electroplating under the resist. It is important to note, however, that the additional delamination in Figure 4(b) may be due to sample preparation.

Further analysis of the cleaned sample was performed using Auger Electron Spectroscopy. Figure 5 shows an image indicating where the data were obtained for comparison. Analysis area (1) is located on the white ring, (2) on the copper field metal, and (3) on the margin of the white ring. Spot sizes for the analyses were 0.5 μm. Auger spectroscopy is a very suitable surface technique for resolving compositional information at areas near the solder bump since the analysis spot size is small and the depth of penetration (4 – 50 Å) is very low.

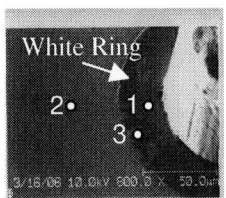

Figure 5. Analysis areas for Auger Spectroscopy.

The survey scans shown in Figure 6 identify all the elements in the sample. The carbon present is adventitious carbon, which is always found in Auger samples, despite the analysis being carried out under high vacuum. The size of each peak in the scan is proportional to the amount of the element that is present in the sample, provided that there is no peak overlap. In this case, the peak at 50 eV cannot be considered diagnostic for tin, copper, or nickel, due to the fact that they all overlap in this region. However, Auger is sensitive to these elements at other binding energies [Sn, ~425 eV; Ag, ~350 eV; Ni, 850 eV; Cu, 920 eV] that do not overlap and these peaks can be used to identify these elements.

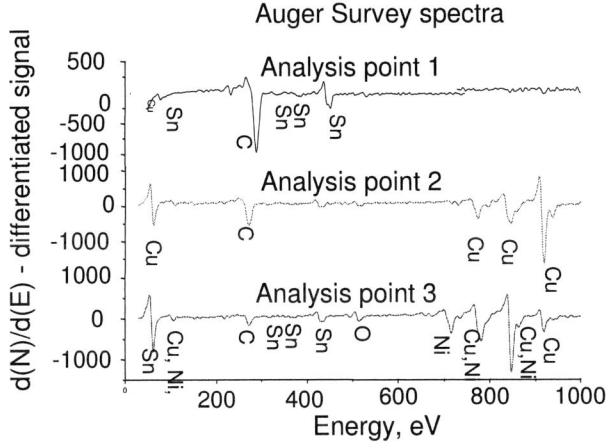

Figure 6. Survey scans of points of analysis.

Since this is a tin-silver lead-free electroplated bump ($Sn_{93}Ag_7$), it was expected that silver would be present in the white ring deposit. The Auger Spectroscopy handbook of reference spectra[ii] shows that Auger spectroscopy is very sensitive to silver and gives the binding energy for the diagnostic peak as 350 eV. However, none of the spectra of the white ring had a peak in the 350 eV region, indicating that silver was not present in the deposit. On review of the metal composition of the electroplating solution, the ratio of tin salt to silver salt was 100:1, meaning that the electroplating solution was very tin rich. Comparison of the standard reduction potentials for tin ($E° = -0.136$ V) and silver ($E° = +0.799$ V) indicated that tin should electroplate preferentially. Because solution is trapped under the laminated photoresist, mass transport effects in the electroplating solution become dominant in determining the composition of the deposit. The trapped solution experiences a different potential and changes in ion concentration from that of the bulk solution, which affect the alloy composition. In this case, the electroplating conditions are changed so much that no silver is electroplated and the deposit is composed of tin only.

As shown in Figure 6, the white ring is composed mainly of tin. Analysis point 2 shows the expected peaks for the copper field metal. Analysis point 3 shows the expected contributions from the copper field metal due to the thinness of the film in this area, as well as contributions from the film itself (tin). An interesting thing to note is that there are contributions from nickel as well which indicates that the delamination occurred in the process step prior to the deposition of the lead-free solder, during the nickel stud deposition.

To further clarify the film components, a depth profile using Auger spectroscopy was recorded on the area of the white ring, (analysis area 1), in Figure 5. The depth profile is shown in Figure 7.

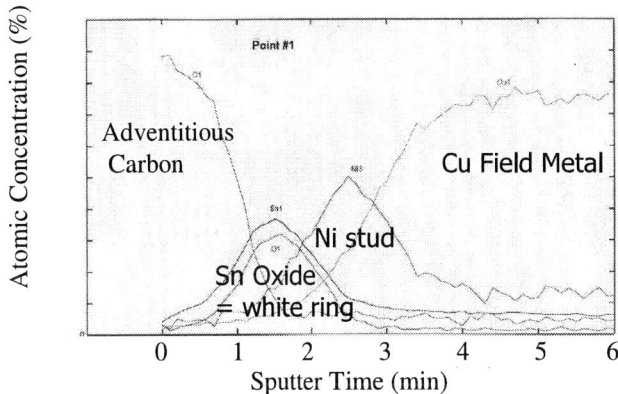

Figure 7. Depth profile through the white ring, the Ni stud and the Cu field metal.

The depth profile (Figure 7) confirms that the photoresist delamination and electroplating started with the nickel stud electroplating step. When it was complete and the sample was electroplated with the lead-free solder (Sn/Ag), the tin deposited on the nickel stud underneath the photoresist. It is believed that the oxidation of the tin was in subsequent processing steps, however, this is still under investigation.

SUMMARY AND CONCLUSIONS

Wafer level packaging relies critically on each process step being carried out successfully to ensure the greatest first pass yield. Formation of white rings is a common phenomenon that may be prevented with a robust photoresist process. The advent of white ring adds significantly to the cost of ownership, especially since the wafer being processed is complete and has the full value added from the front end fabrication facility.

Characterization of success or failure of each process step can be difficult when the wafer surface is not visible to the line operator. Once the thick photoresist is removed, after the solder bumps have been electroplated, defects in resist removal or in previous steps can be viewed for the first time. Use of surface sensitive tools such as SEM and Auger spectroscopy can aid in determining at which step a defect was made.

In the present paper, we have shown that the phenomenon known as "white ring" in lead-free electroplated bumping applications, is caused by photoresist delamination and underplating. Cross-sectional SEM images show that the the ring was present prior to photoresist removal, and Auger spectroscopy results indicate that the resist delamination occurs prior to the nickel stud deposition.

ACKNOWLEDGEMENTS
The authors would like to acknowledge and thank Dr. Nancy Finnegan at the Center for Microanalysis of Materials, the University of Illinois for running the Auger spectra and for many helpful discussions.

REFERENCES

[i] Kietz, K, Circuitree, Tech Talk, "Fine Lines in High Yield (Part LXVIII)", May 1, 2001. Web publication.
[ii] Childs, K. D., Carlson, B. A., LaVanier, L. A., Moulder, J. F., Paul, D. F., Stickle, W. F., Watson, D. G., *Handbook of Auger Electron Spectroscopy, Third Edition*, Physical Electronics, Inc., 1995.

FORMATION OF LEAD-FREE MICROBUMPS BY ELECTROPLATING FOR FLIP-CHIP AND WLP APPLICATIONS

R. Kiumi, F. Kuriyama, N. Saito
Precision Machinery Co., Ebara Corp.
Fujisawa-shi, Kanagawa, Japan
E-mail: kiumi.rei@ebara.com

ABSTRACT

Lead-free micro-bump soldering is an important step toward the removal of hazardous substances from electronics. The bumping processes for the lead-free soldering are required to be compatible with existing machines and to fabricate lead-free micro-bumps without any defects for advanced packaging applications. In our study, the micro-bump formation of lead free electroplating processes for Sn, Sn-Ag, Sn-Cu and Sn-Ag-Cu interconnects was investigated. In order to have a viable lead-free micro-bumping process, we have studied electroplating of lead-free micro-bumps with a dip-type plating bath. The results announce that (1) electroplating technology has been developed to make with high reliability and void-free for lead-free micro-bumps with existing machines; (2) Sn-Cu micro-bumps was much easy to be formed by plating Sn on Cu with a high reliability and without a whisker issue, comparing with other lead-free solder; (3) the non-uniformities of the lead-free micro-bump height were lower than 5% over the whole wafer; (4) the fracture mode was verified being a ductility fracture only in the micro-bumps for the lead-free bumping interconnects between micro-bump and UBM, (5) electroplating processes are suitable for mass production of bumps well-controlled on the morphology, growth shape, thickness and compositions of alloys.

Key Words: micro-bump, lead-free solder bumping, electroplating, flip chip packaging;

INTRODUCTION

The trend of ever increasing I/O density has drastically lowered interconnection pitches. An off-chip pitch of 150 microns is being widely used in the microelectronics industry. 20-micron pitch pads are being projected on the ITRS road map. [1] In fact, a flip-chip technology was practically used by interconnecting a Chip-on-Chip (CoC) with several thousands of micro-solder bumps with a diameter of 30 µm. It made the signal transmission rate between chips comparable to System on Chip (SOC) technology. [2,3] Then, the conventional solder technology may not guarantee the required performance at such fine pitches, due to the higher diffusivity and softening nature of solder. [4] In order to overcome these disadvantages, solder alloys with higher strength and better thermal and microstructure properties are required.

Although lead-free solder plating techniques for electronic applications is driven by the fact that electroplating processes can deposit all the bumps on a wafer simultaneously and at a finer pitch than other technologies, there are still some doubts about fabrication of lead-free micro-bumps by electroplating. For example, can the micro-bumps be perfectly electroplated over a whole wafer with uniform distributions of bump height and alloying compositions; doesn't it increase the probability of Intermetallic compound (IMC) formed through the all bump. According to our research, we believe that electroplated bumping is the most suitable process for fabrication reliability and material quality when fine pitch bumping is required on the flip chip package or wafer level packaging (WLP). Of course, there is a primary issue related to the electric potential differences between the components of lead-free alloys, which will make the composition ratio of the alloys unstable during the electroplating process, and the necessary of adding chelating agent will induce an issue of voiding. In micro-bumping production, there is also a multitude of factors influential to the plated solder composition ratio. In this study, we discussed the micro-bump formation of the lead free electroplating process for Sn, Sn-Ag, Sn-Cu, and Sn-Ag-Cu.

EXPERIMENTS

The different types of lead-free solder bump materials, with Sn constituting a major fraction of the solder, were selected on the basis that these micro-bumps can be fabricated using conventional low cost electroplating techniques, the compatibility with current reflow processes, materials, and surface mount equipment. The lead-free micro-bump media chosen in the study consisted of pure Sn, Ag, Cu, Sn-Ag, Sn-Cu, and Sn-Ag-Cu.

1. Electroplating process

The experiments for micro-bumping were carried out as following, first, to deposit the barrier metal films on the wafers as a plating cathode; then, to make the photo-resist patterns as plating masks, after that, to electroplate bumps, strip resist mask and etch UBM off, finally, to reflow the plated bumps into balls. Each one set of electroplating reactors was used for electroplating Ni, Cu, Ag, Sn, Sn-Ag or Sn-Cu bumps. Two sets of plating reactors were set up for Sn-Ag-Cu plating, with a 2-step electroplating process, which is similar to the 2-step process of Sn-Ag plating introduced before. [5-11] Cu and Sn-Ag reactors, or Sn-Cu and Sn-Ag reactors were used to fabricate Sn-Ag-Cu solder bumps. The plating chemicals mainly comprised of acids, with a nonionic surfactant, were prepared with an appropriate metal concentration. The electroplating of lead free solder bumps was performed with various current densities of 10 to 100mA/cm2 at 20-25°C. The electroplating was carried out by means of a dip-type machine. For the machine, the anode and wafer were vertically set in a plating bath, and the specific wafer holder was used to cover the wafer edge and backside to avoid plating on that areas. [5]

2. Reflow and annealing treatments of lead-free solder bumps

After removing the plating mask from the plated wafers, the plated bumps were reflowed into solder balls. The reflow temperature was tested from 238 to 260°C, and holding time was about 0.5-3min. An annealing treatment was carried out at 125°C for 500 hr after reflow treatment.

3. Evaluation of lead-free solder bumps

The height distributions of the lead-free bumps on wafers was measured before and post reflow by means of step height detection such as a long scan profiler. Composition measurements were performed using ICP spectrometry analysis by dissolving the bumps in strong mixing acidic chemicals. Moreover, to alleviate a concern of whether Sn, Ag, Cu and Ni elements were distributed as uniform compositions in solder alloys, EPMA analyses were performed for the cross-sectional surface of as plated and post reflowed solder bumps. The voiding of bumps was examined by X-ray measurement (Dage, XiDA XD6600).

4. Shear test of lead-free solder bumps

The mechanical property of lead-free solder bumps is examined using a shear tester (Dage, Series 4000P). The failure mode was evaluated with a share test if fractures caused occur inside the solder bump; the reliability of a solder-bump joint is higher, contrary to fractures occurring in the interface between the solder and the pad, which should be a failure mode with lower reliability. So the rate of fracture inside the solder bumps could be simple evaluations for the reliability of solder bump connections.

RESULTS AND DISCUSSION

Formation of Micro-bumps of Lead-free Solder by Electroplating

The electroplating of lead free solder micro-bumps was performed with various current densities 10 to 100 mA/cm2 at 25°C. Figure 1(a) shows the as-plated Sn-Ag micro-bumps, which were electroplated on Cu layer with 30 mA/cm2 at 25°C, and Figure 1(b) shows the Sn-Ag-Cu micro-bumps formed after reflowing the as-plated Sn-Ag micro-bumps on Cu layer at 250°C for 1min. Figure 2(a) shows the as-plated Sn micro-bumps, which were electroplated on Cu layer with 30 mA/cm2 at 25°C, and Figure 2(b) shows the Sn-Cu micro-bumps formed after reflowing the as-plated Sn micro-bumps on Cu layer at 260°C for 1min. The as-plated Sn-Ag micro-bumps are of much finer crystal grains, and smooth top surface, as shown in Figure 1(a), comparing with the as-plated Sn micro-bumps as shown in Figure 2(a). Then, the micro-bumps after reflow treatment, either Sn-Ag or Sn micro-bumps, become into hemisphere with a closed microstructure, a shining and smooth surface, indicating less oxide and impurity on the micro-bump surface. This is a very critical condition to ensure good bond in the assembly.

 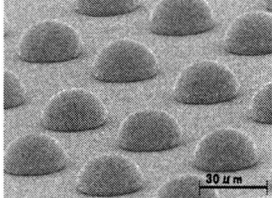

(a) As-plated   (b) Post-reflow

Figure 1 Sn-Ag-Cu micro-bumps

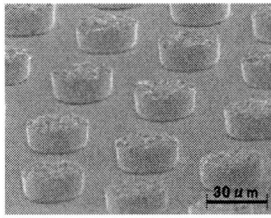 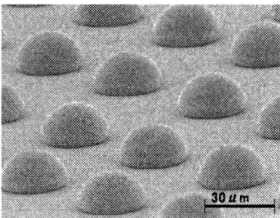

(a) As-plated   (b) Post-reflow

Figure 2 Sn-Cu micro-bumps

The roughness and morphology of micro-bumps were mainly dependent on the additives of electroplating chemicals, although it was also affected to current density and flow rate. In this study, the plated micro-bumps were of a relatively finer microstructure and crystal grains, and there was a good roughness on the top surface. It was observed from Sn, Sn-Pb, Sn-Ag, and Sn-Cu alloy deposition that current density had not a significant influence on the feature scale shape evolution neither on the microscopic surface morphology and roughness. Although Sn bump surface was not formed so even and the as-plated grain size was larger than other lead-free solders, it did not effect on controlling of micro-bump volume so much.

Reflow treatment is to obtain uniform composition within bumps and to get rid of impurity trapped inside during plating and etching. In our study, it was also confirmed whether the reflow temperature of different lead free solder materials was suitable to the existing processes and equipments for meeting the needs of flip chip packaging. Figure 1(b) shows the Sn-Ag solder bumps reflowed with the existing reflowing equipment at 250°C in N_2 atmosphere, and without using any flux. The results

showed that the Sn-Ag bumps could be reflowed into balls, and the reflowed bump balls were arranged with same size and were with smooth and shiny surface, although it was not so critical to performance, and then it would enhance the cosmetic appearance. The above reflow treatments were also an alloying process of different plated layers during reflow, as for reflowing Sn-Ag bumps on Cu, Cu diffused into Sn-Ag bumps and Sn-Ag-Cu bumps were formed during reflow treatment. It was due to liquid and solid phase reaction between melted Sn-Ag bump and its underlayer of Cu, it made Cu diffuse into a melted Sn-Ag bump to form Sn-Ag-Cu solder as reported before. [5] It was verified by Differential Scanning Calorimeter (DSC) analysis. Figure 3 shows the melting points of as-plated lead free solder bumps measured by DSC analysis. The measured melting points of as-plated lead free solders were 221°C for Sn-Ag bumps, 228°C for Sn-Cu bumps and 218°C for Sn-Ag-Cu bumps. The decrease in melting temperatures shows that Sn-Ag-Cu alloy bumps were really formed by reflowing the as-plated Sn-Ag on Cu, that means the as-plated Sn-Ag and Cu did transfer into Sn-Ag-Cu alloy.

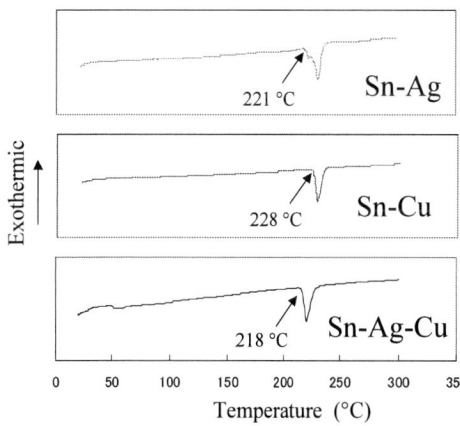

Figure 3 DSC curves of As-plated lead-free bumps

In our study, pure Sn was plated on Cu bump and reflowed at 260°C in N2 atmosphere; the Cu amount dissolved into the Sn solder was measured being not over about 0.5%Cu in bumps, even increasing the thickness of Cu layer much higher than this ratio of alloying elements. There was Sn-Cu IMC layer formed between Sn solder and Cu underlayer after reflow or annealing treatment, which was observed by SEM. About that, there was another report that the Cu amount dissolved into the Sn solder could be up to a higher value with 2.2 to 2.6%Cu. [12] It might be due to different reflow processes or different methods of composition evaluations.

For researching micro-bump formation by electroplating, we tried to electroplate Sn-Ag solder into a test pattern with various size in diameter and 100μm in thickness as shown in Figure 4. Figure 4(a) shows as-plated Sn-Ag bumps with different diameter from 25 to 75μm. Figure 4(b) shows the bump height near about 100μm measured by means of step height detection of a long scan profiler. The test result shows micro-bumps with a high aspect ratio can be formed by electroplating process. From making many kind of microbumps by electroplating, we believe that electroplating is suitable to micro-bumping process for the flip chip and WLP applications.

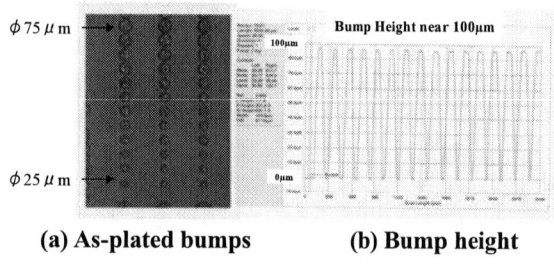

(a) As-plated bumps　　(b) Bump height

Figure 4 As-plated Sn-Ag bumps with diameter 25-75μm and high 100μm

We have put forward, Sn, Sn-Ag, Sn-Cu, Sn-Ag-Cu etc. with various lead-free compositions as replacement for lead for several years. Each metallurgy has its own benefits in term of cost, reliability, yield, environmental impact and process ability. Although there is no clear material of choice for lead-free packaging and one of the main difficulties is finding one solution that meets the reliability requirements for all industry applications. According to our tests, Sn-Cu solder micro-bumps formed by electroplating Sn on Cu, might be a better process to meet the reliability and cost requirements.

Uniformity of the micro-bump height within the wafer
The uniformity of micro-bump height and alloying composition within the chip and across the wafer can significantly affect to reliability of connetion for flip-chip and wafer level packaging applications. In our studies, the nonuniformity of micro-bump height distributed within

the chip and across the wafer was measured being 1-5%, which is almost same with that on other kind of larger size bumps reported before. [7-11] The nonuniformity of 1-5% means the as-plated wafers with good uniformity of bump height, and the values of the uniformity would become much better if it was evaluated after reflow treatment. In this study, the uniformity of Sn-Ag micro-bump height did not show a dependence on current density so clear, although there is generally a tendency for electroplating that the uniformity of bump height decreases with increasing of current density, especially for electroplating of the smaller and higher bumps. It might be considered the electroplating uniformity of Sn-Ag solder was not so sensitive to the change of current density due to the action of additives in its chemicals.

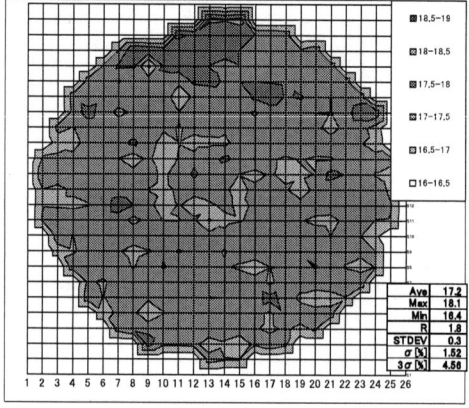

Figure 5 Height map of micro-bumps within the wafer

Systematic height variation within the wafer results from the uneven distribution of the solder bumps across the wafer. The distribution height of micro-bumps over a whole wafer is shown in Figure 5. It shows the height of micro-bumps vs. the position on the wafer measured on all chips over the wafer. Each data point in Figure 5 represents the average height of 5 micro-bumps on the chip. For the Sn-Cu wafer, the minimum bump height is about 16.5, and the maxmum is 18.1 μm and the bump height uniformity of chips across the wafer is pretty good, given that the standard deviation is less to 0.3. The nonuniformity of micro-bump height across the wafer is about σ = 1.52% and 3σ = 4.36%, which is considered due to the uneven current and distribution through the wafer. The uniformity of micro-bump height is mainly dependent on the uniformity of plating current density across the wafer. As we mentioned previously, comparing with Sn-Ag alloy plating, Sn plating was relatively sensitive to the plating current change. Then, it was still a good process to form micro-bumps, as an example shown shown in Figure 5, the bump height of Sn-Cu wafers ranges from 16.5-18.1μm.

Uniformity of the micro-bump composition within the wafer

Concerning the microstructure of these Sn-noble metal eutectic solders as lead-free solders, they are a mixture of Sn and IMC, unlike that of eutectic SnPb that has no IMC. Since Sn has the body-centered tetragonal lattice structure and tends to deform by twinning, its mechanical properties are anisotropic. The mechanical properties of these eutectic solders will be more anisotropic when the dispersion of the IMC is inhomogeneous. There is an issue for the micro-bumps of lead-free solder that IMC might be easier to be formed across the all micro-bump, inducing a worse mechanical properties and reliability.

The change of the metal compositions in as-plated Sn-Ag micro-bumps on 300mm wafer production was evaluated only by ICP spectrometry analysis as shown in Figure 6. The Ag composition in micro-bumps was difficult to evaluate derectly by x-ray, as the diameter of micro-bumps was too small to detect. The evaluation by ICP spectrometry is only a method of average composition of many micro-bumps. The potential difference of metal elements in lead-free materials should be adjusted with controlling the current density, ion concentration and chelating agent amount in plating solution. In our study, there was a possibility of IMC, as for coarse Ag_3Sn needle-like or plate-like crystal shown in Figure 7(a), formed through the all microbump inducing a brittleness of micro-bumps. The imaging of IMC formed through the bump was also observed very well in Figure 7(b) reported by T. Y. Lee. [13] We also analyzed the growth of the compound composition and thickness in the bumps and at the interface between UBM(Cu, Ni) and solder bumps. At the interface between Cu and Sn-Ag, it was also found that there are two kinds of composition, thin Cu_3Sn and nonflat Cu_6Sn_5. Compound growth after the reflow or HTS (High-temperature storage) test of solder was studied detailed including in case of solder bump on Ni layer. The

above detailed results are to be reported in another paper. In the case of lead-free SnAg or SnAgCu solder, it is easer to imagine that this kind of coarse Ag_3Sn needle-like or plate-like crystals would be much easier to occur through the cross-section of micro-bumps which would be easy to induce a early fracture. Comparing with Sn-Cu, the heating process for packaging will tend to make Ag_3Sn coarse needle-like crystals through the cross-section of Sn-Ag solder micro-bumps, which might cause degradation of reliability. The dependence of bump composition on plating chemicals is due to the larger potential difference between metal elements in lead-free materials.

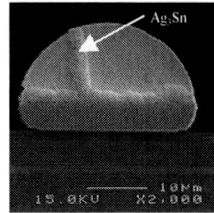 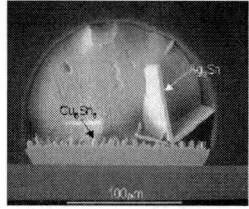

(a) micro-bump   (b) larger than 100μm

Figure 7 Intermetallic compounds formed in the Sn-Ag bump on Cu, (a) after HTS treated 500hr at 125°C; (b) reflowed twice at 260°C and heavy etching. [13]

When comparing the electroplating of lead free alloys with Sn-Pb alloys, it is known that lead free solder is difficult to electroplate due to the large difference in the standard potential of the involved metals [5-10]. Since the potential difference between Sn and Pb is only 11mV, Sn-Pb alloys are rather easy to plate in a wide range of alloy compositions. In contrast, the difference for the Sn-Cu is 473mV, and even much higher, up to 935mV, for the Sn-Ag, which can cause several issues during electroplating the preferred deposition of the more noble metal in the low current density range, immersion plating of the more noble metal on the Sn anode, and redox reaction between divalent Sn and the more noble metal. An anodic reaction is very important for the composition control of lead-free alloy plating. There is no this kind of issue for plating Sn on Cu to form Sn-Cu micro-bumps. Of cause, the issue of whisck was generally worried for Sn electroplating. Then the whisck issue of Sn electroplating did not occur in our many experiments. It might be considered due to reflow treatment or excellent Sn plating chemicals.

Voiding and Shear stress of Lead-free Micro-bumps fabricated by Electroplating

The voiding phenomenon was studeied primarily using electroplating for lead-free solder before. [8] The void issue of micro-bumps were also studied and evaluated by observation of cross-section of micro-bumps and X-ray measurement over wafers. Almost all the micro-bumps being void-free was observed by EPMA on the cross-section of bumps. The results in our study showed that the void-free micro-bumps of lead-free solder could be fabricated by electroplating, and the existing equipments and processes can be used for void-free micro-bumps of lead-free solder. The effective factors on voiding were found mainly related with plating chemicals, especially the chelating agent and its amount in plating chemicals, electroplating and reflow conditions. In our studies, the reflow profile could be utilized with a peak temperature of 238-270°C, according to using of the chemicals and plating condition. From this point, it should be considered that Sn-Cu micro-bumps formed by plating Sn on Cu would be much easier than other lead-free solder on voiding control during mass-production.

It is well known that the mechanical properties, such as strength, ductility, creep and fatigue life in the solder connection of the flip chip packaging will be effected by alloying composition in lead-free solder bumps. In our study, the shear strength of the Sn-Cu, Sn-Ag and Sn-Ag-Cu micro-bumps was much higher than that of Sn-Pb bumps (about 40 N/mm^2), which verified the alloy strengthening effect of Cu or/and Ag addition. As an initial evaluation of the reliability of lead free bumps by shear test, the reliability of a joint would be accepted if fractures occurred inside the bump, on the other hand, it should be considered a failure if fractures occurred on the interface between the solder and the pad. As our observations by microscope on the fracture surfaces as showed in Figure 8, the fractures of the bumps all occurred inside the solder bump. So it can be considered that they are of good reliability for the Sn-Cu, Sn-Ag, and Sn-Ag-Cu solder micro-bumps. Comparison of the fracture surfaces, the Sn-Ag-Cu solder The shear strength

of Sn-Cu, Sn-Ag, Sn-Ag-Cu solder micro-bumps was all higher than 60 N/mm^2, and they were all much higher than the strength of Sn-Pb solder.

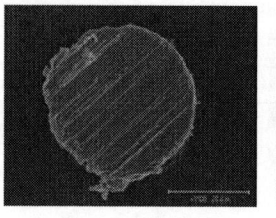 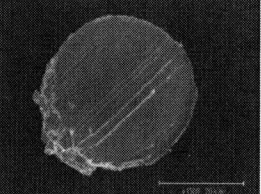

(a) Sn-Ag-Cu   (b) Sn-Cu

Figure 8 Fracture surface showing ductile failure mode for Solder micro-bumps

The results of this study show that lead-free solder micro-bumps can be fabricated with electroplating processes. There is no any fatal issue to limit electroplating of micro-bumps for flip chip and WLP applications, although there might be some issues on amount of side edging for UBM, and perfectly patterning of photoresist. The amount of side edging for UBM after electroplating was commonly about 1-3μm, which was a large issue to limit the application of electroplating for micro-bump fabrication. Fortunately, the top Cu seed layer etching was achieved by IBM internal developed Cu etchant. The total undercut from each etching steps can be limited to less than 0.3 μm. [14] It was successfully etched each seed layers with good uniformity and controlled undercut. For micro-bumping processes, we suggest that the Sn-Cu micro-bumps, which formed by plating Sn solder onto Cu bumps, and alloying to Sn-Cu alloy solder after reflow, because the multi-plating of pure metals will be much suitable to mass-production.

CONCLUSIONS

In this study, the micro-bump processes of lead-free solder Sn-Ag, Sn-Cu, Sn-Ag-Cu have been researched by electroplating for flip-chip and WLP applications. The electroplating processes were tested for fabrication of micro-bumps. It was found the electroplating could form micro-bumps in diameter smaller than 20μm, and it looked like without a clear size limit if the resisit pattern could be formed well. The microstructure of lead-free solder micro-bumps are of fine particles or crystal grains by electroplating. Our processes are suitable for mass-production with well-controlled micro-bump geometry, composition, and uniformity over the entire wafer. The surface of lead-free solder micro-bumps after reflow treament is smooth and shiny and the void issue was suppressed by controlling of plating processes.

REFERENCES

1. International Technology Roadmap for semiconductors-Assembly and Packaging (2003 edition); (http://public.itrs.net/Files/2003 ITRS/ Home 2003.html)

2. T. Ezaki et al., "A 160-Gbps Interface Design Configuration for Multichip LSI" ISSCC 2004.

3. T. Suga, H. Ozaki, and H. Ozawa, "Behavior of Surface Oxide and Intermetallic Compounds in Interconnections of Micro Sn-Ag Solder Bumps" ECTC2006, pp.1141- 1146.

4. C.C. Yeth, W.J. Choi, K.N. Tu. Appl Phys Lett 2002; 80:580.

5. R. Kiumi, J. Yoshioka, F. Kuriyama, N. Saito, "Process Development of Electroplate Bumping for ULSI Flip Chip Technology", 52nd ECTC Symp., pp.711-716 (2002).

6. R. Kiumi, J. Yoshioka, F. Kuriyama, N. Saito, "Processing, Properties, and Reliability of Electroplated Lead-free Solder Bumps", IEEE/ITherm Symp.pp.909-714 (2002).

7. R. Kiumi, J. Yoshioka, F. Kuriyama, N. Saito, "Electroplating Process for Lead-free Bumping in Flip Chip Packaging", Advanced metallization Conference 2003 (AMC2003) pp.669-674.

8. J. Yoshioka, R. Kiumi, S. Takeda, F. Kuriyama, and N. Saito, "VOID-FREE BUMPING PROCESS FOR LEAD-FREE SOLDER", 1st Annual International Wafer-Level Packaging Conference, pp.148-153, (2004)

9. R. Kiumi, J. Yoshioka, F. Kuriyama, N. Saito, S. Takeda, International conference on Lead-Free electronics, Assemblies and Components; IPC/JEDEC 2004, Oct. 20-22, Frankfurt Germany.

10. R. Kiumi, S. Takeda, J. Yoshioka, F. Kuriyama, N. Saito, "Composition Control for Lead-free Alloy Electroplate on Flip Chip Bumping", 55th ECTC Symp., pp.120-126 (2005).

11. R. Kiumi, F. Kuriyama, N. Saito, and K. Kamimura, "Electroplating of Lead-free Bumps on 300mm Wafers for WLP Applications", 2nd Annual International Wafer-Level Packaging Conference, pp.148-153, (2005)

12. Katter U.R. et al., Z. Metallkd. Vol.92, No.7, Jul. 2001, pp740-746

13. T. Y. Lee et al., "Morphology, kinetics, and thermodynamics of solid state aging of eutectic SnPb and Pb-free solders (Sn3.5Ag, Sn3.8Ag0.7Cu, and Sn0.7Cu) on Cu", J. Mater. Res. 17 (2) (2002).

14. H. Gan et al., "Pb-free Micro-joints (50 μm pitch) for the Next Generation Micro-systems: the Fabrication, Assembly and Characterization", 56th ECTC Symp., pp.1210-1215 (2006).

WAFER-LEVEL PACKAGING: EFFECTIVE COST REDUCTION WITH WAFER BONDING

Thorsten Matthias
EV Group Inc.
Tempe, AZ, USA
T.Matthias@EVGroup.com

Markus Wimplinger and Paul Lindner
EV Group
St. Florian/Inn, Austria

ABSTRACT

The market demands for ongoing price reductions in consumer electronics require novel and innovative manufacturing processes. Packaging of the devices is a major cost factor, and therefore continual reduction of the packaging costs is key.

Wafer bonding enables first-level packaging on the wafer level, which has been introduced 15 years ago for automotive devices. Due to the very short product life cycles in consumer electronics, sometimes only a few months, the choice of the appropriate bonding method is driven by the potential for continual cost reduction. Depending on the product, innovative bonding processes using SU-8 and BCB, but also the established MEMS processes like metal-metal and direct bonding, can be the right choice. The decision criteria are:

Highest yield on all manufacturing process is imperative. This involves not only the bonding process, most important pressure and temperature uniformity, but also the pre-processing steps like cleaning and plasma activation. Integration of the pre-processes in the wafer bonding platform enables real-time process control. Due to the accurate timing of the process flow the bonding process time can be optimized and the capacity and throughput are increased.

Very high wafer-to-wafer alignment accuracy allows minimized substrate real estate consumption of the gasket enabling both, more chips per wafer and a higher production capacity. The SmartView® alignment principle reduces the manufacturing steps, as no backside alignment keys are needed, and reduces the material costs as backside polished wafers are not necessary.

Process compatibility between R&D and production equipment significantly reduces cost and time for process qualification. A modular and redundant equipment concept results in high uptime with no trade-offs.

In this paper the technical differences for established and modern wafer bonding processes are reviewed. The impacting factors on cost reduction for current and upcoming product generations are analyzed in detail.

Key Words: wafer-level packaging, wafer bonding, wafer alignment, yield enhancement

INTRODUCTION

The cost of the package is a major price contributor for all electronic devices especially for consumer electronics. Depending on the device costs, the form factor and other specific requirements, the cost of the package can be up to 70% of the device cost. Reducing the cost of the package has the biggest impact on the device cost.

Package innovations through disruptive manufacturing processes are key for successful product development in consumer electronics. It is an acknowledged fact in the semiconductor industry that the move from chip-level packaging to wafer-level packaging is a must. The ability of batch processing of all dies on the wafer as well as the ability for wafer-level testing gives a huge advantage compared to today's established back-end processing schemes.

In addition to the economical advantages of wafer-level backend processes there are technical reasons. As long as the dies are part of a wafer, further processing steps are possible using standard fab equipment. This enables to perform classical front-end processes like lithography or etching after the first packaging steps. Thereby it is possible to add functionality to the package.

Through wafer vias allow to overcome the barriers of conventional packaging technology. Going even further allows vertical stacking of functional entities, a system-in-a-package.

Wafer bonding and wafer-to-wafer alignment are well-established technologies originally coming from the MEMS manufacturing. Wafer-level packaging (WLP) via wafer bonding allows smaller and thinner packages, improves the yield due to higher cleanliness, enables the encapsulation of vacuum or process gas and finally reduces the packaging costs significantly. High precision alignment of device wafer to cap wafer allows real chip size packaging as the required width of the sealing rings is in the low micron range.

The functions of wafer bonding can be summarized as follows:

First level package
The main goal of assembly and packaging operations is to either attach the two halves of a device together (generally used in bulk micro machined sensors) or to provide a protective cover (common for surface machined parts). Sealing a MST device at wafer level provides a high level of cleanliness as it is performed prior to dicing. Particle sensitive MST features could not be diced without a protection – cleaning of the chips after dicing is not very effective. In the case of surface micro machined technology bonding a cap wafer to a sensor wafer provides the required protection for subsequent less clean process steps. Surface micro machined features frequently are first level packaged at the wafer level to protect the sensitive mechanically suspended features from the environment.

Stress Isolation
A bonded wafer pair can allow for a hermetically sealed mounting into the final package. Mechanical stress resulting from the interface between package and board has to be isolated from the sensor Si chip. The wafer level bonding approach can provide sufficient mechanical stability and therefore decouple mechanical stress acting on the package from the sensor chip. Pressure sensors use thick glass pedestals serving as a rigid mechanical base for the silicon sensor chip. The anodically bonded glass pedestals are generated at wafer level and absorb any stress introduced to the final package.

Controlled ambient
A hermetic seal with controlled pressure inside the MST device is essential to adjust the mechanical characteristic of moving features. Dampening properties of an accelerometer or gyroscope are adjusted by encapsulating a defined gas ambient at precisely controlled pressure.

Higher level of integration
With increasing priority wafer bonding is seen as a possibility to increase the functional density of a device. Wafers with different functions like mechanical sensor wafer and an ASIC wafer can be combined at wafer level. For 3D interconnect technologies wafer bonding is an enabling technology.

Whereas the MEMS industry is very open to individual solutions, often described as "one product – one process", the packaging industry is driven by the desire for standardization and ultimately cost reduction.

Cost reduction with wafer bonding
Compared to MEMS devices, where wafer bonding is often the key enabling process, wafer bonding has to compete in the packaging area on the cost level with established and also new backend techniques. In addition to the discussed generic advantages of wafer-level processing versus chip-level processing, there are several important points for the successful implementation of wafer bonding for wafer-level packaging.

Process transfer from R&D to production
Due to the very short product life cycles in consumer electronics, sometimes only a few months, a smooth transfer from research and process development into production is extremely important. The goal is to minimize the time for process qualification. The main requirement is that the equipment used for R&D and production is process compatible, which enables that the process parameters as well as the process control limits can be directly transferred.

Capacity enhancement
In many cases a specific packaging process can be transferred from one product to other similar devices. The wafer bonding platform should be designed in such a way that a capacity enhancement can be easily accommodated. The EVG Gemini® is field upgradeable to up to 4 bond chambers. The individual bond chambers can be tested and qualified prior to installing them on the production system.

However, different applications require different packages and thereby different wafer bonding processes. In consumer electronics the lifetime of semiconductor equipment outlasts the device product life times by magnitudes. A universal bond chamber gives full flexibility to switch between different wafer bonding processes without additional capital equipment expenses. Given a specific device, the choice of the appropriate packaging process can be made upon technical considerations and is not dictated by equipment limitations.

Processes independent of product
In order to minimize the costs for process development and qualification it is desirable to standardize the individual process steps. A novel wafer-to-wafer alignment technique, the SmartView® alignment, enables standardization independent of wafer material, size and properties. Furthermore it is not necessary to perform additional pre-alignment process steps like backside alignment keys or backside polishing. The SmartView® aligner can use any feature on the wafer as alignment key.

A schematic of the SmartView® working principle is shown in Figure 1. Instead of using a single microscope in between the wafers, as is used in other substrate alignment platforms, the SmartView® system employs two microscopes for alignment. One microscope is placed above and the other below the wafer stack. The dual microscopes focus on a common focal plane calibrated for each alignment. Each microscope objective observes one alignment key on the surface of the wafer. The detailed process flow of SmartView alignment is presented below:

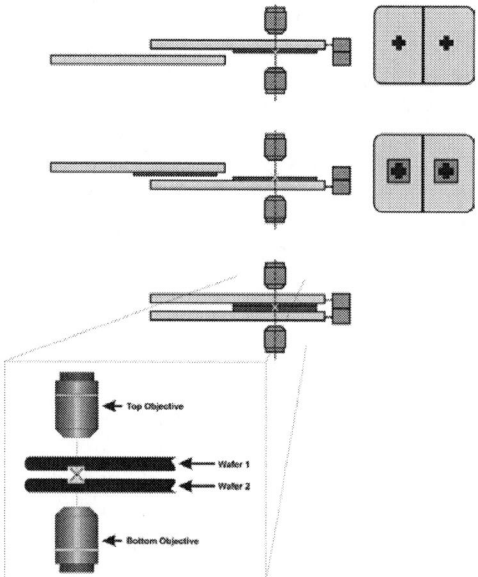

Figure 1: SmartView® alignment principle

Step 1: First the top wafer is loaded face up and is then rotated into face down position. Next it is observed by the bottom optics. The optics are adjusted to bring the alignment keys into view, and then the image is digitized and stored electronically.

Step 2: The top wafer is retracted, allowing the bottom wafer to be brought into position beneath the top optics and then aligned to the existing digitized image of the top wafer. Once complete the two wafers are automatically moved into alignment by calculating the relative X and Y locations of the alignment keys on each wafer and moving the wafers into the final alignment position.

Step 3: Once alignment is complete the wafers are brought into contact and secured for bonding.

Wafer alignment is accomplished using encoded stage motors allowing X and Y movements in increments of 0.1 µm steps and minimized Z-axis travel controlled by three software controlled spindle motors to preserve planarization between the top and bottom wafers.

Yield enhancement
Most wafer bonding applications are very sensitive on the wafer surface parameters. In order to achieve a high yield manufacturing process it is therefore necessary to control the surface properties as tightly as possible. The key pre-processes for wafer bonding are cleaning, plasma actiation and adhesive layer coating. Integration of the pre-processes in the wafer bonding platform enables real-time process control. Due to the accurate timing of the process flow the bonding process time can be optimized and the capacity and throughput are increased.

Wafer cleaning station
For most semiconductor devices a particle can damage the die which it contacts, but only this die. However, in wafer bonding a particle between the wafers prevents the surfaces to get into contact and thereby results in an unbonded area (a "void"). Due to geometrical reasons to unbonded area is much larger than the particle size itself. In the case of Si to SiO_2 bonding using wafer with standard thickness the void can be up to 1000 times larger than the particle itself.

- As the void is much larger than the particle itself, one particle can prevent good wafer bonding over an area of several dies. A single particle can damage multiple dies.

- Furthermore for some subsequent process steps like wafer thinning or wafer dicing a void can result in damage to the whole wafer. Potentially a single particle can result in scrapping the whole wafer stack.

The conservative approach for yield improvement is a cleaning process prior to wafer bonding. Figure 2 shows the cleaning station with spinning chuck, DI water rinse and megasonic nozzle.

Figure 2: Cleaning station with spinning chuck, DI water rinse and megasonic nozzle

Cleaning of the wafers is only the first step, the second step is to prevent that the wafer get contaminated with particles again. The particle build up on a clean wafer depends on

a) The time between cleaning and the next process step.
b) The number of subsequent handling and process steps. Especially cassette loading and unloading operations are very critical.

The most effective way to prevent particle contamination of the clean wafer is to integrate the cleaning module within the wafer bonder platform. This serves both purposes, it minimizes the time between cleaning and bonding, and it minimizes the number of handling and process steps after cleaning.

Furthermore, the integration of the cleaning module in the wafer bonder platform enables the same constant time between cleaning and bonding for every wafer of the cassette. This eliminates a potential wafer-to-wafer process variance and enables easier and more accurate process control.

LowTemp® plasma activation
LowTemp® plasma activation is a wafer surface activation method prior to wafer bonding. By applying a plasma treatment to the wafers prior to bringing them in

contact for bonding, the surface chemistry can be modified. This enables to obtain maximum bond strength even for low temperature thermal annealing schemes.

A major advantage of this process is that it makes possible some bonding applications which are not possible using standard bonding processes due to different materials characteristics (e.g. high thermal mismatch of the two bonding partners, low T_g for polymer bonds, etc.)

Figure 3: Fully automated production wafer bonder EVG Gemini® with SmartView® alignment and LowTemp® plasma activation module

An important aspect of plasma activation is the potential reduction of cycle time. Due to the surface activation the annealing or bonding temperature can be reduced, which reduces the time for heating and cooling. In addition the higher surface reactivity allows reducing the bonding time for many applications.

LowTemp® plasma bonding for WLP
LowTemp® plasma bonding is an enabling technology for direct wafer bonding for wafer-level packaging. Direct wafer bonding provides a lot of ideal bonding properties for packaging applications like hermeticity, no outgassing and very high bond strength. However, so far the need for annealing temperatures up to 1000°C prevented the wide adoption of this technique. LowTemp® plasma activation enables a reduction of the annealing temperature to 200-400°C without compromising the quality of the package.

Furthermore LowTemp® plasma bonding results in a strong pre-bond at room temperature. This pre-bond results in a very high thermal conductivity between the two wafers during the subsequent bonding/ annealing steps. Therefore run-out error can be avoided. This allows faster heating ramps which results in shortened cycle time and thereby increased capacity.

Coating module
Some adhesives used for wafer bonding have a very limited lifetime after coating. Integrating the coating module within the automated wafer bonder allows the use of these adhesives.

Figure 4:. Photograph of an EVG810LT plasma chamber.

Advanced process control
The majority of fab equipment and processes has specifications for the parameters of the incoming wafers. However, in wafer bonding we are dealing with two wafers simultaneously and therefore with the interaction of the wafers during the processes. The interaction of the two wafers results in additive and multiplicative effects, which have to be compensated by the equipment.

State-of-the-art semiconductor production equipment provides multiple safety levels in order to guarantee steady production and uptime.

1. *Field proven design*
   All the components and process modules of the EVG Gemini® have been successfully tested and qualified according to the stringent requirements for reliable production of automotive sensors. The performance and reliability of each single component of hardware and software as well as the interaction of these components is critically reviewed on a regular basis based on EVG's large install base of automated and semi-automated wafer bonders.

2. *Failure tolerant design*
   The critical components of the EVG Gemini® are designed in a failure tolerant way. Wafer bow, warp and thickness variations are compensated in the aligner and the bonder using proprietary wedge compensation mechanisms. The bond chuck components interact in such a way that vibrations and thermal or mechanical shocks do not impact the alignment accuracy of the wafer stack.

3. *Process control*
   Despite of all described measures to avoid problems, it is also necessary to have safety systems in place, which enable to determine problems when they occur. Instantaneous issues like a broken part are usually easy to determine. However, steady degradation or even intermittent errors eventually cost a lot of money, if they are not detected rapidly. Usually the bonded wafer stack cannot be tested directly after bonding and a couple of further process steps are required before testing of the electrical or mechanical properties is possible. Therefore tight process control is mandatory. It is necessary to be able to analyze the impact of each equipment module. Wafer ID

and bondtool ID tracking are enabling key components of process control on production wafer bonding equipment.

Review: Wafer bonding processes

Anodic Bonding is the most widely used packaging method. Electric fields assist in the thermal diffusion of ions across the bond interface to achieve solid state mixing of the glass and the silicon. Anodic bonding is used in the assembly of many fluidic devices and packaging of pressure sensors. Anodic bonds are also quite common in MOEMS (optical MEMS) components because the transparent cover is needed. Anodic bonding is considered the workhorse of MEMS packaging and accounts for majority of all packaging applications. Advancements in wafer clamping and independent thermal processes for top and bottom chucks are key to managing thermal expansion differences and maintaining submicron alignment during whole wafer processing.

Silicon Direct Bonding became wide spread during the 1990's. Numerous materials can be bonded using this technique provided the surfaces meet the rigid roughness and flatness standards. Hydrophilic surfaces created by wet chemistry or plasma activation will immediately bond upon contact via van der Waals attractions between adsorbed water groups. Following the bond, batch thermal annealing is used to transition the bonds to covalent Si-O-Si bonds and achieve bond strengths equivalent to bulk silicon. Because silicon direct bonding requires atomic intimacy at the interface, point of use cleaning methods are essential in high volume production. Modern cluster tools capable of cleaning, activation, aligning and bonding can achieve 25-40 wafer/hour throughputs and are contained in class 1 mini-environments.

Thermal Compression Bonding includes three main subcategories; glass and glass frit, eutectic, and diffusion. During glass frit and glass bonding the intermediate layer at the interface begins to flow under the influence of pressure when heated above the glass softening temperature. The glasses can be applied via extrusion, screen-printing, spraying, or sedimentation methods. Bonding techniques using metals as intermediate layers typically form a hermetic seal and are high vacuum compatible (low out gassing materials with low permeabilities):

Eutectic bonding takes advantages of a metallurgical phase transformation in which the binary phase formed from constituents has a lower melting point than either solute. In essence, eutectic bonding is a special case of a diffusion bond that allows for very strong intermetallic bonds to be formed at relatively low temperatures. When two materials diffuse a mixture is formed. This mixture has a very low melting point at the eutectic composition. Once the eutectic forms and becomes liquid the reaction accelerates under the influence of liquid phase diffusion at the liquid solid interface. Figure shows a schematic of this process for the Au-Si system. The reaction begins with inter-diffusion and the eutectic is formed after resolidification from the melt.

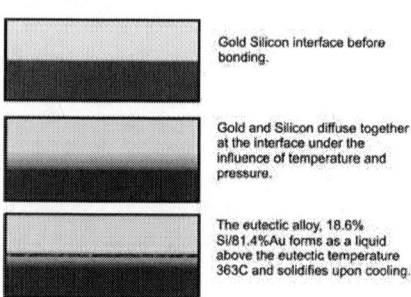

Figure 5: Schematic representation of the gold-silicon eutectic reaction. Pressure is applied to maintain intimate contact during diffusion and temperature is regulated to achieve alloy formation.

Solid-state thermo compression bonding is similar to eutectic bonding because an alloy is formed. However, these reactions do not involve melting of the diffused interface layer. In solid state bonding, the key is to identify systems with low temperature solid-state phase transformations and rapid diffusion coefficients. The phase that forms is most often an intermetallic compound that is very tough and gives structural stability to the assembly. Formation is via diffusion and realignment of the atoms into stoichiometric structures.

Diffusion bonding is generally applicable to systems in which the diffusion coefficient is rapid at relatively low temperature. This occurs for some FCC metals such as Au and Cu. Thus it is possible to create Au-Au and Cu-Cu bonds or even Cu-Au bonds at low temperature using diffusion driven kinetics. In these cases, no alloys are formed and the interface is a mixture of the two solutes, like sugar water. In some applications, a diffusion bond is preferable to intermetallic or eutectic alloy formation because of the generally brittle nature of those alloys.

Adhesive bonding with low-k dielectrics like Benzocyclobutene (BCB) gains more and more attention since they allow for the creation of electrical interconnects between different functional modules.

CONCLUSIONS

Reducing the package costs is the most efficient approach to reduce the total device costs for many consumer electronics devices. Disruptive manufacturing innovations like wafer-level-packaging instead of chip-level packaging give a huge competitive advantage.

Wafer bonding is used for first-level packaging, for stress isolation and for the encapsulation of controlled ambient. Due to the stacking of individual device layers a higher degree of integration can be achieved.

However, as wafer bonding is competing with other backend methods, cost control and cost reduction is of highest importance. In this paper the critical procedures and process steps are reviewed and analyzed with special emphasis on cost reduction. Integration of pre-bond processing steps into the wafer bonding platform has been identified as key for cost reduction and reduced cost-of-ownership.

Surface Mount Technology Association (SMTA)
6600 City West Parkway
Eden Prairie, MN 55344
USA

ISBN 978-1-5108-3302-9